Advances in Computational Approaches in Biomechanics

Pritam Pain
Haldia Institute of Technology, India

Sreerup Banerjee
Haldia Institute of Technology, India

Goutam Kumar Bose
Haldia Institute of Technology, India

A volume in the Advances in Mechatronics and Mechanical Engineering (AMME) Book Series

IGI Global
PUBLISHER of TIMELY KNOWLEDGE

Published in the United States of America by
IGI Global
Engineering Science Reference (an imprint of IGI Global)
701 E. Chocolate Avenue
Hershey PA, USA 17033
Tel: 717-533-8845
Fax: 717-533-8661
E-mail: cust@igi-global.com
Web site: http://www.igi-global.com

Library of Congress Cataloging-in-Publication Data

Names: Pain, Pritam, 1993- editor. | Banerjee, Sreerup, 1981- editor. | Bose, Goutam Kumar, 1970- editor.
Title: Advances in computational approaches in biomechanics / Pritam Pain, Sreerup Banerjee, Goutam Kumar Bose, editors.
Description: Hershey, PA : Engineering Science Reference, [2022] | Includes bibliographical references and index. | Summary: "This book gives an insight into the current trends of application of intelligent computational techniques used to analyze a multitude of phenomena in the field of biomechanics, offering a series of sophisticated techniques used for computer simulation in both solid mechanics, fluid mechanics and fluid-solid interface across different domain of biological world and across various dimensional scales along with relevant case studies"-- Provided by publisher.
Identifiers: LCCN 2021050136 (print) | LCCN 2021050137 (ebook) | ISBN 9781799890782 (hardcover) | ISBN 9781799890799 (paperback) | ISBN 9781799890805 (ebook)
Subjects: LCSH: Biomechanics. | Bioengineering. | Computational biology.
Classification: LCC QH513 .A35 2022 (print) | LCC QH513 (ebook) | DDC 570.285--dc23/eng/20211115
LC record available at https://lccn.loc.gov/2021050136
LC ebook record available at https://lccn.loc.gov/2021050137

This book is published in the IGI Global book series Advances in Mechatronics and Mechanical Engineering (AMME) (ISSN: 2328-8205; eISSN: 2328-823X)

British Cataloguing in Publication Data
A Cataloguing in Publication record for this book is available from the British Library.

For electronic access to this publication, please contact: eresources@igi-global.com.

Advances in Mechatronics and Mechanical Engineering (AMME) Book Series

J. Paulo Davim
University of Aveiro, Portugal

ISSN:2328-8205
EISSN:2328-823X

Mission

With its aid in the creation of smartphones, cars, medical imaging devices, and manufacturing tools, the mechatronics engineering field is in high demand. Mechatronics aims to combine the principles of mechanical, computer, and electrical engineering together to bridge the gap of communication between the different disciplines.

The **Advances in Mechatronics and Mechanical Engineering (AMME) Book Series** provides innovative research and practical developments in the field of mechatronics and mechanical engineering. This series covers a wide variety of application areas in electrical engineering, mechanical engineering, computer and software engineering; essential for academics, practitioners, researchers, and industry leaders.

Coverage

- Manufacturing Methodologies
- Bioengineering Materials
- Intelligent Navigation
- Autonomous Systems
- Design and Manufacture
- Medical Robotics
- Intelligent Sensing
- Micro and nanomechanics
- Tribology and surface engineering
- Control Methodologies

The Advances in Mechatronics and Mechanical Engineering (AMME) Book Series (ISSN 2328-8205) is published by IGI Global, 701 E. Chocolate Avenue, Hershey, PA 17033-1240, USA, www.igi-global.com. This series is composed of titles available for purchase individually; each title is edited to be contextually exclusive from any other title within the series. For pricing and ordering information please visit http://www.igi-global.com/book-series/advances-mechatronics-mechanical-engineering/73808. Postmaster: Send all address changes to above address.

Titles in this Series

For a list of additional titles in this series, please visit: www.igi-global.com/book-series

Developing Charging Infrastructure and Technologies for Electric Vehicles
Mohammad Saad Alam (Aligarh Muslim University, Aligarh, India) Reji Kumar Pillai (India Smart Grid Forum, India) and N. Murugesan (India Smart Grid Forum, ndia)
Engineering Science Reference • © 2022 • 343pp • H/C (ISBN: 9781799868583) • US $215.00

Multidisciplinary Perspectives on Green Electromobility and Charging Stations
Yu Miao (East China University of Sciencer and Technology, hina)
Engineering Science Reference • © 2022 • 305pp • H/C (ISBN: 9781799839972) • US $225.00

Electric Vehicles and the Future of Energy Efficient Transportation
Umashankar Subramaniam (Prince Sultan University, Saudi Arabia) Sheldon S. Williamson (Ontario Tech University, Canada) Mohan Krishna S. (Alliance University, India) and Febin Daya J. L. (Vellore Institute of Technology, India)
Engineering Science Reference • © 2021 • 293pp • H/C (ISBN: 9781799876267) • US $215.00

Modeling and Optimization of Solar Thermal Systems Emerging Research and Opportunities
Jagadish (National Institute of Technology, Raipur, India) and Agnimitra Biswas (National Institute of Technology, Silchar, India)
Engineering Science Reference • © 2021 • 215pp • H/C (ISBN: 9781799835233) • US $185.00

Recent Trends on Electromagnetic Environmental Effects for Aeronautics and Space Applications
Christos D. Nikolopoulos (Department of Electronic Engineering, School of Engineering, Hellenic Mediterranean University, Greece)
Engineering Science Reference • © 2021 • 285pp • H/C (ISBN: 9781799848790) • US $265.00

Modern Machining Processes and Techniques for Aerospace Materials
Anand Pandey (Manipal University Jaipur, India)
Engineering Science Reference • © 2021 • 300pp • H/C (ISBN: 9781799845102) • US $215.00

Evaluation and Applications for Ultrasonic Methods of Non-Destructive Testing
Ashish Khaira (Samrat Ashok Technological Institute, India)
Engineering Science Reference • © 2021 • 215pp • H/C (ISBN: 9781799845645) • US $195.00

Handbook of Research on Advancements in Manufacturing, Materials, and Mechanical Engineering
Leonid Burstein (Independent Researcher, Israel)
Engineering Science Reference • © 2021 • 462pp • H/C (ISBN: 9781799849391) • US $295.00

IGI Global
PUBLISHER of TIMELY KNOWLEDGE

701 East Chocolate Avenue, Hershey, PA 17033, USA
Tel: 717-533-8845 x100 • Fax: 717-533-8661
E-Mail: cust@igi-global.com • www.igi-global.com

Editorial Advisory Board

Table of Contents

Preface xv

Chapter 1
Development of Computational Approaches in Biomechanics: A Historical Perspective 1
Hrijuta Datta, National Institute of Technology, Agartala, India
Moutoshi Singha Roy, National Institute of Technology, Agartala, India

Chapter 2
Finite Element Analysis in Biomechanics 16
Subhomoy Chatterjee, Schieffelin Institute of Health Research and Leprosy Centre, India

Chapter 3
Design for Additive Manufacturing in Medical Devices 48
Shammodip Roy, Stryker Corporation, USA

Chapter 4
Gait Analysis: An Effective Tool to Measure Human Performance 65
Monisha Gowri S., National Institute of Technology, Rourkela, India
Ravi Kant Avvari, National Institute of Technology, Rourkela, India
Mirza Khalid Baig, National Institute of Technology, Rourkela, India
Thirugnanam Arunachalam, National Institute of Technology, Rourkela, India

Chapter 5
Biomechanical Analysis on Vancouver Periprosthetic Fracture in Femur Using the Finite Element Modeling 88
Raja Dhason, SRM Institute of Science and Technology, India
Sandipan Roy, SRM Institute of Science and Technology, India
Shubhabrata Datta, SRM Institute of Science and Technology, India

Chapter 6
Recent Advancements in Design and Material Perspective of the Implants in the Biomechanics of the Cervical Spine 94
Pechimuthu Susai Manickam, SRM Institute of Science and Technology, India
Balamurugan S., SRM Institute of Science and Technology, India
Sandipan Roy, SRM Institute of Science and Technology, India

Chapter 7
Injury Prevention and Improving the Performance of Athletes 100
Tharani Kumaran, National Institute of Technology, Rourkela, India
Mirza Khalid Baig, National Institute of Technology, Rourkela, India
Ravi Kant Avvari, National Institute of Technology, Rourkela, India
Thirugnanam Arunachalam, National Institute of Technology, Rourkela, India

Chapter 8
Computational Modeling of the Mechanics of Tissue Engineering Fibrous Scaffolds 121
Dhruba Jyoti Mech, Indian Institute of Technology, Hyderabad, India
Mohd Suhail Rizvi, Indian Institute of Technology, Hyderabad, India

Chapter 9
Biomechanics of the Aortic Valve in Health and Disease 137
Thirumalai Deepak, National Institute of Technology, Rourkela, India
Patina Yamini, National Institute of Technology, Rourkela, India
Anju R. Babu, National Institute of Technology, Rourkela, India

Chapter 10
Fluid Mechanics in Arterial Diseases: Computational Study 153
Dipak Kumar Mandal, Department of Mechanical Engineering, College of Engineering and Management, Kolaghat, India
Partha Goswami, Kolaghat Thermal Power Station, West Bengal Power Development Corporation, India
Nirmalendu Biswas, Department of Power Engineering, Jadavpur University, Salt Lake, India
Nirmal K. Manna, Department of Mechanical Engineering, Jadavpur University, Kolkata, India

Chapter 11
Small Intestinal Peristalsis: Biomechanics and Clinical Prominence of Digestion 179
Jetal Bhanarkar, National Institute of Technology, Rourkela, India
Rashi Singh, National Institute of Technology, Rourkela, India
Anupama Usha Rani Siddhanathi, CWS Hospital, India
Ravi Kant Avvari, National Institute of Technology, Rourkela, India

Chapter 12
Computational Study of In-Vitro Ureter Urine Flow in DJ Stent 198
Ranjit Barua, CHST, Indian Institute of Engineering Science and Technology, Shibpur, India
Pallab Datta, National Institute of Pharmaceutical Education and Research, Kolkata, India
Amit Roy Chowdhury, Indian Institute of Engineering Science and Technology, Shibpur, India
Surajit Das, R. G. Kar Medical College and Hospital, India

Chapter 13
Computational FEM Application on Percutaneous Nephrolithotomy (PCNL) Minimum Invasive Surgery Through Needle Insertion Process 210
Ranjit Barua, CHST, Indian Institute of Engineering Science and Technology, Shibpur, India
Surajit Das, R. G. Kar Medical College and Hospital, India
Pallab Datta, National Institute of Pharmaceutical Education and Research, Kolkata, India
Amit RoyChowdhury, Indian Institute of Engineering Science and Technology, Shibpur, India

Chapter 14
Biomechanics of an Avian Flight: Exploration of the Complex Interactions Between the Musculoskeletal, Sensory, and Neural Systems 223
Yugesh Ramdhun, National Institute of Technology, Rourkela, India
Priyobroto Basu, National Institute of Technology, Rourkela, India
Ravi Kant A., National Institute of Technology, Rourkela, India
Thirugnanam A., National Institute of Technology, Rourkela, India

Chapter 15
Viromechanics of Fluid: A Novel Computational Method to Model the COVID-19 Transmission 244
Nima Norouzi, Chongqing Medical University, China

Compilation of References 260

About the Contributors 299

Index 304

Detailed Table of Contents

Preface xv

Chapter 1
Development of Computational Approaches in Biomechanics: A Historical Perspective 1
Hrijuta Datta, National Institute of Technology, Agartala, India
Moutoshi Singha Roy, National Institute of Technology, Agartala, India

Studies involving applications of mechanics in the biological systems to understand different physiological processes date back to ancient times. The development of the subject gained momentum in the medieval and renaissance periods by different stalwarts in the field of science, like Galileo, da Vinci, von Helmholtz, to name a few. However, it is not very long ago that the computational approach was recognized and accepted as the potentially important avenue for exploring biomechanical phenomena. However, as the computers became more and more powerful, it was possible for the scientists and engineers to simulate more complex phenomena having complex three-dimensional geometries, dynamically changing loading conditions, and interaction between multiple phases, including fluid-solid interactions. This chapter gives a detailed description of the evolution of different computational approaches in the field of biomechanics, along with their comparative benefits and shortcomings from a historical perspective with future directions in the research.

Chapter 2
Finite Element Analysis in Biomechanics 16
Subhomoy Chatterjee, Schieffelin Institute of Health Research and Leprosy Centre, India

Finite element method (FEM) is among the methods of approximation, which is abundantly used to solve the real-life problems that cannot be solved with the analytical methods. The biological systems are complex in nature in terms of geometry, material properties, functioning, and case-to-case deviation. Thus, the standard analytical methods suffer catastrophic limitations in this area. Mainly for this reason, biomechanics is highly enriched with the integration of FEM, which opens a new window with extended capabilities against a logically acceptable margin of approximation. This chapter focuses on designing customized medical devices (orthopaedic and dental implants) and correlating their design with the presented clinical condition and, thus, leading to a criteria of design prescription based on clinical conditions.

Chapter 3
Design for Additive Manufacturing in Medical Devices 48
Shammodip Roy, Stryker Corporation, USA

The chapter will cover the fundamentals of product design principles for additive manufacturing in medical devices, with a key focus on orthopedic implants and related devices. The chapter will describe commonly used additive manufacturing processes for medical devices and orthopedics, key design considerations, and how they have opened up possibilities in healthcare. The chapter will detail the fundamentals of design principles, which are expanding the boundaries of rapid, meaningful innovation and positively impacting the innovation cycles of a wide range of industries. It will describe how topology optimization, with the help of computing power, is providing design engineers with the tools to accurately understand structure-functional relationships of designs and in the process re-imagine the biomedical designs of tomorrow.

Chapter 4
Gait Analysis: An Effective Tool to Measure Human Performance 65
Monisha Gowri S., National Institute of Technology, Rourkela, India
Ravi Kant Avvari, National Institute of Technology, Rourkela, India
Mirza Khalid Baig, National Institute of Technology, Rourkela, India
Thirugnanam Arunachalam, National Institute of Technology, Rourkela, India

The impact of the musculoskeletal system on human locomotion is acquired using force platforms, motion analysis systems, and EMG markers. The kinetic and kinematic parameters can be obtained using the force platform, EMG markers, and motion analysis system. The videos are captured using the motion analysis system and analyzed using suitable software. If the typical gait pattern is irregular, it indicates a disorder or abnormality. An abnormal gait cycle or pattern may rise due to joint pain, muscle strain, deformities of bone, weakness, and other impairments in limbs. The gait analysis is used to study various gait abnormalities. The obtained pattern of abnormal subjects is clinically correlated for the assessment of gait disorders. Gait analysis has a wide range of applications like sports for enduring athletes' performance and injury prevention, in rehabilitation, post-surgery analysis, design of orthotics and shoes, and biomechanics studies of astronauts.

Chapter 5
Biomechanical Analysis on Vancouver Periprosthetic Fracture in Femur Using the Finite Element Modeling 88
Raja Dhason, SRM Institute of Science and Technology, India
Sandipan Roy, SRM Institute of Science and Technology, India
Shubhabrata Datta, SRM Institute of Science and Technology, India

After total hip arthroplasty (THA), high demands occur in the femoral fixation, especially for elderly patients due to periprosthetic fractures. Fractures happened in the femur after THA was classified based on Vancouver periprosthetic femoral fracture. Choosing the fixation for these types of specific fracture types are challenging due to pattern and orientation of the fracture in the THA prosthesis. Several researchers have reported the clinical, experimental, and computational studies about the failure of fixation methods, and they highlighted the remedies and scope in the further studies. Most of the authors recommended the computational studies having the advantage to predict the inner behaviour of the bone because invivo/invitro study availability of the specimen are limited and it needs ethical clearance. The current study focussed on computational studies in periprosthetic fractures and aims to discuss the three dimensional model creation, mesh generation, material properties, boundary conditions/loading, limitation, opinions, and future thoughts in implant design.

Chapter 6
Recent Advancements in Design and Material Perspective of the Implants in the Biomechanics of the Cervical Spine 94
Pechimuthu Susai Manickam, SRM Institute of Science and Technology, India
Balamurugan S., SRM Institute of Science and Technology, India
Sandipan Roy, SRM Institute of Science and Technology, India

Understanding biomechanics helps us to understand the problems involved in the cervical spine. The most common problems in the cervical spine are cervical spondylosis, herniated disc, degenerative disc disease, and spinal stenosis. The surgeries to treat these problems are anterior cervical discectomy and fusion surgery, cervical spinal fusion, laminectomy, laminoplasty, and artificial disc replacements. The post-surgery effects of the implants and the surgical studies can be simulated using numerical simulation. In the numerical study, by varying the design of the implants and by varying the material models, the simulation can be conducted and the best comes can be used further for in vitro study. In the in vitro study, the local stress and strains cannot be calculated but the finite element study provides a clear in sight in the stress and strain distribution across the full structure. In summary, the presurgical clinical evaluation can be conducted using the numerical study and the in vitro experiments to understand the biomechanics of cervical spine with implants.

Chapter 7
Injury Prevention and Improving the Performance of Athletes 100
Tharani Kumaran, National Institute of Technology, Rourkela, India
Mirza Khalid Baig, National Institute of Technology, Rourkela, India
Ravi Kant Avvari, National Institute of Technology, Rourkela, India
Thirugnanam Arunachalam, National Institute of Technology, Rourkela, India

Sports biomechanics helps one to understand the sport movements and helps increase the performance of individuals and prevent injury. Understanding the mechanics involved in every sport is crucial for kinesiologists help patients to recover from traumatic or overuse injuries. Each sport activity has different phases, and in each phase, the kinetics and kinematics are studied for better understanding. Computational technologies help to understand sports movements and forces developed during motion and provide proper guidance that can be followed to avoid injuries and enhance performance. The aspects of biomechanical analysis of sports, injuries and their causes in sports, various techniques for recording and analysis of sports movements, and lastly, performance improvement and preventing injury using the analysis are discussed in this chapter.

Chapter 8
Computational Modeling of the Mechanics of Tissue Engineering Fibrous Scaffolds 121
Dhruba Jyoti Mech, Indian Institute of Technology, Hyderabad, India
Mohd Suhail Rizvi, Indian Institute of Technology, Hyderabad, India

Tissue engineering fibrous scaffolds play a crucial role in regenerative medicine. This chapter discusses the underlying mechanics and various approaches presently available to model these scaffolds and their limitations. The fibrous scaffolds are subjected to forces or deformation at two very different length scales (i.e., macroscopic forces during mechanical characterization and microscopic cellular forces). The scaffold behaves very differently under these two loading conditions, and very few computational frameworks capture the true nature of the scaffold under microscopic loading. The authors have also briefly discussed the two different ways a cell can sense the stiffness of the underlying scaffold and how it differs from the macroscopic stiffness.

Chapter 9
Biomechanics of the Aortic Valve in Health and Disease 137
Thirumalai Deepak, National Institute of Technology, Rourkela, India
Patina Yamini, National Institute of Technology, Rourkela, India
Anju R. Babu, National Institute of Technology, Rourkela, India

The aortic valve is composed of collagen, elastin, proteoglycan, valvular interstitial cells (VIC), and valvular endothelial cells (VEC). In the open condition, the aorta valve allows blood to leave the heart, and in the closed condition, it prevents the backflow of the blood to the left ventricle. However, when the aortic valve cups become narrow or thickened, cusp motion is impaired and obstructs the blood flow. This chapter investigates the structure and composition of the aortic valve cusp and the role of VIC, VEC, and cross-talk of VEC-VIC. In addition, biomechanical characterization of the aortic cusps such as uniaxial, biaxial, flexure, three-point bending, cantilever bending, and viscoelasticity was discussed. Furthermore, etiology, in vitro cell culture and in vivo animal models, and ex vivo models mimicking aortic stenosis and regurgitation were summarized.

Chapter 10
Fluid Mechanics in Arterial Diseases: Computational Study 153
Dipak Kumar Mandal, Department of Mechanical Engineering, College of Engineering and Management, Kolaghat, India
Partha Goswami, Kolaghat Thermal Power Station, West Bengal Power Development Corporation, India
Nirmalendu Biswas, Department of Power Engineering, Jadavpur University, Salt Lake, India
Nirmal K. Manna, Department of Mechanical Engineering, Jadavpur University, Kolkata, India

With the advancement of computer power, computational solution is considered as a useful predictive tool for study of the blood flow through a diseased artery. The blood flow is governed by the continuity and Navier-Stokes equations. Womersley number and Reynolds number are physiologically important non-dimensional flow parameters and have existence in the non-dimensional governing equations. In the computational solution, the mathematical flow modelled is established considering some boundary conditions, and then it is solved by numerical simulations. The governing equations are converted to discretised form and then solved numerically by readily available commercial CFD software and/or by in-house developed CFD code using appropriate algorithms. Grid independence test and validation of CFD model are the crucial parts of computational solutions. The chapter delivers the knowledge on impact of fluid mechanics on arterial diseases and computational solution techniques.

Chapter 11
Small Intestinal Peristalsis: Biomechanics and Clinical Prominence of Digestion 179
Jetal Bhanarkar, National Institute of Technology, Rourkela, India
Rashi Singh, National Institute of Technology, Rourkela, India
Anupama Usha Rani Siddhanathi, CWS Hospital, India
Ravi Kant Avvari, National Institute of Technology, Rourkela, India

Digestion of food forms an essential process in a living organism. Digestion of food can be classified into two categories: (1) mechanical digestion and (2) chemical digestion. Whereas mechanical digestion contributes to the physical breakdown of food using mechanical forces, chemical digestion contributes via breaking down the chemical bonds present in the food. The process is manifested by muscular contraction. The intestine develops undulations via contraction of the muscle fiber that involves contraction and relaxation of muscle fibers (referred to as the peristalsis). As a consequence, the momentum generated by the muscular contraction serves to develop fluid motion in the lumen, which eventually leads to mixing and transport. Whereas the bodily secretions of the digestive enzymes participate in innate digestion, the gut flora, on the other hand, also influence digestion to an extent by means of fermentation. Small intestinal peristalsis serves as the key mediator, whether the digestion proceeds normally during physiology or abnormally leading to pathology.

Chapter 12
Computational Study of In-Vitro Ureter Urine Flow in DJ Stent 198
Ranjit Barua, CHST, Indian Institute of Engineering Science and Technology, Shibpur, India
Pallab Datta, National Institute of Pharmaceutical Education and Research, Kolkata, India
Amit Roy Chowdhury, Indian Institute of Engineering Science and Technology, Shibpur, India
Surajit Das, R. G. Kar Medical College and Hospital, India

Many researchers and urologists are presently studying different designs of ureteral stents to advance the feature of their surgeries and the succeeding recovery of the patient. With the aim of help during this design procedure, several computational models have been established to simulate the performance of various biological tissues and deliver an accurate computational environment to estimate the stents. As a result of the high difficulty of the complicated issues, they generally introduce interpretations to create these simulations a smaller amount computationally trying. A DJ stent (double J) is used to improve the blocking of urine in the upper urinary tract while there is ureteral stenosis, which causes the disturbance of normal urine flow and affects renal or kidney failure. The intention of employing a DJ stent is to confirm enough urine flow in the ureter, but the DJ stent performs as a foreign body in the urinary tract and sometimes acts as a difficulty in achieving satisfactory urine flow.

Chapter 13
Computational FEM Application on Percutaneous Nephrolithotomy (PCNL) Minimum Invasive Surgery Through Needle Insertion Process 210
Ranjit Barua, CHST, Indian Institute of Engineering Science and Technology, Shibpur, India
Surajit Das, R. G. Kar Medical College and Hospital, India
Pallab Datta, National Institute of Pharmaceutical Education and Research, Kolkata, India
Amit RoyChowdhury, Indian Institute of Engineering Science and Technology, Shibpur, India

PCNL or percutaneous nephrolithotripsy is one of the foremost interventional surgical treatment modalities for big kidney stones, which are more than two centimeters in diameter. With the application of miniaturized procedures, the signs for percutaneous nephrolithotripsy have been increased to smaller renal stones. Particularly for urologist surgeons without an accent on endourology, it is challenging to indicate the developing multitude of existing methods and procedures. Several makers have established different percutaneous nephrolithotripsy methods with changing diameters and different features. The suggestions for the dissimilar methods are intersecting. Reflective studies presented decreased complication rates. This chapter defines the presently available methods for percutaneous stone surgical treatment with their particularities and suggestions and studies the steering of a surgical flexible needle into the kidney to take out the stone by this procedure and finite element model analysis force of surgical needle and deformation of kidney tissue model.

Chapter 14
Biomechanics of an Avian Flight: Exploration of the Complex Interactions Between the Musculoskeletal, Sensory, and Neural Systems 223
Yugesh Ramdhun, National Institute of Technology, Rourkela, India
Priyobroto Basu, National Institute of Technology, Rourkela, India
Ravi Kant A., National Institute of Technology, Rourkela, India
Thirugnanam A., National Institute of Technology, Rourkela, India

Birds have always fascinated scientists and opened their eyes to new areas of flight mechanisms through biomimicry of these flyers. These flyers can sustain and control the flight through clever maneuvering by flapping their wings. It involves an intricate aerodynamic force production to generate sufficient lift force to overcome the drag and perform useful maneuvers. Development of the forces differs widely among the species due to natural selection of the flyers. As far as the flapping mechanism is considered, it is very efficient. The differences in flight mechanisms may be explored by determining the kinematics, kinetics, and aerodynamics of the flyers. Wing kinematics determines the aerodynamic forces, which vary with flapping speed. The maneuverability and stability are regulated by complex muscle action and neural control allowing the flyer to perform specific tasks. Computational models have emerged as powerful tools to predict the flow around the flyers with potential exploration of the complex interactions between the skeletal system, sensory system, and neural control of the flight.

Chapter 15
Viromechanics of Fluid: A Novel Computational Method to Model the COVID-19 Transmission 244
Nima Norouzi, Chongqing Medical University, China

This study deals with estimating the trajectory of COVID-19 coronavirus adhering to horizontally projected respiratory droplets, considering the geographic altitude. The size of the viruses and respiratory droplets are the factors that determine the trajectory of the microparticles in a viscous medium such as air. For this purpose, a graphic comparison of the diameters and masses of the microparticles that are produced in respiratory activity has been made. The estimation of the vertical movement of the microparticles through the air is based on Stokes' law, and it was determined that respiratory droplets smaller than 10μm in diameter have very small terminal velocities; in practice, they are floating for brief seconds before evaporating in the air. Regarding the horizontal displacement of respiratory droplets, frames from Beggs determine its scope.

Compilation of References 260

About the Contributors 299

Index 304

Preface

Mechanics is a subject which is pretty close to my heart and I am enjoying teaching this subject to the budding engineers for more than last two decades. My passion towards this fundamental subject, common to almost all engineering disciplines, grew with time and it made me inclined towards its minute intricacies, manifested in subjects like fracture mechanics or biomechanics.

Out of them, Biomechanics is a very special one where I had the chance to learn a lot of fascinating things from practical experience and through various experiments. I really found it interesting and challenging to correlate the results found in the experiments with the theories. But, with the passage of time, I felt that the laboratory simulation in experiments was often too difficult and complicated. And despite being an experimentalist, I had to take the help of computational approach to solve lot of problems. To be honest, be it solid Biomechanics or fluid, the efficacy of the computational techniques has often proved to be handy, realistic and it enhanced the theoretical understanding as well. But challenges are still there. The use of standard codes or writing new codes are still the most preferred approaches, rather than addressing the complicated and real-life bio-physics puzzles. Here lies the scope of new venture in the domain of Biomechanics.

The present compilation is a special one where special care has been taken to strike a proper balance between the two approaches. It presents a well thought of sequence of articles catering the academic interest of a wide variety of people, students as well as practising professionals.

The organisation the book, the presentation of the problem and solution, the correlation with real life scenario is very good. I have no doubt that it is going to be a very useful book with a great potential to offer exceptional exposure for all concerned in the domain of biomechanics.

Chapter 1: Studies involving applications of mechanics in the biological systems to understand different physiological processes dates back to ancient period. The development of the subject gained momentum in the medieval and renaissance period by different stalwarts in the field of science, like Galileo, da Vinci, von Helmholtz to name a few. However, it is not very long ago when the computational approach, was recognized and accepted as the potentially important avenue for exploring biomechanical phenomena. However, as the computers became more and more powerful, it was possible for the scientists and engineers to simulate more complex phenomena having complex three-dimensional geometries, dynamically changing loading conditions and interaction between multiple phases, including fluid-solid interactions. This book chapter gives a detailed description of the evolution of different computational approaches in the field of biomechanics, along with their comparative benefits and shortcomings from a historical perspective with future direction in the research.

Chapter 2: Finite element method (FEM) is among the methods of approximation which is abundantly used to solve the real life problems those cannot be solved with the analytical methods. The biological systems are complex in nature in terms of geometry, material properties, functioning and case to case deviation. Thus the standard analytical methods suffer catastrophic limitations in this area. Mainly for this reason Biomechanics is highly enriched with the integration of FEM which opens a new window with extended capabilities against a logically acceptable margin of approximation. This chapter focuses on designing of customized medical devices (orthopaedic and dental implants) and correlating their design with the presented clinical condition and, thus, leading to a criteria of design prescription based on clinical condition.

Chapter 3: The proposed chapter will cover the fundamentals of product design principles for additive manufacturing in medical devices, with a key focus on orthopedic implants and related devices. The chapter will describe commonly used additive manufacturing processes for medical devices and orthopedics, key design considerations, and how that has opened up possibilities in healthcare. The chapter will detail the fundamentals of design principles, which are expanding the boundaries of rapid, meaningful innovation and positively impacting the innovation cycles of a wide range of industries. It will describe how topology optimization, with the help of computing power, is providing design engineers the tools to accurately understand structure-functional relationships of designs, and in the process, re-imagine the biomedical designs of tomorrow.

Chapter 4: The impact of the musculoskeletal system on human locomotion is acquired using force platforms, motion analysis systems and EMG markers. The kinetic and kinematic parameters can be obtained using the force platform, EMG markers and motion analysis system. The videos are captured using the motion analysis system and analyzed using suitable software. If the typical gait pattern is irregular, it indicates a disorder or abnormality. An abnormal gait cycle or pattern may rise due to joint pain, muscle strain, deformities of bone, weakness and other impairments in limbs. The gait analysis is used to study various gait abnormalities. The obtained pattern of abnormal subjects is clinically correlated for the assessment of gait disorder. Gait analysis has a wide range of applications like sports for enduring athletes' performance and injury prevention, in rehabilitation, post-surgery analysis, design of orthotics and shoes, and biomechanics studies of astronauts.

Chapter 5: After Total Hip Arthroplasty (THA), high demands of problem occurs in the femoral fixation especially for elderly patients due to periprosthetic fractures. Fractures happened in the femur after THA was classified based on vancouver periprosthetic femoral fracture. Choosing the fixation for these types of specific fracture types are challenging due to pattern and orientation of the fracture in the THA prosthesis. Several researchers are reported the clinical, experimental and computational studies about the failure of fixation methods and they highlighted the remedies and scope in the further studies. Most of the authors recommended the computational studies having the advantage to predict the inner behaviour of the bone because invivo/invitro studies availability of the specimen are limited and it needs ethical clearance. The current study focussed on computational studies in periprosthetic fractures aims to discuss the three dimensional model creation, mesh generation, material properties, boundary conditions/loading, limitation, opinions and future thoughts in implant design

Chapter 6: The understanding in the biomechanics helps us to understand the problems involved in the cervical spine. The most common problems in the cervical spine are cervical spondylosis, herniated disc, degenerative disc disease and spinal stenosis. The surgeries to treat these problems are anterior cervical discectomy and fusion surgery, cervical spinal fusion, laminectomy, laminoplasty and artificial disc replacements. The post-surgery effects of the implants and the surgical studies can be simulated using numerical simulation. In the numerical study by varying the design of the implants and by varying the material models the simulation can be conducted and the best comes can be used further for in vitro study. In the in vitro study the local stress and strains can not be calculated but the finite element study provides a clear in sight in the stress and strain distribution across the full structure. In summary the pre surgical clinical evaluation can be conducted using the numerical study and the in vitro experiments to understand the biomechanics of cervical spine with implants.

Chapter 7: Sports biomechanics helps to understand the sport movements and helps to increase the performance of individuals, and prevents injury. Understanding the mechanics involved in every sport is crucial for kinesiologists to recover from traumatic or overuse injuries. Each sport activity has different phases, and in each phase, the kinetics and kinematics are studied for better understanding. Computational technologies help to understand sports movements and forces developed during motion and provide proper guidance that can be followed to avoid injuries and enhance performance. The aspects of biomechanical analysis of sports, injuries and their causes in sports, various techniques for recording and analysis of sports movements and lastly, performance improvement and preventing injury using the analysis are discussed in this chapter.

Chapter 8: Tissue engineering fibrous scaffolds play a crucial role in regenerative medicine. This chapter discusses the underlying mechanics and various approaches presently available to model these scaffolds and their limitations. The fibrous scaffolds are subjected to forces or deformation at two very different length scales, i.e., macroscopic forces during mechanical characterization and microscopic cellular forces. The scaffold behaves very differently under these two loading conditions and very few computational frameworks capture the true nature of the scaffold under microscopic loading. The authors have also briefly discussed about the two different ways how a cell can sense the stiffness of the underlying scaffold and how it differs from the macroscopic stiffness.

Chapter 9: The aortic valve is composed of collagen, elastin, proteoglycan, valvular interstitial cells (VIC), and valvular endothelial cells (VEC). In the open condition, the aorta valve allows blood to leave the heart, and in the closed condition, it prevents the backflow of the blood to the left ventricle. However, when the aortic valve cups become narrow or thickened, cusp motion is impaired and obstructs the blood flow. This chapter investigates the structure and composition of the aortic valve cusp and the role of VIC, VEC, and cross-talk of VEC-VIC. In addition, biomechanical characterization of the aortic cusps such as uniaxial, biaxial, flexure, three-point bending, cantilever bending, and viscoelasticity was discussed. Furthermore, etiology, in vitro cell culture and in vivo animal models, and ex vivo models mimicking aortic stenosis and regurgitation were summarized.

Chapter 10: With the advancement of computer power, computational solution is considered as a useful predictive tool for study of the blood flow through diseased artery. The blood flow is governed by the continuity and Navier-Stokes equations. Womersley number and Reynolds number are physiologically important non-dimensional flow parameters, have existence in the non-dimensional governing equations. In the computational solution, the mathematical flow modelled is established considering some boundary conditions, and then it is solved by numerical simulations. The governing equations are converted to discretised form and then solved numerically by readily available commercial CFD software and or by in-house developed CFD code using different algorithms. Grid independence test and validation of CFD model are the crucial parts of computational solution. The present chapter delivers the knowledge on impact of fluid mechanics on arterial diseases and computational solution techniques.

Chapter 11: Digestion of food forms an essential process in a living organism. Digestion of food can be classified into two categories; (i) mechanical digestion and (ii) chemical digestion. Whereas mechanical digestion contributes to the physical breakdown of food using mechanical forces, chemical digestion contributes via breaking down the chemical bonds present in the food. The process is manifested by muscular contraction. The intestine develops undulations via contraction of the muscle fiber that involves contraction and relaxation of muscle fibers (referred to as the peristalsis). As a consequence, the momentum generated by the muscular contraction serves to develop fluid motion in the lumen, which eventually leads to mixing and transport. Whereas the bodily secretions of the digestive enzymes participate in innate digestion, the gut flora, on the other hand, also influence digestion to an extent by means of fermentation. Small intestinal peristalsis serves as the key mediator, whether the digestion proceeds normally during physiology or abnormal leading to pathology.

Chapter 12: Many researchers and urologists are presently studying different designs of ureteral stents to advance the feature of their surgeries and the succeeding recovery of the patient. With the aim of help during this design procedure, several computational models have been established to simulate the performance of various biological tissues and deliver an accurate computational environment to estimate the stents. Though, as a result of the high difficulty of the complicated issues, they generally introduce interpretations to create these simulations a smaller amount computationally trying. A DJ stent (double J) is used to improve the blocking of urine in the upper urinary tract while there is ureteral stenosis, which causes the disturbance of normal urine flow and affects renal or kidney failure. The intention of employing a DJ stent is to confirm enough urine flow in the ureter, but the DJ stent performs as a foreign body in the urinary tract and sometimes acts as a difficulty in achieving satisfactory urine flow.

Chapter 13: PCNL or Percutaneous nephrolithotripsy is one of the foremost interventional surgical treatment modalities for big kidney stones, which is more than two centimeters in diameter. With the application of miniaturized procedures, the signs for percutaneous nephrolithotripsy have been increased to smaller renal stones. Particularly for urologists' surgeons without an accent on endourology, it is challenging to indicate the developing multitude of existing methods and procedures. Several makers have established different percutaneous nephrolithotripsy methods with changing diameters and different features. The suggestions for the dissimilar methods are intersecting. Reflective studies presented decreased complication rates. This chapter defines the presently available methods for percutaneous stone surgical treatment with their particularities and suggestions and studies the steering of a surgical flexible needle into the kidney to take out the stone by this procedure and finite element model analysis force of surgical needle and deformation of kidney tissue model.

Chapter 14: Birds have always fascinated scientists to a whole new area of flight mechanism; through bio-mimicry of these flyers. These flyers can sustain and control the flight through clever maneuvering by flapping their wings. It involves an intricate aerodynamic force production to generate sufficient lift force to overcome the drag and perform useful maneuvers. Development of the forces differs widely among the species due to natural selection of the flyers. As far as the flapping mechanism is considered, it is very efficient. The differences in flight mechanisms may be explored by determining the kinematics, kinetics and aerodynamics of the flyers. Wing kinematics determines the aerodynamic forces which vary with flapping speed. The maneuverability and stability are regulated by complex muscle action and neural control; allowing the flyer to perform specific tasks. Computational models have emerged as a powerful tool to predict the flow around the flyers with potential exploration of the complex interactions between the skeletal system, sensory system and neural control of the flight.

Chapter 15: This study deals with estimating the trajectory of COVID-19 coronavirus adhering to horizontally projected respiratory droplets, considering the geographic altitude. The size of the viruses and respiratory droplets is the factor that determines the trajectory of the microparticles in a viscous medium such as air; For this purpose, a graphic comparison of the diameters and masses of the microparticles that are produced in respiratory activity has been made. The estimation of the vertical movement of the microparticles through the air is based on Stokes' Law, and it was determined that respiratory droplets smaller than 10μm in diameter have very small terminal velocities; in practice, they are floating for brief seconds before evaporating in the air; Regarding the horizontal displacement of respiratory droplets, frames from Beggs to determine its scope.

I feel honoured to write few words about this valuable endeavour. I wish all sorts of success
Best wishes

Abhijit Chanda
Jadavpur University, India

Chapter 1
Development of Computational Approaches in Biomechanics:
A Historical Perspective

Hrijuta Datta
National Institute of Technology, Agartala, India

Moutoshi Singha Roy
National Institute of Technology, Agartala, India

ABSTRACT

Studies involving applications of mechanics in the biological systems to understand different physiological processes date back to ancient times. The development of the subject gained momentum in the medieval and renaissance periods by different stalwarts in the field of science, like Galileo, da Vinci, von Helmholtz, to name a few. However, it is not very long ago that the computational approach was recognized and accepted as the potentially important avenue for exploring biomechanical phenomena. However, as the computers became more and more powerful, it was possible for the scientists and engineers to simulate more complex phenomena having complex three-dimensional geometries, dynamically changing loading conditions, and interaction between multiple phases, including fluid-solid interactions. This chapter gives a detailed description of the evolution of different computational approaches in the field of biomechanics, along with their comparative benefits and shortcomings from a historical perspective with future directions in the research.

INTRODUCTION

Biomechanics is the study of why and how a living thing moves through a certain environment in order to complete a task. It is focused on the laws of mechanics applied to different biological systems including human body. Humans, like all other life forms on the earth, are constantly subjected to the universal force of gravitation, as well as internal and external influences. From a very general perspective of life every human being desires to improve their quality of lifestyle. The structure, function, and mobility of

DOI: 10.4018/978-1-7998-9078-2.ch001

the human body may be analyzed, and the resultant knowledge can be utilized to assess and improve quality of life, by studying the interplay of these forces and their consequences in health and disease. The musculoskeletal system is responsible for the weight bearing and mobility of the human body, which in turn is regulated by a sophisticated neurological system that maintains a highly coordinated mechanical contact that exists between the bones, muscles, ligaments, and joints of the musculoskeletal system. Any kind of injury, bruises, lesion, degeneration or deformity, caused to any part of the musculoskeletal system will change the entire mechanical interaction, leading to inefficient movement, instability, and loss of normal ranges of motion for the joints. Proper understanding of the mechanical environment and its modification, manipulation, and control, on the other hand, can help prevent injury, correct irregularities, and speed up healing and rehabilitation and improve performance in normal and sporting activities. As a result, using different analysis techniques to understand the biomechanics of human movement is useful for determining root cause of a disease, formulating therapy recommendations, and assessing treatment outcomes and performance of the sportspersons.

Biomechanics is broadly divided as

1. Kinematics
   a. Quantitative analysis
   b. Qualitative analysis
2. Kinetics
   a. Statics
   b. Dynamics
3. Anthropometry

The human body consists of numerous joints and muscles that are necessary for performing daily routine functions. As a result, numerous musculoskeletal movements might be used in coordination to complete a particular type of movement. A cognitive process used to compensate for a cognitive impairment. Someone, who is worse in terms of spatial than that of verbal abilities, for example, may employ compensating methods to address spatial challenges, such as mentally rotating a geometric figure using linguistic processes. While this compensatory mechanism is crucial for dealing with the effects of musculoskeletal injuries or diseases, it makes non-invasive assessment of internal forces challenging. Anatomical modelling and noninvasive movement data, such as segment position and strain on force-measuring devices, are being combined to produce reliable findings for estimating the physiological movements.

Computational biomechanics is a promising topic of study that analyzes different body parts ranging from the orthopedic ones like the foot, ankle, knee, hip, lower limb, spine, head, and teeth at the tissue level, as well as soft tissue organs. (Sweeting & Mock, 2021) One of the key areas where computational methods where employed is orthopedic biomechanics. It provides the information on the complex biomechanical characteristics of normal and abnormal human joints in order to bring innovative orthopedic therapy and rehabilitation strategies to light. The insight thus obtained allows the surgeons to map and perform surgical operations with more precision and with better outcome. Computer-assisted surgical systems have the potential to enhance clinical outcomes and healthcare delivery efficiency. It also aids in the establishment of a foundation for computer-integrated medicine by retrieving clinically relevant data on the physical condition of the underlying biology at the cell, tissue, organ, and system levels.

HISTORY OF BIOMECHANICS

The study of human motion using the laws of mechanics started since the ancient time. Biomechanics has its origins in a scientific concern with the human body and its anatomy. Papyri from Egypt dated from 1700-1600 BC, such as the Edwin Smith and the Ebers papyri, show this kind of curiosity. Embalming procedures needed knowledge of anatomical structures, although this field was only emerging at the time. Hippocrates (460-377 BC) founded modern medicine in the 4th century BC, despite the fact that anatomical and physiological information was limited due to the human body's inability to be dissected. Hippocrates, on the other hand, treated diseases and joint problems using logic and reasoning. (Innocenti, 2018) To Aristotle, projectile motion and trajectory attracted him, making him recognized as the "Father of Kinesiology". He has also written a book named 'About the movement of Animals'. Kinesiology is the scientific study of movement of human beings in a safe and secure mannered in order to accomplish the task effectively. Late 1800s to 1900s is designated as the kinesiology era. The concept of biomechanics emerged from physical education as a specialized area of research in the mid-1960s and 1970s. The most fascinating era of biomechanics is in the mid 20th century and since then till now there is continuous development in the field of biomechanics at a rapid pace never seen before.

Some of the key scientists and their contribution in the field of biomechanics is briefed below:

1. In the second century, Galen of Pergamon (130-201 or 216), the Roman emperor Marc Aurèle's doctor, circumvented the prohibition on human dissection by examining animal anatomy, particularly the anatomy of the maggot monkey. This frequently erroneous anatomical information would remain constant for over 1300 years, until the Flemish doctor Andreas Vesalius (1514-1564) performed human dissections, which would revolutionize human anatomy and science.
2. Leonardo da Vinci (1452-1519) was one of the first to dissect human beings. He is responsible for many anatomical descriptions of bones, joints, and muscles. In his Codex Atlanticus essay, Leonardo da Vinci detailed the mechanics of human movement in various planes of space for the first time.
3. Galileo elevates mechanics to the status of a science (1564-1642). Indeed, he was enthralled by mechanics and movement. He studied medicine and science, and his work on the center of gravity resulted in the validation of several theorems. He was also enthralled by the pendulum and the fall of bodies.
4. Giovanni Alfonso Borelli (1608-1679) made essential attempts at scientific analysis of living beings' movement in space i.e. locomotion using Galileo's theory of mechanics. He compared a man's movement to that of a miniature boat and its rower in his work De motu animalium (1679), noticing parallels between the foot's support on the ground and the support of the shovel paddle in the water. In the second phase of his study, he attempts to characterize the internal forces, or muscle contractions. Borelli studied mobility on land, in the water, and in the air in walking animals, swimming fish, and flying birds.
5. When Eadweard Muybridge (1830-1904) took the first series of images of a movement in space, the scholarly interest in human mobility has developed. At the time, there was debate over whether the horse's four legs could be in the air at the same moment during a gallop. Muybridge had 12 cameras in a line in 1878. Each camera is fired past by a galloping horse. This initial set of images demonstrated that the horse's four legs are all in the air at the same time.

6. In 1881, Muybridge was approached by the French scientist Étienne-Jules Marey (1830-1904), who was interested in investigating the mechanics of flying birds. At the time, France was the epicentre of scientific research impact in different fields. The group was led by Professor E.J. Marey of the Collège de France. Among the scientists who participated in this group were Carlet, Demeny, Pages, Marey (1872), Carlet (1872) and Mare (1873, 1874).
7. Guillaume-Benjamin Duchenne (1867, 1873) and Adolf Fick (1860, 1866) have made substantial contributions to a better knowledge of the muscular and articular systems.
8. Christian Wilhelm Braune (1831-1892) and Otto Fischer were influenced by Marey's work (1861-1917). After Braune's demise, Fischer continued the research. He used four chronophotographic devices to improve Marey's movement study approach. While researching walking, experiments and data analysis were more precise, and the results were more significant. He arrived to the conclusion that when walking, the lower limb does not behave like a pure pendulum, but rather is influenced by the muscle forces.
9. Rudolf Fick, the son of Adolf Fick and a pupil of Otto Fischer, is the author of the early twentieth-century anatomy book Manual of Anatomy and Mechanics of the Joints. In the three parts that make up the book, each muscle and joint is methodically covered. At this time, Jules Amar's (1879-1935) work, which related articular motion theories to human physiology for the rehabilitation of amputee patients in need of prosthesis, was very crucial. Amar invented the "dynamographic sidewalk," which measures patient pressures on the ground and allows him to personalize prosthesis for them. This equipment gave rise to the force plate, which can now be found in almost every biomechanics laboratory. During First World War, the number of cases of lower and upper limbs amputations increased considerably. As a result, a number of scholars (including Mommsen (1918), Shede (1918), Bloch (1919), Schmetz (1921), Verth (1927), and others) focused their efforts on the study of movement and the development of prostheses.
10. Ivan Sechenov (1829-1905), a physiologist, and Peter Lesgaft, a physicist and anatomist, were among the first to establish biomechanics in Russia (1837-1909). The international face of Russian biomechanics is Nikolai Bernstein (1896-1966), a neurophysiologist with a background in mechanics and mathematics. The purpose of this scientist and his colleagues' research was to improve workers' performance by studying human mobility. The term "biomechanics" was coined by Bernstein to characterize the application of mechanical principles to the study of motion.
11. The Movements of the Human Body, a landmark study on sports biomechanics, was published in 1938 by K. Michael Ivanitski (1895-1969). Over 100 research publications on the functional anatomy of movement in relation to physical education and sports have been published by him. Among the Russian scientists of the mid-twentieth century, we must highlight Lev Nikolaev (1898-1954), whose work Guide of Biomechanics Applied to Orthopedics, Traumatology, and Prosthesis (1947-1950) proved his expertise gained during WWII. Following WWII, experimental biomechanics research in Germany was effectively ceased, while the rest of Europe was badly damaged. For obvious reasons, only work on assisting millions of invalids through the development of prostheses, orthoses, and orthopedic research was supported. After that, the scientific epicenter migrated to North America. Near the close of the twentieth century, movement analysis sciences resurfaced in Europe and Asia.
12. Basler, a German physicist, focused on mobility in the 1930s. He was particularly interested in the human body's center of gravity. He created a specific dynamometer to examine the response forces of the foot on the ground.

13. Various scientific studies undertaken by American scientist Elftman in a Colombian institution took place between 1938 and 1943. He investigated the distribution of mass at the feet, the role of the arms while walking, body rotations, and ground reaction forces, among other things. His name is most known for designing the first force plate, which was published in the prestigious scientific magazine Science in 1938.
14. Scherb, a Swiss scientist, published his studies on muscular activation in the 1940s. When he walked on a treadmill, several muscles' electrical potential was recorded by him. Myokinesiology was the name he gave to his technique. He used his findings to rule out potential muscle issues and undertake post-transplantation controls. He was one of the first scientists to promote the ;'theory that the neuromuscular approach for autonomic actions like walking is acquired through experience and is deeply recorded for the rest of one's life.

APPROACHES IN BIOMECHANICS

Conceptual, physical, and mathematical models have all been demonstrated to be successful in biomechanics. Only a few occasions have conceptual models been used to elucidate a notion without requiring physical construction or mathematical investigation. Models that represent many body segments and muscles, as well as sophisticated bone shapes, range in complexity from the extreme simplicity of some walking and running models to the complexity of models that represent multiple body segments and muscles. The more straightforward the model, the more evident which features are required for the estimated outcome. Computer models have been increasingly popular in the field of biomechanics during the last decade, thanks to fast expanding computer power. Multi-body models and numerical models are the two most common types of biomechanical computer models. As a medical imaging tool, mesh creation and material modeling are important to accomplish the purpose of individualizing the model. Possibly on parallel computer architectures, this chain of tools is necessary. (Alexander, 2003)

Biomechanics has one fundamental limitation: due to ethical issues, the forces and pressures inside the human body are notoriously difficult to quantify. Since the body and environment are intertwined, biomechanics measurement technology tends to concentrate on this interaction. Electromyography (EMG) measures the action potentials of contracting muscles using electrodes attached to the skin, whereas force platforms dynamically calculate reaction forces when a person walks or runs across the sensors. Different models in computational biomechanics are as follows:

Conceptual Models

In biomechanics, conceptual models that do not require mathematical analysis are utilized only infrequently. It could be a system for assisting individuals in learning about, understanding, or simulating the subject the model represents. Conceptual models are representations of real-world objects.

Physical Models

In biomechanics, physical models have been proved to be useful for a variety of applications. For starters, a physical model can demonstrate how a suggested method actually works. Observations that would have been difficult to make on live animals or plants in the past have been made using physical models. Stress patterns in irregularly formed bones were studied using photoelastic models.

The Finite Element Analysis

The finite element method (FEM) is a prominent numerical analysis approach widely used for obtaining approximate solutions to a wide variety of engineering problems by solving systems of partial differential equations (PDEs) on a computer in different fields of engineering etc. It has been emerged as a sophisticated computer methodology for structural stress analysis established in engineering mechanics. It is a recent method for numerical stress analysis that has the benefit of being adaptable to irregularly shaped objects with heterogeneous material characteristics. Such numerical tools may lead to a better understanding of particular tissues' responses and interactions.

In 1972, the FEM was used for the first time in biomechanics. The basic idea behind FEM is that a solution region has to be designed analytically with the aid of some discrete elements, the method popularly known as discretization. In other way, this is of utmost use where the user needs to choose a physical situation that may be described by a system of partial differential equations with complex geometry (e.g. the human head or knee), an analytical (mathematically correct) solution is impossible because of the complex geometry.

The general procedure of FEM can be summarized to have the following steps:

- First of all, the object of interest has to be chosen, and a discrete representation or discretization of the domain or the object of interest has to be generated.
- The physical behavior of the object of interest is to be defined by selecting the appropriate interpolation functions.
- All the systems of equations are formulated accordingly, and stiffness matrix (The mechanical qualities of the underlying structure are essentially represented by a stiffness matrix) is set up.
- Solutions to all the system of equations are computed for the nodes and are interpolated across the elements. It makes the problem's numerical solution ready.

The application of FEM in the orthopedic as well as dentistry is undeniably vast. In 1972, the finite element method (FEM) was brought to orthopedic biomechanics to assess stresses in human bones. Since then, this approach has been used more often for stress assessments of bone and bone-prosthesis systems, fracture fixation devices, and numerous tissues other than bone.

Although the FEM yielded a limited number of noteworthy results and helpful concepts during its first decade of application in this sector, numerous articles served to demonstrate its possibilities and limits. The approach is now well established as a tool for fundamental research and design analysis in orthopedic biomechanics, and the number of publications that employ it is fast growing. Meanwhile, as advanced computing system evolved, the method's capabilities are expanded, along with the rising sophistication of computers, ensuring exciting future possibilities. However, as the biological structures involved, as well as the clinical difficulties, are complicated, scientific development in this domain necessitates a solid grasp of engineering mechanics on the one hand, a great comprehension of mathematics on the other, coupled with a sound knowledge of the biological systems where the result of the analysis belongs to.

Stress analysis of dental structures has been a popular topic in recent years, with the goal of identifying stresses in dental structures and improving their mechanical strength. The mouth cavity, as we all know, is a complicated biomechanical system with restricted access. As a result, the majority of biomechanical research in the oral environment, such as orthodontics, restorative dentistry, endodontics, prosthodontics, and so on, has been done in vitro. Apart from the orthodontics, FEM is successfully applied to different weight bearing joints of the human body, such as hip, knee and shoulder joints as well as different segments of spine.

THE EVOLUTIONARY FRAMEWORK IN COMPUTATIONAL BIOMECHANICS

Computational methods and tools in biomechanics are used for the analysis of mechanical interactions that have naturally transferred from traditional engineering fields to the study of the cell, tissue, and organ biomechanics. Models that describe and predict behavior have become increasingly complicated, despite the availability of current computational tools, as it aims to replicate the complex mechanical behavior of the biological systems. Despite the fact that these models have yielded sufficiently good results in predicting the behaviour of the system under inspection, their utility is occasionally questioned. The validation and verification of models are insufficient, which has led to criticism. So, different strategies are employed to examine and validate the model and explain the principles of verification, validation, and sensitivity studies. In computational biomechanics, model analyses and interpretations are the key. Real-world examples are provided for a better understanding of evolution. This chapter provides a drone view of the computational methods that has been evolved with the growth of digitization. The evolution of computational biomechanics has been strategized on the basis of physiological parameters and biological systems. Here we have categorized it as follows:

MAJOR AREAS OF RESEARCH ASSOCIATED WITH COMPUTATIONAL BIOMECHANICS

- Musculoskeletal Biomechanics
- Orthopaedic Biomechanics
- Cardiovascular Biomechanics
- Injury Biomechanics
- Sports Biomechanics
- Cellular Biomechanics
- Rehabilitation Biomechanics

Section I: Musculoskeletal Biomechanics

Musculoskeletal biomechanics study involving computational approaches is one of the prominent areas of biomechanical studies. The computational techniques can be used to assess the gait patterns of different subjects, joint forces and range of motions in health and diseased conditions, as well as understanding the internal mechanical environments of the musculoskeletal systems using numerical techniques like FEA.

1. **Computational Modeling of Trabecular Bone Mechanics:** The porous structure of the trabecular bone contains bone marrow, which provides a particular habitat for a range of cell types, including

mesenchymal stem cells (MSCs). The fact that marrow possesses the properties of a fluid while being surrounded by bone that is prone to deformation has prevented prior experimental and computational investigations from accurately representing the complex mechanical environment imposed on MSCs. As a result, past experimental and computational studies have been unable to adequately depict the complex mechanical environment in the trabecular structures. A recent achievement in this research was the construction of a fluid structure interaction (FSI) model of trabecular bone and marrow to predict the mechanical environment of MSCs in vivo and analyze how this environment varies throughout osteoporosis. (Birmingham et al, 2010, Lycke et al, 2019) Lower bone mass and a variety of marrow viscosities, both of which are associated with osteoporosis, were utilized to study the impact of lower bone mass and a variety of marrow viscosities on the shear stress induced inside bone marrow utilizing these methods. The findings show that shear stresses generated inside bone marrow under physiological loading circumstances are within the range that has been shown to cause a mechanobiological response in MSCs in vitro. Furthermore, a decrease in bone marrow viscosity reduces marrow shear stress, whereas a decrease in bone mass increases marrow shear stress.

2. **Computational Modeling of Extravascular Flow in Bone:** The resident cells of bone tissues are the micromachines in charge of maintaining and modifying tissue shape to meet the demands of bone's dynamic, physiological activities. Mechanical strain causes extravascular fluid flow, which transmits mechanical and chemical signals that affect bone cell function. On the other hand, the mechanisms by which cell-scale processes are transmitted to functional adaptation at the organ scale are unknown. Researchers can use predictive multi-scale models to digitally analyze the effects of certain model parameters, increasing efficiency and speeding up the discovery of functional adaptation processes. The creation of computer modeling tools is meant to predict the relationship between mechanical stress of bone, load-driven fluid flow, and molecular transport augmentation inside bone. Recent research has found that common idealizations in geometry, geographical distribution and material features of cells and tissues reduce the accuracy of extravascular flow projections. Fluid drag-induced stresses on cell surfaces, for example, are underpredicted by orders of magnitude when pericellular fluid space geometries are idealized. (Steck et al, 2010) New bottom-up methodologies will help researchers decode the mechanical and chemical signals that make up the mechanophysiological environment of bone at several length scales, which is crucial for understanding mechanotransduction and how cells adapt bone tissue in health and disease. (Terkawi et al, 2022)
3. **Computational Modeling of Cell Mechanics in Cartilage:** Mechanical interactions between cells and their extracellular matrix influence mechanotransduction in mammalian tissues. It is now possible to estimate material characteristics and their variations with critical parameters such as location, age, and disease thanks to the discovery and implementation of solid-based computational techniques for multiphasic continuum models of articular cartilage cells (chondrocytes). Models for a variety of in vitro micromechanical test designs for evaluating cell characteristics, as well as simulations of mechanical cell-matrix interactions under physiologically acceptable stress conditions, are among the uses. The computational approaches, which include applications to contact concerns and multiscale modelling, use the boundary element method (BEM) and the finite element method (FEM) (Haider et al, 2010)

Section II: Evolution of Computational Biomechanics Approaches for Physiological Fluids

Computational methodologies and tools for analyzing mechanical events in biological systems have readily moved to the study of cell, tissue, and organ biomechanics from traditional engineering domains. Despite the availability of modern computer capabilities, models that explain and forecast the future have gotten increasingly sophisticated as biological systems' mechanical activity has become more complex. Despite the fact that these models have produced positive research outcomes, their usefulness is occasionally called into doubt. Model validation and verification are insufficient, which has resulted in criticism. The link between the constructs and the concepts of verification, validation, and sensitivity studies are explained. In computational biomechanics, model analysis and interpretation are critical. For a better understanding of evolution, real-life examples are offered. Some of the applications of the computational biomechanics in the area of physiological fluids are given below.

1. **Immersed Boundary/Continuum Methods:** In the study of biological systems, the relationship between fluid and solid structure is crucial. Accessible finite element codes and computational fluid dynamics codes have historically been linked via staggered iterations. Despite this, intricate dynamical system behaviours are usually lost in the process, despite the simplicity of the approach. For this reason, the researcher focused on the fluid-solid interaction (FSI) systems as a whole in order to capture the dynamical aspects of the system and develop efficient methods for system model reduction. (Wang, 2006) Since its inception, the immersed boundary technique has been applied to a wide range of problems. The first application of this strategy was for the structures that are very flexible and have a less stringent time step constraint. Both elastic fibre and beam (rod) networks can be utilized to mimic complicated nonlinear systems in modern versions of immersed boundary techniques. Immersed finite element formulations have also incorporated advanced nonlinear solid models. The implicit compressible immersed continuum approach's preliminary results demonstrate that correct time steps can be used for stiff FSI systems. Furthermore, immersed boundary/continuum approaches can be used to tackle compressible fluid flow issues.
2. **Computational Modeling of ATP/ADP Concentration:** The ability of vascular endothelial cells (ECs) to adapt to mechanical stresses resulting from blood flow is crucial for proper vascular function, and abnormalities in this area can result in major problems. EC mechanotransduction is important in vascular pathology. Intracellular calcium mobilization by ECs is one of the first and most important reactions to flow. Calcium mobilization is stimulated in ECs because the adenine nucleotides ATP and ADP are present outside the cell, indicating how flow modifies the EC-surface. Knowledge of ATP/ADP levels is required to comprehend the calcium flow response. (Choi & Barakat, 2010) Because quantifying ATP/ADP concentration at the EC surface experimentally is difficult, researchers have concentrated their efforts on building mathematical models that quantify the effect of flow on ATP/ADP concentration at the EC surface.
3. **Multiscale Transport and Absorption with Intestinal Function Using a Lattice-Boltzmann Method:** When dealing with related multiscale flow problems, scalar transport is used. A two-dimensional model is obtained using a lattice-Boltzmann model with a multigrid approach and changing boundaries, but the method is directly generalizable to three dimensions, and the approaches were used to investigate nutrient intake at the small intestine epithelium. The fine grid is nested within a coarse grid in this design, having an overlap region between the two grids. Due to

the information transfer between the two grids, mass and momentum are conserved while stress is maintained. The scalar is subjected to a modified moment propagation strategy, which allows for greater Schmidt numbers than competing methods. The shifting borders were treated using second-order boundary conditions, which interpolated to the exact wall position. Scalar transport from the border to a temporary lattice node was constructed for scalar and scalar flux boundary conditions, and it was then projected back to the node adjacent to the boundary. (Wang et al, 2010) A macro-scale cavity flow was simulated on the lower surface with micro-scale finger-like protuberances moving in a pendular motion as a model of macro-micro scale interactions in fluid motions, scalar mixing, and scalar uptake at the surface of the villi that line the epithelium of the human intestines to demonstrate the method's utility.

Section III: Computational Evolution of Cardiovascular Biomechanics

In cardiovascular development, the interaction of genetic and epigenetic (environmental) factors is dynamic. Despite the fact that the majority of current research focuses on genetic factors, it is clear that mechanical plays a vital role in the development of the heart and blood vessels. Understanding the interplay between mechanical and developmental processes is essential for detecting changes that can lead to congenital cardiovascular problems. Tissue engineering, infarct repair, and other fields could benefit from this information. Engineers are investigating the impact of mechanical pressures on cardiovascular morphogenesis, which is a relatively new subject in developmental biology. The prior idea included volumetric increase and active contraction. The next sections explore the use of this and other ideas to cardiovascular development difficulties.

1. **Vascular Hemodynamics Computational Modeling:** In addition to biochemical considerations, hemodynamic parameters governed by lumenal geometry and blood flow rates are likely to have a role in the development of cardiovascular disease. Although the function of flow variables in aneurysmal illness is still contested, several computational and experimental studies have been related certain hemodynamic characteristics to the onset and progression of atherosclerotic plaques. The approaches used in applying Computational Fluid Dynamics (CFD) to model patient-specific flow have also been demonstrated. Typical modelling assumptions and boundary conditions are used to define post-processing and visualization techniques. The current improvements in medical imaging and numerical approaches (Rayz & Berger, 2010) lead to further development of the computational modelling will grow into a therapeutic tool that will provide patient-by-patient guidance for cardiovascular disease treatment.
2. **Computational Modeling of Coronary Stents:** Both medical device design optimization and cardiovascular surgery pre-operative simulation can benefit from realistic computer modelling. The study of fluid and solid stress can reveal important information on tissue interaction and response mechanisms. In approaches for fluid and solid modelling connected to stents, the hemodynamic and mechanical data generated by simulations are discussed in terms of their consequences on vascular health and mechanotransduction. (Chen & Kassab, 2010)
3. **Aortic Heart Valves Computational Modeling:** The mechanics of the aortic heart valve can be studied through computational modelling. In the heart valve, solid components interact with a fluid domain, resulting in complicated dynamical and mechanical behaviour. There is no standard way to modelling the heart valve at the moment, and numerous strategies have been utilized to address

the various components of heart valve modelling. The imposition of the load, the simulated segment of the cardiac cycle, the inclusion of the fluid component of the problem, the intricacy of anatomical components, and the description of material properties are all examples of differences in technique. Simplicity saves time and money, but it may compromise accuracy. More physiologically appropriate models will be possible when modelling techniques progress and are used. (Croft & Mofrad, 2010) Computational research on the aortic valve has resulted in a better knowledge of normal valve mechanics, new insights on sick valve evolution, and predictions of the durability and efficacy of surgical repairs and valve replacements.

Section IV: Computational Evolution of Soft Tissue Biomechanics

Computational biomechanics of soft biological tissue is boosting our ability to address transdisciplinary difficulties in academia, industry, and medicine. Soft biological tissue such as the artery wall, the heart wall with the heart valves, and the ligament are studied using some of the available computational methodologies. The inherent complexities of the biological microstructure and functions of each researched soft tissue are briefly reviewed, research effort related to constitutive modelling is catalogued, and a sample constitutive model for each analyzed soft tissue is explored in further depth. Residual stresses, as well as biological processes like growth and remodeling, are described. (Usyk & McCulloch, 2003)

1. **Mathematical and Computational Foundations of Biomechanical Imaging:** The term "biomechanical imaging" refers to the measuring of mechanical properties of tissues in situ and in vivo. (Barbone & Oberai, 2010) The mechanical property distributions can be visualized to create the tissue pictures. The capacity to scan the tissue while it is being deformed by external pressures is critical to the method. Image processing is used to infer the displacement (or, in some circumstances, velocity) field across the region of interest. An inverse problem for the appropriate mechanical features is presented using the measured displacement fields, an assumed form of the tissue's constitutive equation (e.g., linear elastic), and the rule of conservation of momentum. (Barbone & Oberai, 2010)
2. **Meshfree Computational Method:** In the late 1970s, meshless methods (MMs) were first utilized to solve astrophysics problems. It is not necessary to explicitly link nodes in MMs since the spatial domain is represented by a collection of nodes (cloud of points) rather than by elements as in most mesh-based techniques (e.g., finite difference method, finite element method, finite volume method). A brief discussion of the applications, advantages, and disadvantages of many MMs developed and applied in the context of computational biomechanics. (Garg & Pant, 2017) The meshless total Lagrangian explicit dynamics methodology operate alongside strong and weak formulations, as well as novel interpolation algorithms such as modified moving least squares and the discretization correction particle strength exchange method. The approaches' usefulness in multiscale issues, as well as their inherent parallelization and advantages over classic mesh-based numerical methods, can be analyzed through diverse applications. MMs are common numerical approaches that can be used to solve difficult engineering problems. MMs will be sturdy enough to be deployed by industry after extensive and rigorous research in the field.
3. **Computational Biomechanics of the Human Cornea:** The increased use of corrective refractive surgery has piqued scientists' curiosity in the human cornea's biomechanical activity. To improve the success of surgical treatments, it is critical to understand the link between mechanical and

optical performance. This entails the creation of trustworthy computer imaging, modelling, and virtual training methods. A good biomechanical model can accurately replicate refractive surgery in both healthy and sick corneas, which lowers the risk of surgery failure (exploring the possibility of successful surgery), how a computer model of corneal biomechanics can be used to aid and improve clinical practice can be possibly shown. Photorefractive keratectomy (PRK), laser in situ keratomileusis (LASIK), and laser subepithelialkeratomileusis are the most popular refractive error correction methods (LASEK). More clinical outcome research and the utilization of new technology, techniques, and procedures have been done as a result of the technique's effectiveness. The scientific interest in the biomechanical activity of the human cornea is further explained by the growing popularity of corrective refractive surgery. (Lycke et al, 2019)

The majority of the computational studies focus on testing method modelling, material characterization to account for the stroma's underlying microstructure, standard surgery technique outcome prediction, and intraocular pressure measurement accuracy. Unfortunately, basic science research has lagged behind clinical advances, and the anatomic, biologic, and physiologic aspects of the wound healing process following refractive surgery remain unknown and unmolded. Improved predictability of refractive surgery outcomes, novel surgical options for damaged corneas, and preoperative detection of eyes at risk of developing ectasia following refractive surgery could all be aided by a better understanding of corneal biomechanics. It could be crucial in the development of a corneal replacement in the near future.

Section V: New Scope in Dentistry

The application of biomechanics in dentistry is a topic less discussed. This is because it is often assumed that its application in orthodontics and implantology relates to the movement of the mandible or the application of forces. As a result, the basic information and essential ideas necessary in this field for study and comprehension of the nature of tissues have not been particularly effective. (Manea et al, 2019) Biomechanics is an important branch of study in dentistry. It enables complete understanding of the nature and mechanical characteristics of oral tissues, mandibular and temporomandibular joint movement, and external agents to which the entire stomatognathic system is exposed.

COMPARATIVE ANALYSIS OF DIFFERENT METHODS

For everyday computational biomechanics applications, current computational models have the potential to become commodities in the future. Thanks to the availability of models and related resources, others will be able to grasp their capability, assess their quality, and repurpose them for other use cases without having to rebuild them. The biomechanics community has recognized this opportunity and has begun to lay the framework for a computational modeling and simulation sharing culture, with some success. The community became aware of the issues in model sharing as many groups sought to include model exchange strategies into the workflow of academic research. There are initiatives and rules that are synergistic, yet they are sometimes redundant or contradictory. The architecture and attitude for sharing models is improving, and simulation software is expanding to support open science and hence more data for computer simulations. There are differing views on what should be shared, why it should be shared, how it should be shared, and when it should be shared. All of this looks to be creating a positive dynamic inside the community and among diverse stakeholders, which will aid in driving the model sharing culture toward a unified understanding backed by a long-term platform. It is possible to have a greater knowledge of how this system operates, particularly the symbiotic interaction that exists amongst oral tissues, through it. This offers up the prospect of better diagnosis, treatments, and the identification of more appropriate repair materials. One of the probable reasons why it is not viewed as important anymore is because it is deemed impracticable to utilize owing to the sophisticated handling required for its experimentation.

FUTURE ANALYSIS

Computational biomechanics employs variety of scientific disciplines, including computer science, mathematics, mechanics, and biology. As a result, the expansion of this field depends on the development in technology to enable faster implementation of the base disciplines. The effectiveness of computational biomechanics must be judged in terms of how accurately it can predict material behaviour and as a consequence, how successfully it helps to augment quality of life.

The progress of computational biomechanics both in terms of societal as well as natural occurrence has been visualized over the past few years in solving various biological or biomechanical problems. We can explain a lot more today than we could before when the computational biomechanics was a vague idea and was solely associated with engineering. In the realm of medical research, computational biomechanics is projected to have one of the most significant impacts. The human body is a complicated system in which various mechanical and fluid processes occur at all times. Computational biomechanics may aid in the accurate modeling of processes such as blood flow through the human body, organ and limb function, and much more. Medical practitioners and researchers will be able to develop more advanced treatment procedures as a result of this. It also helped in the field of prosthetics, orthodontics and rehabilitation.

It is apparent that models of physical and biological systems have given computational biomechanics a bright future Integration of biological, quantum, and molecular mechanics has made this possible through the use of computer simulations. As a result, a variety of systems, including communication, medical, military, and industrial, to mention a few, have seen significant improvements.

REFERENCES

Alexander, R. M. (2003). Modelling approaches in biomechanics. *Philosophical Transactions of the Royal Society of London. Series B, Biological Sciences*, *358*(1437), 1429–1435. doi:10.1098/rstb.2003.1336 PMID:14561333

Barbone, P., &Oberai, A. (2010). *A Review of the Mathematical and Computational Foundations of Biomechanical Imaging*. . doi:10.1007/978-90-481-3575-2_13

Birmingham, E., Grogan, J. A., Niebur, G. L., McNamara, L. M., & McHugh, P. E. (2013, April). Computational modelling of the mechanics of trabecular bone and marrow using fluid structure interaction techniques. *Annals of Biomedical Engineering*, *41*(4), 814–826. doi:10.100710439-012-0714-1 PMID:23519534

Chen, H., & Kassab, G. (2010). *Computational Modeling of Coronary Stents.* . doi:10.1007/978-90-481-3575-2_6

Choi, H. W., & Barakat, A. (2010). *Computational Modeling of ATP/ADP Concentration at the Vascular Surface.* . doi:10.1007/978-90-481-3575-2_2

Croft, L., & Mofrad, M. (2010). *Computational Modeling of Aortic Heart Valves.* . doi:10.1007/978-90-481-3575-2_7

Garg, S., & Pant, M. (2017). Meshfree Methods: A Comprehensive Review of Applications. *International Journal of Computational Methods*, *15*(04), 1830001. Advance online publication. doi:10.1142/S0219876218300015

Haider, M., Benedict, B., Kim, E., & Guilak, F. (2010). *Computational Modeling of Cell Mechanics in Articular Cartilage.* . doi:10.1007/978-90-481-3575-2_11

Innocenti, B. (2018). Biomechanics: A fundamental tool with a long history (and even longer future!). *Muscles, Ligaments and Tendons Journal*, *7*(4), 491–492. doi:10.32098/mltj.04.2017.02 PMID:29721449

Lycke, R., Walls, M., & Calve, S. (2019). Computational Modeling of Developing Cartilage Using Experimentally Derived Geometries and Compressive Moduli. *Journal of Biomechanical Engineering*, *141*(8), 0810021–0810028. Advance online publication. doi:10.1115/1.4043208 PMID:30874718

Manea, A., Bran, S., Dinu, C., Rotaru, H., Barbur, I., Crisan, B., Armencea, G., Onisor, F., Lazar, M., Ostas, D., Baciut, M., Vacaras, S., Mitre, I., Crisan, L., Muresan, O., Roman, R., & Baciut, G. (2019). Principles of biomechanics in oral implantology. *Medicine and Pharmacy Reports*, *92*(Suppl No 3), S14–S19. doi:10.15386/mpr-1512 PMID:31989104

Rayz, V., & Berger, S. (2010). *Computational Modeling of Vascular Hemodynamics.* . doi:10.1007/978-90-481-3575-2_5

Steck, R., Tami, A., Sidler, H-J., Anderson, E., & Niederer, P. (2010). *Computational Modeling of Extravascular Flow in Bone.* . doi:10.1007/978-90-481-3575-2_10

Sweeting, K., & Mock, M. (2007, June). Gait and posture - assessment in general practice. *Australian Family Physician*, *36*(6), 398–401, 404–405. PMID:17565395

Terkawi,, M.A., Matsumae,, G., Shimizu,, T., Takahashi,, D., Kadoya,, K., & Iwasaki,, N. (2022). Interplay between Inflammation and Pathological Bone Resorption: Insights into Recent Mechanisms and Pathways in Related Diseases for Future Perspectives. Int J Mol Sci., 23(3). doi:10.3390/ijms23031786 PMID:35163708

Usyk, T., & McCulloch, A. (2003). Computational Methods for Soft Tissue Biomechanics. *Biomech. Soft Tissue Cardiovasc. Syst.*, *441*, 273–342. Advance online publication. doi:10.1007/978-3-7091-2736-0_7

Wang, X. (2006). From Immersed Boundary Method to Immersed Continuum Methods. *International Journal for Multiscale Computational Engineering*, *4*(1), 127–146. doi:10.1615/IntJMultCompEng.v4.i1.90

Wang, Y., Brasseur, J., Banco, G., Webb, A., Ailiani, A., & Neuberger, T. (2010). *Development of a Lattice-Boltzmann Method for Multiscale Transport and Absorption with Application to Intestinal Function*. doi:10.1007/978-90-481-3575-2_3

Chapter 2
Finite Element Analysis in Biomechanics

Subhomoy Chatterjee
Schieffelin Institute of Health Research and Leprosy Centre, India

ABSTRACT

Finite element method (FEM) is among the methods of approximation, which is abundantly used to solve the real-life problems that cannot be solved with the analytical methods. The biological systems are complex in nature in terms of geometry, material properties, functioning, and case-to-case deviation. Thus, the standard analytical methods suffer catastrophic limitations in this area. Mainly for this reason, biomechanics is highly enriched with the integration of FEM, which opens a new window with extended capabilities against a logically acceptable margin of approximation. This chapter focuses on designing customized medical devices (orthopaedic and dental implants) and correlating their design with the presented clinical condition and, thus, leading to a criteria of design prescription based on clinical conditions.

INTRODUCTION

The standard analytical methods of solution have got only limited scope to solve the real life laws of physics. The space and time dependent problems are often expressed as partial differential equations (PDE)s whose solution can be addressed through methods of approximation. There are few established numerical methods of approximation like finite element method (FEM), finite volume method (FVM), finite difference method (FDM), etc. These methods and their differences are precisely mentioned by Bjorn Sjodin in Machine Design (https://www.machinedesign.com)(Sjodin, 2016). This chapter gives an overall idea of the working mechanism of solid structural analysis using FEM. It is focused on the biomechanical problem of design customization of orthopaedic and dental implants.

DOI: 10.4018/978-1-7998-9078-2.ch002

BACKGROUND

Designing of medical devices has been greatly benefited by the integration of FEM in the product development process over the last decade (Driscoll, 2019). FEM is a numerical tool, which is used to solve boundary value problems. In this method, a complex entity is considered to be a combination of finite number of small entities of regular configuration, known as **Elements**. As for example, in case of solid structural analysis, a complex geometry is considered to be composed of an aggregation of small geometrical elements of regular shape (Fig. 1a). We can take the example of a complex brick structure (Fig. 1b). It's quite difficult to identify the status of stress or deformation of any particular point within the irregular structure. But we may calculate these entities for each brick, which is a regular structure. Each brick passes its force output to the next brick in the sequence, and there this force acts as the input force. Once the mechanical parameters of all the bricks are calculated, we can aggregate the individual outcomes and obtain the overall results for the whole structure or for any particular zone of interest. Unlike the bricks which are cemented to each other through the interfacing surface, the elements in the FE system are connected to each other at certain defined points, known as **Nodes**, which can pass information from one element to another. A system of field equations is considered, expressed as PDEs, that represent the particular aspect of physics we are concerned with. This is calculated for each element and the individual contribution of each element is assembled to obtain the system output. With this consideration, mechanical analyses of complex and irregular structures are possible which is otherwise impossible through analytical methods. This clearly indicates that the analytical outcome will be approximated to an extent dependent on the structural complexity and the number of elements we break it to. The process of breaking down the complex structure into finite number of elements is known as **Meshing**. The more the number of elements we will mesh the structure into (resulting in smaller elements), more closer we will be able to reach to the real outcome (Fig. 1c). But solution with higher mesh size is achieved against higher computational demand. Generally, this method generates stiffness matrices and solves them. The size of these matrices increases with increase in number of elements, imposing higher computational load. This is further elaborated under the section 'FE Analysis'.

This chapter focuses on use of FEM in design customization of prosthetic replacements in orthopaedic and dental sciences. Design customization of the prosthetic components has been tried to maximize the biomechanical outcome. There are documented cases of failure of total hip replacement (THR) owing to stress shielding (Kraaij et al., 2014; Yamako et al., 2014a; Yamako et al., 2014b). Stress shielding occurs as a result of difference in stiffness of the implant material with that of the portion of the bone it replaces and the bone tissue that surrounds it. This issue can be partially addressed by using isoelastic stems where the incompatibility of material stiffness is reduced by proper material selection (Kärrholm et al., 2002; Yamako et al., 2014a). But achieving a material of appropriately required stiffness at required location is often a fantasy. Moreover, biocompatibility is certainly a concern while choosing materials for *in vivo* use, reducing our choice of materials. Alternatively, isoelasticity has been realized by introducing different grades of porosity into the implant design while using the conventional Ti alloys as implant material (Boobalan and Shankar, 2013; Fernandes et al., 2004; Mattheck et al., 1990; Ruben et al., 2012; Ruben et al., 2007; Schmidt and Hackenbroch, 1994; Virulsri et al., 2015; Yang et al., 2009). At this juncture, it may be understood that introduction of porosity parameters should be customized according to the particular case presented by each subject. FEM is an essential tool for carrying out research on design customization owing to the complexity and non-uniformity in the models involved.

MAIN FOCUS OF THE CHAPTER

The problems in biomechanics are associated with complexity and irregularity of structure, materials, etc. Solution of such problems is particularly improved by incorporation of FEM. This chapter is focused on design customization of orthopaedic and dental implants. Damage of peri-implant bone, leading to implant loosening over time, is a major concern in this area. Among the leading causes of implant loosening is aseptic resorption of bone due to stress-shielding. Here, an approach to solve this issue is described that involves incorporation of internal porosity and design customization, worked out using FE analyses.

Figure 1. Meshing of a complex irregular geometry of biomechanics into finite number of elements, each having a regular geometry. (a) The irregular geometry of the mandible has been meshed into elements of regular tetrahedral shape. (b) The concept is similar as building a complex brick structure, where each of the bricks is having a regular geometry. (c) More the number of elements more closer will be the outcome to reality.

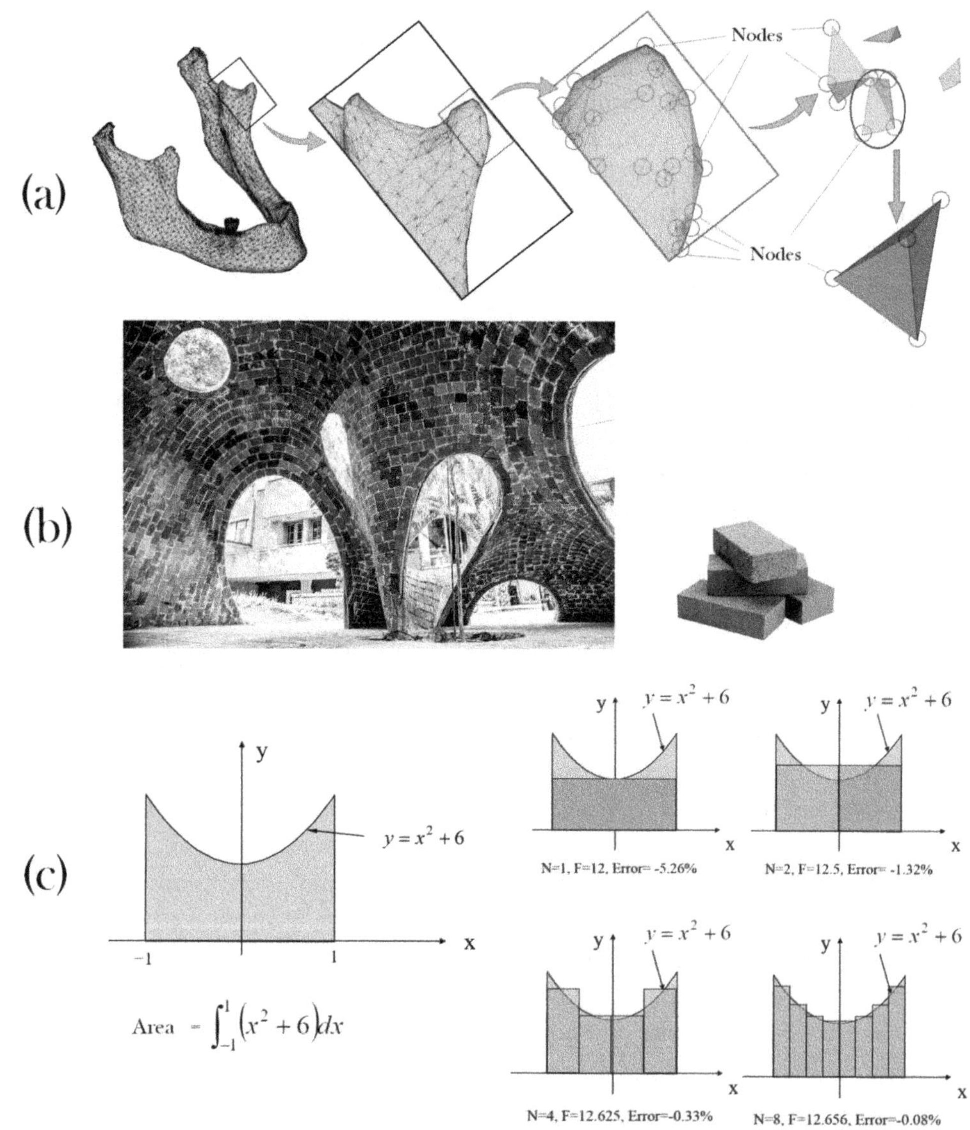

FINITE ELEMENT MODELLING IN BIOMECHANICS

The mechanical environment at each point within a biomechanical system can be analyzed and interpreted through FEM which is otherwise impossible to obtain experimentally. Moreover requirement of surgical placement of sensors *in-vivo* for data collection, which is a hazardous and expensive process, is not required. Here only a reconstructed anatomical model is required. Thus it is widely used in the field of biomechanics and has become a popular method for biomechanical analysis.

Orthopaedics and Dental sciences form a significant area of biomechanical research and this chapter is focused on these areas. Orthopaedic implantation is generally done for artificial replacement of a diseased joint or for fracture fixation. In Dental sciences implantation is done to replace one or multiple damaged teeth. Generally the implant materials are stiffer than normal bone. Consequently, when it is placed longitudinally along the bone axis as in femoral stem in case of THR, the load is mainly borne by the implant, reliving the surrounding bone from load bearing. Thus the implant gets stressed more while the surrounding bone is shielded from being stressed. Hence, the difference in stiffness between the bone and the implant material results in a condition known as **stress-shielding** in the surrounding interfacial bone. Bone is a smart living material which senses the external mechanical stimulus and remodels itself accordingly. The concept of bone remodelling is elaborated under the Case study 1. The mineral content in the bone is physiologically regulated by the activities of special cells within the bone, known as **osteoblasts** (mineral depositors) and the **osteoclasts** (mineral removers). Stress-shielding is a debilitating situation as the bone senses lack of load bearing and thus, its requirement. Then osteoclastic activity takes over the osteoblastic activity, leading to overall bone resorption and damage. This is among the major causes of loosening and failure of load-bearing implants. In this backdrop, it is anticipated that the stiffness reduction of the implant can address the issue of stress shielding in many cases. Moreover the stiffness of the implant component should be in accordance to the stiffness (in turn, density) of the surrounding bone and the portion of the bone replaced by the implant volume. Thus the process of stiffness reduction of an implant component must take into account the bone physiology and other subject characteristics such as body weight.

It has been reported that the clinical outcomes of several new materials and designs of joint replacement component are below expectation (Cipriano et al., 2008; Massoud et al., 1997). This boils down to the need for preclinical evaluation based on the variation of subject parameters that the components would experience clinically (Fisher, 2011; Harris, 2012; Hua et al., 2014; Langton et al., 2011; Wagner et al., 2012). In order to come down to an implant design, that would reduce stress shielding, there is requirement for rigorous analysis of stress/strain environment in every point of the peri-implant tissues and within the implant over time in different load cases and finally, customization of the implant design for the particular clinical case. But such study in physiological relevance is almost beyond possibility using classical experimental or analytical procedures. FEM provides a computational simulation of the real life affairs which is capable for analyzing the biomechanical environment in each element and node at different points of time according to the loading configuration, though with some degree of approximation. It is becoming indispensable in the preclinical evaluation, owing to its specific advantages over experimental studies like the provision of obtaining site specific *in vivo* results (Stops et al., 2012).

STEPS OF GENERATION OF FINITE ELEMENT MODEL IN BIOMECHANICS

Generally the steps involved in a complete FE analysis in biomechanics are reconstruction of the anatomical model, meshing each structure into elements, identification of material properties of each object and assignment of the same, application of surface behaviours at different interfaces within different anatomical objects, provision of the boundary conditions (this may include several load cases), analysis, observation and interpretation of the outcome post processing. If the study contains some *in-vivo* / *in-vitro* devices (e.g. orthopaedic/dental implants, prosthetic heart valves, intravenous stents, plates and screws to address bone fracture, external prosthetic devices for amputated subjects, external orthopaedic / dental orthotic devices, etc.) then some further steps may be appended to the process. This includes modelling of the device in CAD, importing the CAD model into the FE interface, positioning of the *in-vivo* / *in-vitro* device to assemble them with the anatomical models to simulate the clinical condition, identification of material properties of the object(s) of such devices, its surface behaviour at the human-device and inter-device interfaces, etc. Such steps need to be completed before the analysis.

Reconstruction of Anatomy

As an initial phase of virtual simulative study in biomechanics, the body part which is under the purview of the study has to be modelled. The procedure for generating the most near-natural anatomical model is through reconstruction from imaged data of real life. Generally this is the scanned data from magnetic resonance imaging (MRI) or computed tomography (CT) procedures. The reconstruction is done on the basis of the gray values, resulting from the X-ray opacity. This is expressed in terms of Hounsfield Units (HU).

Hounsfield Units (HU)

It is a scalar quantification of X-ray opacity. It is calculated based on the values of linear attenuation of X-rays by a particular substance. As the X-ray beam passes through a medium, it gets attenuated on its way. Its intensity after it travelled (w) cm from the reference point within the medium is obtained from Eq. 1

$$I_w = I_0 \times e^{-\mu w} \tag{1}$$

Here I_0 and I_w are the X-ray intensities at the reference point and at a distance of (w) cm from it respectively. (μ) is the linear attenuation coefficient.

From Eq. 1, we can obtain the expression for the linear attenuation coefficient (μ) as shown in Eq. 2.

$$\mu = \ln(I_0 / I_w) / w \tag{2}$$

The HU scale is a linear transformation of (μ) taking distilled water at standard pressure and temperature (STP) as the reference. HU of distilled water at STP is considered as zero. Materials having X-ray opacity higher than water have positive values of HU whereas materials having X-ray opacity lower than water have negative values of HU. The HU of air is considered as −1000. With these considerations, HU of a material (m) is calculated using Eq. 3.

$$HU_m = 1000 \times \frac{\mu_m - \mu_{water}}{\mu_{water} - \mu_{air}} \tag{3}$$

Here μ_m, μ_{water} and μ_{air} are the linear attenuation coefficients of the material (m), water and air respectively.

It is quite relevant that the X-ray opacity of one particular material is directly related with its mass density. But different materials will have different X-ray opacity for one particular mass density. There are various empirical equations to relate the mass density with the HU value for different materials.

Figure 2. Reconstruction of femur. The first part shows the reconstruction after putting the threshold of HU values and slice-wise deleting the selected areas outside AOI. The second part shows reconstruction after further slice-wise editing to create the femur model as a single solid body and smoothing the manual slice-wise editing errors.

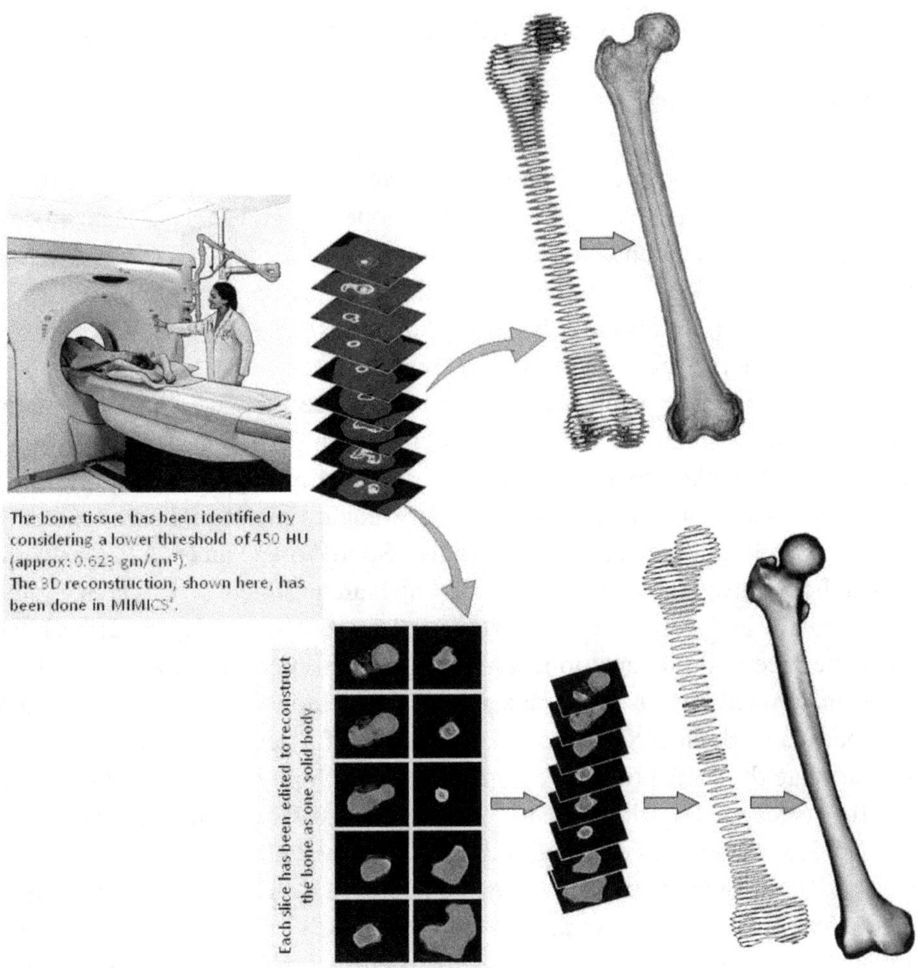

The area of interest (AOI) in each of the scanned slices, i.e. the area corresponding to the image of the structure which we are going to model, can be identified through proper area selection operation. These patches can be considered to be stacked in a pile with their location in the plane of the slice that each of the patches belongs to. The perpendicular gap in between them can be interpolated to reconstruct the anatomical geometry (Fig. 2). Thus, it is obvious that less the gap between the slices, more accurate would be the reconstructed model

The author used the medical reconstruction software MIMICS® (Materialise NV, Leuven, Belgium) for this purpose. Here, the scanned data is imported in Digital Imaging and Communications in Medicine (DICOM) format and a separate file is formed for each scanned case. The selection of area in each slice of a file is initially done on the basis of HU values of each pixel. It is created tough proper selection of the upper and lower thresholds of HU and based on this a '**Mask**' is created. All the pixels, whose HU values (calculated from the gray values) falls between the applied upper and the lower thresholds, are selected. This is applied to all slices simultaneously. The shortcoming of this method is that areas outside our AOI, having HU within our applied range, would be selected simultaneously while some area within our AOI may fall out of the range and, thus, would remain unselected. This problem is addressed by slice-wise editing the Mask. This is done manually, this time not on the basis of HU but simply like graphical painting and erasing operation. In this way a clean selection of the patches corresponding to a particular anatomical object can be obtained manually. Even after creation of the 3D model by interpolating these patches, the model can be smoothened to remove the errors of manual boundary selection in each slice.

Preparation of the FE Model

After reconstruction of the 3D model, the volume is imported into the FE software interface. The author used ANSYS® (ANSYS Inc., Pennsylvania, USA) for FE modelling and analysis. The procedure adopted by the author for transferring the reconstructed model from MIMICS® to ANSYS® is described hereafter.

First the surface of the 3D model is meshed into multiple triangular areas in MAGICS® (Materialise NV, Leuven, Belgium) which, in this case, is an ancillary package with MIMICS®. These areas were imported in ANSYS® where the solid 3D mode was recreated.

If the model contains multiple reconstructions (e.g. the knee joint which will contain portions of femur, tibia and fibula and the patella and may also contain the articular cartilages, ligaments, etc.), each of them needs to be brought into one ANSYS® interface following the same procedure. The model may contain artificial devices such as implants, fracture fixation plates and screws, etc. Such devices may be designed separately in other CAD software like SolidWorks® and CATIA® (Computer Aided Three-Dimensional Interactive Application), both of which are products of Dassault Systèmes, Vélizy-Villacoublay, France. These devices should be designed in real (non-magnified) scale and imported into the same FE interface with the anatomical models. These objects are then properly placed in accordance with the anatomical models. This placement may include virtual removal certain portions of the anatomical models and Boolean operations for insertion of a device within the anatomical model. As for example, to create the model (one side) simulating total hip replacement (THR), the hemipelvis and the femur of the corresponding side have to be reconstructed. The implant components; like the cement layer (in case it is simulating a cemented fixation), the acetabular shell and liner, the femoral head and the femoral neck-stem component; need to be modelled in a CAD package. While assembling these objects, the anatomical femoral head and neck parts need to be virtually removed, followed by placement of the implant components and the cement layers (if present) in their respective position and using

Boolean operation to simulate implantation. Additional removal of portions of the anatomical models will be required to simulate the broached areas of the femur and the hemipelvis. The entire model is then meshed in the FE software. Figure 3 shows FE modelling of femoral stem implant into femur, considering simplified physiological loading. The author used 10 noded tetrahedral elements for this purpose owing to its compatibility with MIMICS® in the further procedures. The used element is SOLID187 which has three degrees of freedom at each of the 10 nodes, i.e. translation in *x, y,* and *z* directions.

Figure 3. FE modelling and post-processed results. (a) reconstructed femur and model of implant, virtual osteotomy of femoral head and implant fitment through Boolean operation; (b) meshed model with boundary conditions and (c) contour plot of von Mises strain over the longitudinal interfacial surface of femur.

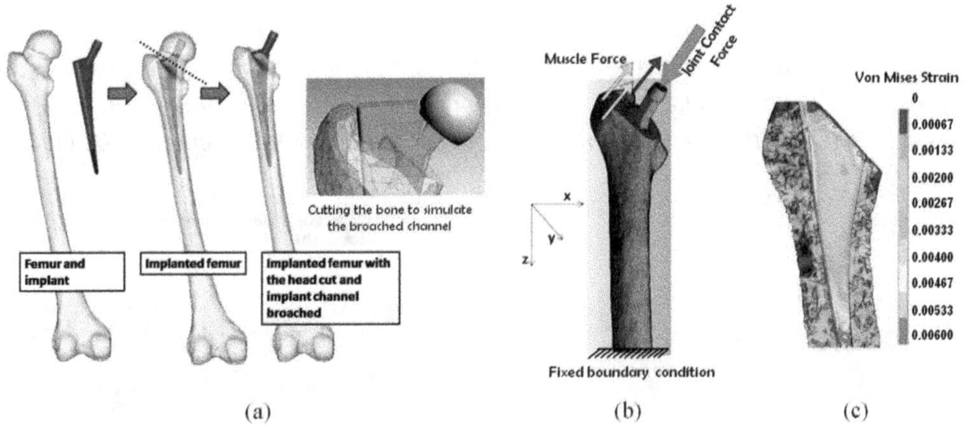

The surface behaviours at the various interfaces are to be defined. These interfaces are those between the various anatomical structures (e.g. femoral head and acetabular cartilage in a FE model of natural hip joint), between anatomical structure and device (e.g. between femoral stem and surrounding bone) or between the various components of a device (e.g. between acetabular shell and liner). The surface behaviours can be frictional (the coefficient of friction, i.e. μ, has to be defined), bonded (i.e. $\mu = 1$), frictionless (i.e. $\mu = 0$), etc. Such definition has to be done even when the surfaces are initially non-contact, but they are expected to come in contact during operation. For example, there may be a gap between the acetabular shell (inner diameter 48mm) and the acetabular liner (outer diameter 47.88mm) for engineering clearance. If such surface characteristics are not defined, the FE program would be unable to identify the restrictions of motion of one component due to the existence of another component. Considering the previous example, if the surface behaviour between the non contacting acetabular shell and liner is undefined, the liner volume would go through the shell volume without recognising its existence.

It has been already discussed that finer mesh will reduce the errors of approximation in the FE analysis. But it demands higher computational power of the system and consumes much more time for each analysis. In this backdrop, it is clear that an optimization of the mesh size is required based on the demands of the model and the computational power available.

Selection of Mesh Size

If we reduce the size of the elements (i.e. generate finer mesh), the computational demand will increase. We need to stay within the limits of the computational power available with us. While doing so, we may miss the critical delicate contours of certain areas of the model which might be of importance to us. This is shown on a non biomechanical example model in figure 4(a). In order to avoid this problem, different mesh sizes may be used at different zones of the model (Fig. 4b). This can be self adaptive, such that finer mesh would be automatically generated in the locations of curvatures or delicacies, while keeping the rest of the volume grossly meshed. Else it can be due through manual selection of the zones of interest. In the second case, we may keep the general element size bigger, while denote certain AOI where a different size (finer) or even type of mesh will be applicable.

Figure 4. Controlling the mesh size: (a) systemic application of mesh refinement (b) application of mesh refinement on selected locations. The pictures are taken from Finite element mesh refinement, multiphysics cyclopedia, COMSOL (https://www.comsol.com, accessed on 15th September, 2021).

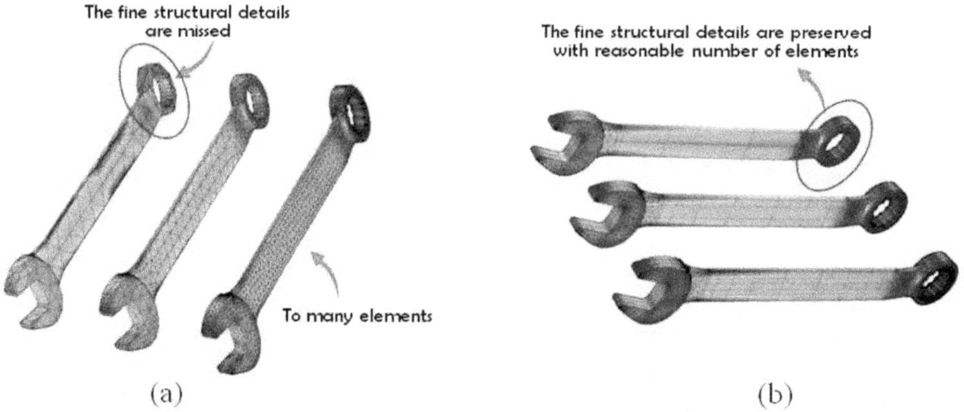

In this way, the design intricacies in the particular AOI are maintained while the number of elements in the overall model can be kept reasonable.

Assignment of Material Properties

After the meshing is done, the material properties of each of the objects are to be defined. For solid models, if the operational stress at every location is anticipated to be within the elastic limits of the corresponding material, then the material properties are provided in terms of elastic modulus, Poisson's ratio and density. But if plastic deformation at any point is anticipated, then the non-linear stress–strain characteristics of such materials has to be defined considering multiple points of the stress-strain curve of the specific material. The material properties of the device volumes are considered as uniform. But the anatomical components are heterogeneous in nature. For example in the mentioned case of THR, the anatomical structures are the femur and the hemipelvis. We know that bone is a thoroughly heterogeneous material. Its property varies from point to point and thus assignment of its material properties is done on an element-wise manner. It is done based on voxel-wise X-ray opacity (in terms of HU) as on the scanned data from which they have been reconstructed. For this purpose the bone elements are again transported back to the MIMICS® interface. The average of the HU values of all the voxels within the volume of each element is assigned to the particular element. The density of the element is calculated from this average HU value using some established empirical equations which are defined for some particular materials (e.g. Eq. 12). The value of elastic modulus of bone is also calculated similarly in an element-wise manner using other established empirical equation (e.g. Eq. 12). The Poison's ratio of bone is to be provided in addition to this. The author obtained this value from the existing literature. Thereafter, the elements to the anatomical structures are transferred back to the FE interface. The device volumes are assigned with their material properties in the FE software itself. The procedure of assignment of material properties to the anatomical structures can also be done in a simplified manner in the FE software as well. But in his case it wouldn't be done in element-wise manner. For example, the femur can be splitted into two volumes, one corresponding to the cortical bone and the other the cancellous bone. Average properties of these two types of bones can be assigned to them respectively.

Provision of Boundary Conditions

After the FE model is prepared, the working loads and the location where it is held fixed are to be defined (Fig. 3b). In biomechanics, these working loads are the physiological loads to which the system is exposed to. It includes the body weight, muscle tension, ligamentous constrains, etc. While application of these loads the area over which it is applicable is identified and in most cases it is applied as a distributed load over this area. Bodyweights may, sometimes, be considered to be working at some nodes inside some bone (e.g. the head of thee femur) for simplification (Ghosh et al., 2015). The loads are to be applied maintaining their appropriate directions. Loads can be applied to the nodes in the Cartesian '*x*, '*y* and '*z* directions. In order to ensure proper direction, the load vector needs to be resolved according to the Cartesian directions. An alternative procedure is to orient the nodes, on which the load would be applied, angularly, without shifting their location, so that any of the reoriented Cartesian axes of those nodes will point exactly towards the loading direction. Then the load may be applied as a single vector in that reoriented axis.

The FE model must have some movement restriction at some point. Otherwise it would be a case of unconstrained motion in space and will have no solution. Some location(s) within the model have to be considered as fixed. These locations can be one or more surfaces or portion(s) of one or more volumes and the movement restriction is applied on the respective nodes. These locations would belong to one or few of the modelled objects and the restrictions would be imposed on the rest of the objects through surface definitions at interfaces with these objects and so on. Based on the loading directions, we may need to restrict the movement of the nodes in all the three Cartesian directions, or any two directions or, even in some cases, any one of the directions.

FE Analysis

In FE analysis, a series of algebraic equations are formed regarding the displacements of the nodes considering the stiffness of the material, the applied forces and the boundary constraints. The equations are formed automatically behind the interface of the FE software. This set of equations may be represented in the form of matrix as mentioned in Eqs. (4).

$$\mathbf{K}_{ij}\,\mathbf{u}_j = \mathbf{f}_j \tag{4a}$$

$\mathbf{K}_{ij}$ is the stiffness matrix, $\mathbf{u}_j$ is the displacement matrix of the individual nodes and $\mathbf{f}_j$ is the matrix of force applied on each node. These can be more expanded as:

$$\begin{bmatrix} K_{11} & K_{12} & \cdots & K_{1n} \\ K_{21} & K_{22} & \cdots & K_{2n} \\ \vdots & \vdots & \ddots & \vdots \\ K_{n1} & K_{n2} & \cdots & K_{nn} \end{bmatrix} \begin{Bmatrix} u_1 \\ u_2 \\ \vdots \\ u_n \end{Bmatrix} = \begin{Bmatrix} f_1 \\ f_2 \\ \vdots \\ f_n \end{Bmatrix} \tag{4b}$$

$\mathbf{u}_j$ is the displacement of the jth node, $\mathbf{f}_j$ is the force applied on the jth node and $\mathbf{K}_{ij}$ is a member of the stiffness matrix which is the influence of jth displacement on the ith force.

Different geometries of the elements are considered in FE method which includes line element (one dimensional), shell element (two dimensional), volume element (three dimensional), etc. There are different types of elements within each of these categories. For example the volume elements can be tetrahedral, hexahedral, etc. A tetrahedral volume element can also be either 4-noded or 10-noded. The line element is discussed here because of its simplicity.

Line Element

The force vectors of a line element under tension is shown in figure 5(a). The displacement of the two nodes *i* and *j* under tension are resolved into the horizontal (u_i and u_j and the vertical (v_i and v_j components.

Figure 5. Line element (a) a single line element with force vectors, (b) a truss containing two bars, (c) consideration of the truss to be composed of two line elements and (d) consideration of the forces on each element.

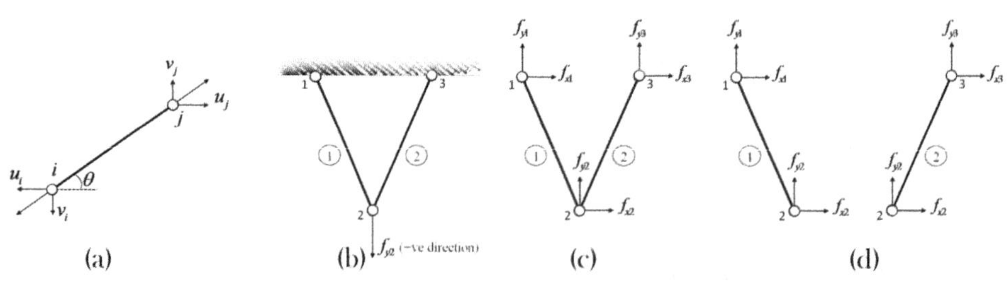

The elongation of the element can be expressed as

$$\delta = \left(u_j cos\theta + v_j sin\theta\right) - \left(u_i cos\theta + v_i sin\theta\right) \text{ or } \delta = \left[-cos\theta - sin\theta \; cos\theta \; sin\theta\right]\begin{Bmatrix} u_i \\ v_i \\ u_j \\ v_j \end{Bmatrix} \tag{5}$$

Now we know that $\delta = \dfrac{PL}{AE}$ or $P = \dfrac{AE}{L}\delta$ where P is the applied load (in N), L is the initial length of the element (in m), E is the Elastic Modulus of the material (in N/m^2) and A is the cross-sectional area (in m^2) perpendicular to the line of application of P. The load P can also be resolved in vertical and horizontal directions considering both the nodes i and j. Thus:

$$f_{xi} = -PCos\theta \;\; f_{yi} = -PSin\theta \;\; f_{xj} = PCos\theta \;\; f_{yj} = PSin\theta$$

Thus we can write:

$$\begin{Bmatrix} f_{xi} \\ f_{yi} \\ f_{xj} \\ f_{yj} \end{Bmatrix} = \begin{Bmatrix} -cos\theta \\ -sin\theta \\ cos\theta \\ sin\theta \end{Bmatrix} P = \begin{Bmatrix} -cos\theta \\ -sin\theta \\ cos\theta \\ sin\theta \end{Bmatrix} \frac{AE}{L}\delta \tag{6}$$

Substituting δ from Eq. 5, we can write:

$$\begin{Bmatrix} f_{xi} \\ f_{yi} \\ f_{xj} \\ f_{yj} \end{Bmatrix} = \begin{Bmatrix} -cos\theta \\ -sin\theta \\ cos\theta \\ sin\theta \end{Bmatrix} \frac{AE}{L}\left[-cos\theta - sin\theta \; cos\theta \; sin\theta\right]\begin{Bmatrix} u_i \\ v_i \\ u_j \\ v_j \end{Bmatrix} \tag{7}$$

Or

$$\underset{[F]}{\begin{Bmatrix} f_{xi} \\ f_{yi} \\ f_{xj} \\ f_{yj} \end{Bmatrix}} = \frac{AE}{L}\underset{[K]}{\begin{bmatrix} Cos^2\theta & Cos\theta Sin\theta & -Cos^2\theta & -Cos\theta Sin\theta \\ Cos\theta Sin\theta & Sin^2\theta & -Cos\theta Sin\theta & -Sin^2\theta \\ -Cos^2\theta & -Cos\theta Sin\theta & Cos^2\theta & Cos\theta Sin\theta \\ -Cos\theta Sin\theta & -Sin^2\theta & Cos\theta Sin\theta & Sin^2\theta \end{bmatrix}}\underset{[D]}{\begin{Bmatrix} u_i \\ v_i \\ u_j \\ v_j \end{Bmatrix}}.$$

In equation 7, [F] denotes the 'Force Matrix', [K] denotes the 'Stiffness Matrix' and [D] denotes the 'Displacement Matrix'.

Considering the truss structure (Fig. 5b), initially it needs to be meshed into multiple elements. Since here the structure is extremely simple, we have considered it to be meshed into just two elements (Fig. 5c). Here we have 3 nodes as 1, 2 and 3. We have to consider each of these elements as separate entities. Nodes (1) and (2) are associated with element (1) while nodes (2) and (3) are associated with element (2). Each element is exposed to a load condition (Fig. 5d). Equation 7 applies for each element. Let us consider the 'Stiffness Matrix' for elements (1) and (2) as $k^{(1)}$ and $k^{(2)}$ respectively. Thus,

For element (1):

$$\begin{Bmatrix} f_{x1} \\ f_{y1} \\ f_{x2} \\ f_{y2} \end{Bmatrix} = \frac{AE}{L} \begin{bmatrix} k_{11}^{(1)} & k_{12}^{(1)} & k_{13}^{(1)} & k_{14}^{(1)} \\ k_{21}^{(1)} & k_{22}^{(1)} & k_{23}^{(1)} & k_{24}^{(1)} \\ k_{31}^{(1)} & k_{32}^{(1)} & k_{33}^{(1)} & k_{34}^{(1)} \\ k_{41}^{(1)} & k_{42}^{(1)} & k_{43}^{(1)} & k_{44}^{(1)} \end{bmatrix} \begin{Bmatrix} u_1 \\ v_1 \\ u_2 \\ v_2 \end{Bmatrix} \tag{8a}$$

For element (2):

$$\begin{Bmatrix} f_{x2} \\ f_{y2} \\ f_{x3} \\ f_{y3} \end{Bmatrix} = \frac{AE}{L} \begin{bmatrix} k_{11}^{(2)} & k_{12}^{(2)} & k_{13}^{(2)} & k_{14}^{(2)} \\ k_{21}^{(2)} & k_{22}^{(2)} & k_{23}^{(2)} & k_{24}^{(2)} \\ k_{31}^{(2)} & k_{32}^{(2)} & k_{33}^{(2)} & k_{34}^{(2)} \\ k_{41}^{(2)} & k_{42}^{(2)} & k_{43}^{(2)} & k_{44}^{(2)} \end{bmatrix} \begin{Bmatrix} u_2 \\ v_2 \\ u_3 \\ v_3 \end{Bmatrix} \tag{8b}$$

For analysis of the entire structure, these individual Stiffness Matrices are to be assembled to form the equation for the system.

$$\begin{Bmatrix} f_{x1} \\ f_{y1} \\ f_{x2} \\ f_{y2} \\ f_{x3} \\ f_{y3} \end{Bmatrix} = \frac{AE}{L} \begin{bmatrix} k_{11}^{(1)} & k_{12}^{(1)} & k_{13}^{(1)} & k_{14}^{(1)} & 0 & 0 \\ k_{21}^{(1)} & k_{22}^{(1)} & k_{23}^{(1)} & k_{24}^{(1)} & 0 & 0 \\ k_{31}^{(1)} & k_{32}^{(1)} & k_{33}^{(1)} + k_{11}^{(2)} & k_{34}^{(1)} + k_{12}^{(2)} & k_{13}^{(2)} & k_{14}^{(2)} \\ k_{41}^{(1)} & k_{42}^{(1)} & k_{43}^{(1)} + k_{21}^{(2)} & k_{44}^{(1)} + k_{22}^{(2)} & k_{23}^{(2)} & k_{24}^{(2)} \\ 0 & 0 & k_{31}^{(2)} & k_{32}^{(2)} & k_{33}^{(2)} & k_{34}^{(2)} \\ 0 & 0 & k_{41}^{(2)} & k_{42}^{(2)} & k_{43}^{(2)} & k_{44}^{(2)} \end{bmatrix} \begin{Bmatrix} u_1 \\ v_1 \\ u_2 \\ v_2 \\ u_3 \\ v_3 \end{Bmatrix} \tag{9}$$

Equation (9) gives an idea about how the computational demand increases to solve bigger Stiffness Matrix as number of elements increases. Hereafter comes the consideration of fixed boundary conditions. In the given example, nodes (1) and (3) are fixed. Thus the values of u_1, v_1, u_3, and v_3 are zero. This reduces Equation (9) as:

$$\begin{Bmatrix} f_{x2} \\ f_{y2} \end{Bmatrix} = \frac{AE}{L} \begin{bmatrix} k_{33}^{(1)} + k_{11}^{(2)} & k_{34}^{(1)} + k_{12}^{(2)} \\ k_{43}^{(1)} + k_{21}^{(2)} & k_{44}^{(1)} + k_{22}^{(2)} \end{bmatrix} \begin{Bmatrix} u_2 \\ v_2 \end{Bmatrix} \tag{10}$$

Further, if we consider only vertical displacement of node (2) then value of u_2 will also become zero. Thus Equation (10) can be further simplified as:

$$f_{y2} = \frac{AE}{L}\left(k_{44}^{(1)} + k_{22}^{(2)}\right)v_2 \tag{11}$$

Or

$$v_2 = \frac{f_{y2}L}{\left(k_{44}^{(1)} + k_{22}^{(2)}\right)AE}$$

Post Processing of the Outcome

After the analysis, the generated results need to be seen and retrieved up to the required extent and in the required format. The results of our interest may be viewed by plotting their contour over the object geometry (in the form of a continuous color plot) (Fig. 3c) or the results of the nodes or elements of the zone of interest may be listed for achieving their exact value. We may want to retrieve the results of stress (directional, von Mises, etc.), strain, displacement, etc.

CASE STUDIES

In different research studies, the author tried to analyze the biomechanical environment of the implant and peri-implant bone in cases of Orthopaedic and Dental implantation, through FE analysis, in order to suggest some design features that may enhance the post-implantation biomechanical outcome. The scenario of Orthopaedic implantation that the author considered were related to THR. The stress/strain contour over the interfacial bone and the stress/strain distribution within the implant volume were studied. The implant stability and biomechanical response of periprosthetic bone around THR devices depend on a plethora of parameters, including design of articulating materials, gait cycle and physiological parameters. These parameters were considered individually and their effect on the biomechanical response was analyzed. The bone models were reconstructed from CT data using MIMICS®. The implant geometries were modelled using CAD softwares: CATIA® and SolidWorks®. FE modelling and analyses were done in ANSYS®. Assignment of the material properties for the bone was done element-wise based on the attenuation of X-ray. The average attenuation of X-ray in the voxels within the periphery a particular element is used to assign the material properties to that element. This is done in MIMICS® when the meshed models of the bone parts were imported from ANSYS® for this purpose. Virtual osteotomy and implantation was done through slice and Boolean operation. After FE analyses, the contour of the desired outcome were viewed through pictographic display of results (the contour plot) whereas the exact values from were extracted from the results in the text form (the list result).

CASE STUDY 1

Study to estimate the role of pre-surgical bone condition and bodyweight in customization of stiffness of femoral stem, in case of THR, to improve the periprosthetic biomechanical response (Chatterjee et al., 2020)

Background

Customization of the component designs was generally based on the anatomical geometry. This study demonstrated for the first time the importance of other subject parameters, such as bone condition and body weight, to be considered while such design customization. Here internal hollowness has been introduced as the modality of stiffness reduction. The external morphology of the femoral stem was customized according to the femoral geometry (including the geometry of the marrow lumen) while the internal hollowness architecture was matched to the stiffness requirement according to bone condition and body weight. The aim was to attain close to natural strain distribution, and thus stress distribution, at peri-implant bone, to reduce interfacial bone loss due to stress-shielding over time. The biomechanical outcomes were obtained through FE analyses. Preservation of interfacial bone over time has been studied here through analysis of bone remodelling, carried out through manual coding in ANSYS®.

Figure 6. The overall work flow.

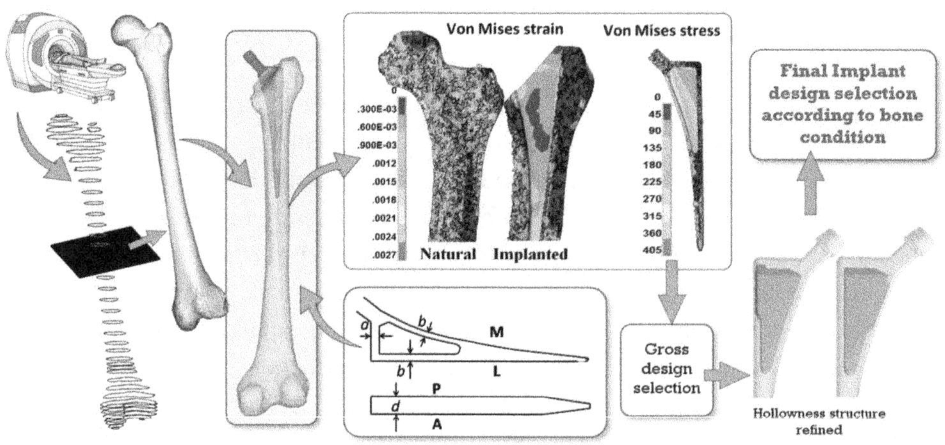

Methods

The customized model of femoral stem was generated in CATIA®. Virtual osteotomy and implantation was done in ANSYS®. The broached channel for implant insertion was simulated by cutting the bone over the implant by a surface generated by extruding the boundary edges of the implant roof. 10-noded tetrahedral elements (SOLID187) were used for meshing (Aversa et al., 2016; Chatterjee, 2013; ten Broeke et al., 2014). Equation 12 shows the empirical relation of density (ρ in g.cm^{-3}) and Elastic Modulus (E in MPa) with HU values (Anderson and Madigan, 2013; Majumder et al., 2007; Norman et al., 2013; Rieger et al., 2013; ten Broeke et al., 2014; Yamako et al., 2014b).

$$\rho = 0.2389 + (0.0008531 \times HU) \tag{12}$$

$$E = 6850 \times \rho^{1.49}$$

The solid implant was considered as the standard based on which the advantages of incorporated hollowness was studied. The un-implanted femur was modelled for comparison for studying the effect of implantation. All models were provided with same constraints and interfacial characteristics. They were loaded with joint load and muscle loads, whose magnitude depended on the body weight. The parameters that were varied in this study were categorized as Subject parameters and Implant parameters. The Subject parameters were 'Bone condition (density)' and 'Body weight'. The scanned data, which belonged to a subject with average bone density (in this case, female), was chosen for FE modelling. Four other bone conditions were simulated by manual alteration of the bone density as 80%, 90%, 110% and 120% of this average case. Three cases of body weights were considered as: 70kg (~686.47N), 90kg (~882.6N) and 110kg (~1078.73N). The Implant parameters were: 'Type of hollowness' and 'Extent of hollowness'.

The criterion for selecting the design of the implant was least deviation of strain in the periprosthetic bone as a result of implantation. The overall strain deviation from the natural femur was calculated by a scalar parameter defined in this study and designated as 'M' such that:

$$\mathrm{M} = |(\epsilon_x - \epsilon_x^{'})| + |(\epsilon_y - \epsilon_y^{'})| + |(\epsilon_z - \epsilon_z^{'})| \tag{13}$$

ϵ_x, ϵ_y, and ϵ_z are the average strains within each zone of the interfacial bone in x, y, and z directions, respectively, while that of the corresponding zones in natural (in-implanted) models are $\epsilon_x^{'}$, $\epsilon_y^{'}$, and $\epsilon_z^{'}$.

Obviously, lower value of 'M' is desirable for an implant design. Thus the option of the values of the implant parameters was chosen for a particular subject parameter considering the lowest value of 'M'. Thereafter, its perpetual effect on the bone formation/resorption over the surrounding bone in the cancellous epiphyseal bone region was estimated through custom coded Bone remodelling algorithm (BRA)

Bone Remodelling Algorithm (BRA)

This algorithm was prepared through coding in Ansys Parametric Design Language (APDL). The objective is to study the perpetual effect of the alteration of loading environment, created due to implantation, on the bone density over time. This is an iterative FE solution procedure based on the concept that the bone density changes according to the external mechanical stimulus. The osteocytes have canaliculus structures which sense the mechanical stimulation and send feedback to some specialized surface lining cells (osteoblasts and osteoclasts). The osteoblasts and osteoclasts are capable of laying down bone mineral matrix and resorping them respectively. Thus the bone density is controlled according to the applied loading. Since any external loading scenario will produce different loading environment at different points within the volume, the extent of bone remodelling is different for zone. Thus BRA was applied in element-wise manner in the FE model. The BRA prepared by the author is meant to register the alteration of bone densities over time in response to the implantation. Thus here, a simplification is imposed in the assumption that the natural bone is having most optimized loading scenario and is not subjected to remodelling. Though this assumption is not strictly true, but it can be adopted to study the bone remodelling which is due to deviation of loading because of implantation. Thus the remodelling stimulus is calculated here based on the deviation of the biomechanical environment due to implantation.

For each element, the change in the density in each iteration ($\Delta\rho$ is obtained according to Equation 14.

$$\Delta\rho = \begin{cases} v(\rho)\{S-(1\pm s)k\}\psi\Delta t, & \text{if } S \le (1-s)k \text{ or } S \ge (1+s)k \\ 0 & \text{if } (1-s)k < S < (1+s)k \end{cases} \tag{14a}$$

'S is the mechanical stimulus for remodelling. 'k' is the reference stimulus. 'ψ' is the adaptation rate, taken as 129.6 g/[mm^2Jg^{-1} month] (Ghosh and Gupta, 2014; Weinans et al., 1993). $v(\rho)$ (mm^2mm^{-3}) is the internal free surface area per unit volume of bone.

'S was obtained for each element from the averaged elastic strain energy density (U for unit bone mass, over a loading history (n, (Eqn. 14b (Ghosh and Gupta, 2014; Ghosh et al., 2013a):

$$S = \frac{1}{n}\sum_{i=1}^{n}\left(\frac{U_i}{vol.\rho}\right) \tag{14b}$$

'k' was obtained in a similar manner from the natural (intact) bone.

In equation 14a we see that when we subtract the reference stimulus 'k' from the obtained stimulus 'S, there is a factor of threshold (sk). We consider that there is no change in bone density if the value of 'S is within $(1\pm s)k$. This is termed as the 'Dead zone' (Fig. 7). In this study, the value of (s) has been taken to be 0.75.

Figure 7. Change of bone density according to the applied mechanical stimulus. The position of the reference stimulus (k), i.e. the stimulus existing in the natural bone, the 'Dead zone', the zones of bone densification and bone resorption, the zone where the cancellous bone turns cortical and the zone of total bone loss are clearly shown in this graph.

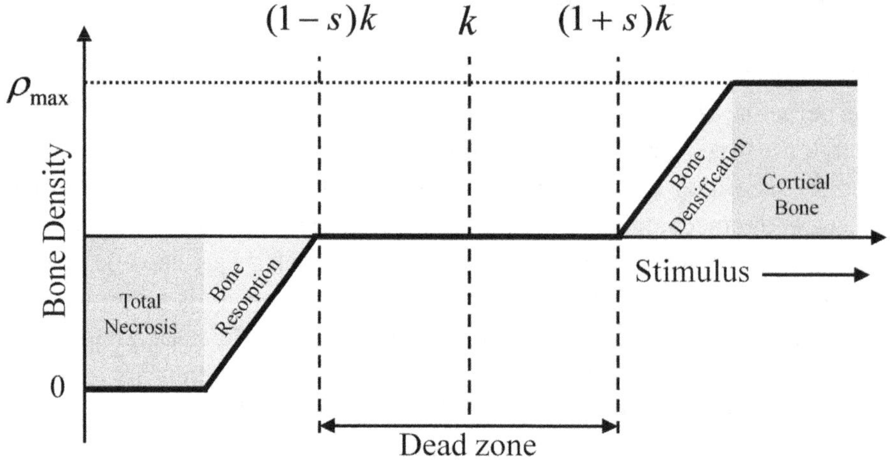

Moreover, it can be anticipated that $v(\rho)$ cannot be measured. We can predict that higher density of cancallous bone is associated with for number of internal filamentous boney structures, leading to higher value of $v(\rho)$. Thus, here also an empirical equation (Eqn. 14c) (Ghosh et al., 2013a; Martin, 1972, 1983) was used to formulate it as a function of the bone density, as clearly indicated by its symbol [$v(\rho)$].

$$v(\rho) = -0.0293 + 8.512\rho - 4.887\rho^2 - 1.568\rho^3 + 3.718\rho^4 - 1.6352\rho^5 \quad (14c)$$

The bone density can increase up to that of the cortical bone, denoted in Fig. 7 as (ρ_{max}). Here it is taken as 1.84 g/cm^3. After a particular iteration, the altered density of each element is obtained from which their altered Elastic Modulus is calculated as according to Eq. 12. With these altered material properties, solution is again done in the next iteration. If the density of a particular bone element (e) reaches ρ_{max} it is excluded from the remodelling iterations. If ρ_e is less than 0.13 g/cm^3, it is considered as complete bone resorption, as hence also excluded from the remodelling iterations. Each iteration can have multiple sub-steps with different loading scenarios considering different phases of an action (e.g. gait cycle). These sub-steps are performed one after the other without altering the material properties of bone. In this case, after the full iteration is completed, we will consider the averaged stimulus (S) of all the sub-steps for Eq. 13(a) for preparing the material properties element-wise for the next iteration. If the change in bone density for the concerned element 'e', (i.e. $\Delta\rho_e$ is negligible for a particular element ($\leq$0.005), we assume that it has attended a condition of optimized density required for the situation and will not remodel further. So this element will excluded from this iterative procedure further. The overall iterative process is terminated after a decided number of iterations. Figure 8 shows the entire working flowchart for the iterative bone remodelling procedure.

Figure 8. The working flowchart for the iterative procedure of bone remodelling.

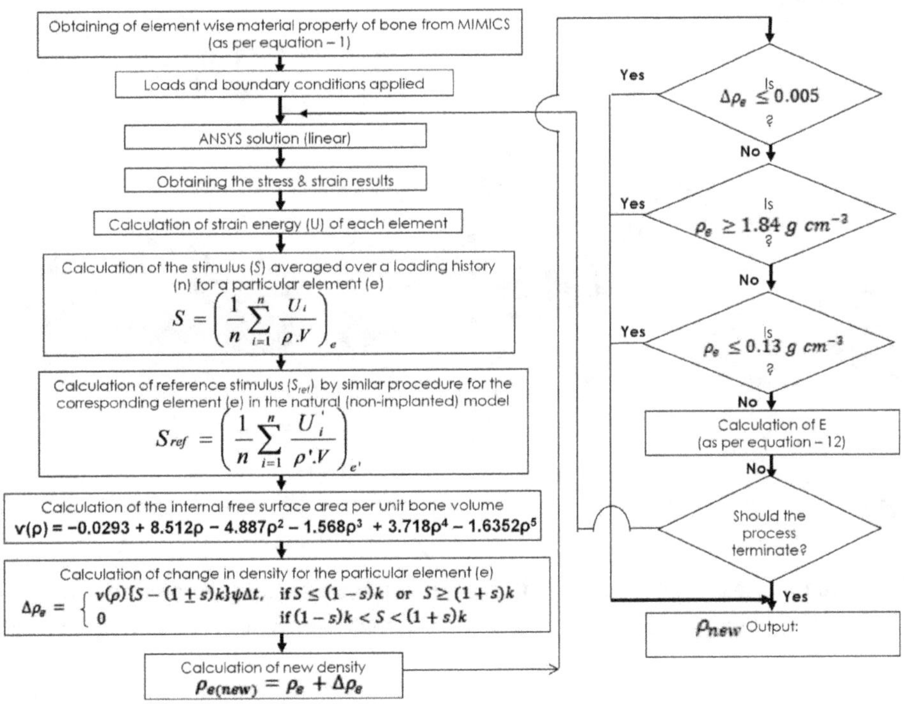

Solutions and Recommendations

Based on the FE analyses, internal hollowness within the femoral stems has been tuned according to the existing bone condition (bone density) and bodyweight in order to prevent stress-shielding and finally implant loosening. The importance of these subject parameters has been established for implant design customization for achieving required biomechanical environment. These parameters should be used to customize the stiffness of the implant, while its overall shape has to be customized according to the bone geometry. Special attention is required for higher body weights since in this case, if we use higher internal hollowness, implant tends to fail if the bone is stronger while damage of bone tends to occur for it is weaker. Thus the extent of hollowness needs to be reduced. A special finding of this FE study is that a gradual shift of the orientation of the internal hollowness from medial to lateral direction has been advised as the bone conditions varies from stronger to weaker. This ensures favourable interfacial bone remodelling over time.

The findings of this study are likely to facilitate designing of femoral stems for achieving better physiological outcomes and enhancement of the quality of life of patients undergoing THR.

CASE STUDY 2

Study on the effect of functional stiffness gradation of dental implants on the biomechanical response, considering single or multiple loss of molar teeth (Chatterjee et al., 2019).

Figure 9. The overall work flow for this case study.

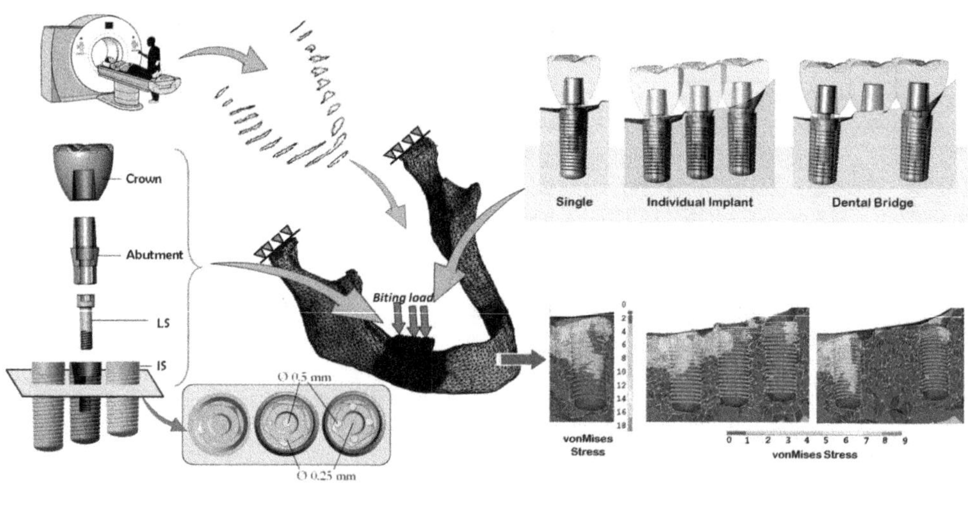

Background

Here also hollow channels were introduced within the implant screws (IS) designs for overall stiffness reduction. Extensive FE analyses were involved to associate the porosity type and approach of implantation with the type of dental loss presented. Two contrasting radial gradients of diameters of the hollow channel were designed. Two clinical situations of edentulism of one side of the mandible i.e. a) loss of one molar (first), and b) loss of all the three molars, were considered in this study. Absence of single/ multiple teeth can be addressed either by dental implantation (Turkyilmaz et al., 2007) or by using a dental bridge supported over the adjoining natural crowns on both sides. Additionally, some studies report use of dental bridge supported over Osseo integrated implants (Fernandez-Redondo et al., 1998; Li et al., 2006; Li et al., 2004a, b; Li et al., 2005; Pietrabissa et al., 2000; Quinn et al., 2010; Shi and Fok, 2009). Figure 10 describes few among the approaches of implantation for the loss of a single/multiple teeth. The implant supported dental bridge (ISDB) designs are used in cases where there is a loss of at least three consecutive teeth (Fig. 10b). In these designs, all of the prosthetic crowns are not supported by IS. Some crowns in the middle are supported by the surrounding prosthetic crowns. The terminal crowns are associated with IS. This design reduces the bone invasion and also risk of fracture in between two consecutive IS. In case of consecutive loss of many teeth (as in case of total edentulism) there are more implant screws in between to support the entire bridge architecture (Fig. 10c).

Figure 10. Few among the approaches of implantation in different clinical situations. (a) Implantation for a single dental loss (https://www.perioperio.com), (b) implant supported dental bridge (ISDB) configuration for loss of three consecutive teeth (https://www.karpovichdental.com) and (c) ISDB for complete mandibular edentulism (https://www.sanfranciscodentalcare.com).

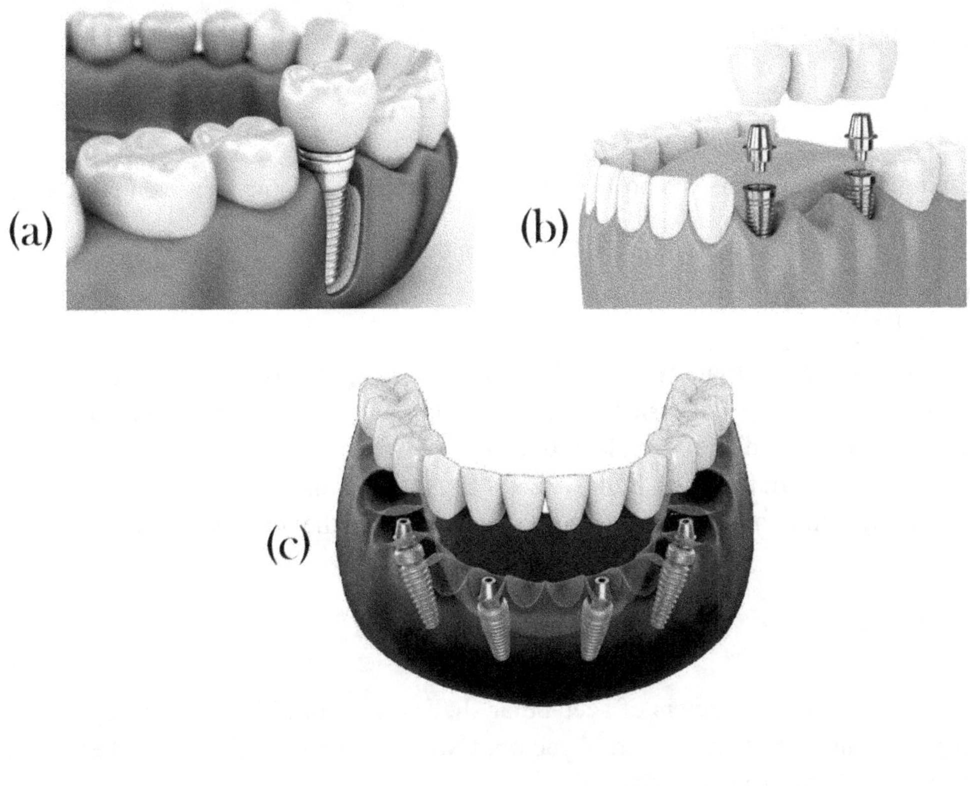

FE analyses have been extensively utilized to study the dental implant biomechanics (Lin et al., 2008; Natali et al., 2010; Turkyilmaz et al., 2007). These studies included areas like analysis of design misfits

(Pietrabissa et al., 2000) or incorporation of a specific design feature like Morse taper (Toniollo et al., 2012) and so on. Chen et al. (2012) adopted 3D FE analysis to study the biomechanics for 3 missing posterior teeth in the mandible, either implanted individually or treated with fixed partial denture supported by two implants.

In this study, both the implantation approaches were simulated for multiple mandibular dental loss (loss of three molars on the right side), i.e. individual implantation and implant supported dental bridge. Loss of a single molar was also considered as the other clinical scenario. FE was used to model and analyze the combined effect of radial gradation of porosity of IS, length of IS and the implantation approach on the biomechanical environment over the peri-prosthetic bone, under physiological masticatory loading.

Methods

The modelling and analytical aspects here are similar to the Case Study 1. Here the dental implant models were generated in SolidWorks®. The design of the individual implant, that was modelled here, was composed of 4 components: the IS (Ti-6Al-4V), the abutment (Ti-6Al-4V), the locking screw (Ti-6Al-4V) and the crown (porcelain). In case of the bridge construction, the complete component architecture was modelled for the 1st and the 3rd molar areas, while only the crown (supported by the adjacent prosthetic crowns) was modelled for the 2nd molar.

Empirical equations similar to Eq. 12 were used to calculate the density and Elastic Modulus from HU, mentioned here as Eq. 15 (Lin et al., 2009; Roy et al., 2017).

$$\rho = 1.028 + (0.000769 \times HU) \tag{15}$$

$$E = 2349 \times \rho^{2.15}$$

The meshing, the FE analysis and post-processing of the FE outcome were done in a similar manner as in Case Study 1.

Solutions and Recommendations

It has been found that the functionally graded IS with increasing density from periphery to core provides desirable interfacial bone stress. Considering this situation, the IS length of 11mm ensures best biomechanical environment. Regarding the implantation approach for multiple molar loss, similar biomechanical condition has been found with both individual implantation and dental bridge configuration. Thus dental bridge configuration may be considered as a choice owing the other benefits described previously.

CASE STUDY 3

Study on the influence of parameters of acetabular shell in case of THR (shell design and liner material) and subject parameters (bone condition and body weight) on the biomechanical environment in the peri-implant bone (Chatterjee et al., 2018).

Background

This study establishes the biomechanical response of periprosthetic bone in the acetabulum with pre-clinically tested liner materials together with new shell design for different subject conditions. Here, the impact of shell design and liner material on the biomechanical response of periprosthetic bone has been analysed using FE method.

The conventional material for joint replacement was metal. Ceramic materials have been used for bearing surfaces in THR for their increased wear resistance (Cilingir, 2010; Slonaker and Goswami, 2004; Zhang et al., 2010). Moreover, the ceramic wear particles are reported to be less hazardous to the body than metals (Essner et al., 2005; Warashina et al., 2003). Polymers can also be used in this case. The conventional polymer for this purpose is ultra high molecular weight polyethylene (UHMWPE). Its wear resistance can be reduced through cross-linking and reducing the free radical concentration (Muratoglu, 2001; Muratoglu et al., 2001; Muratoglu et al., 1999). It is then known as highly crosslinked UHMWPE (HC−UHMWPE). High density polyethylene composite with hydroxyapatite and alumina (HDPE−HA−Al_2O_3) has been developed as an alternative polymer and its osseointegration property has been established (Bodhak et al., 2009; Nath et al., 2007, 2009; Saha et al., 2012; Saha et al., 2010; Tripathi and Basu, 2014; Tripathi et al., 2012; Tripathi et al., 2013).

Developments in design of the acetabular shell put forward some altered designs. This included, among many others, a design with twelve fins and another design with three spikes(Baleani et al., 2001; Perona et al., 1992). An increase of the axial and frontal stability was found with the finned design (Baleani et al., 2001). The overall component motion was mostly decreased with the spiked design (Perona et al., 1992). This is especially required for bone ingrowth.

Figure 11. The overall work flow for this case study.

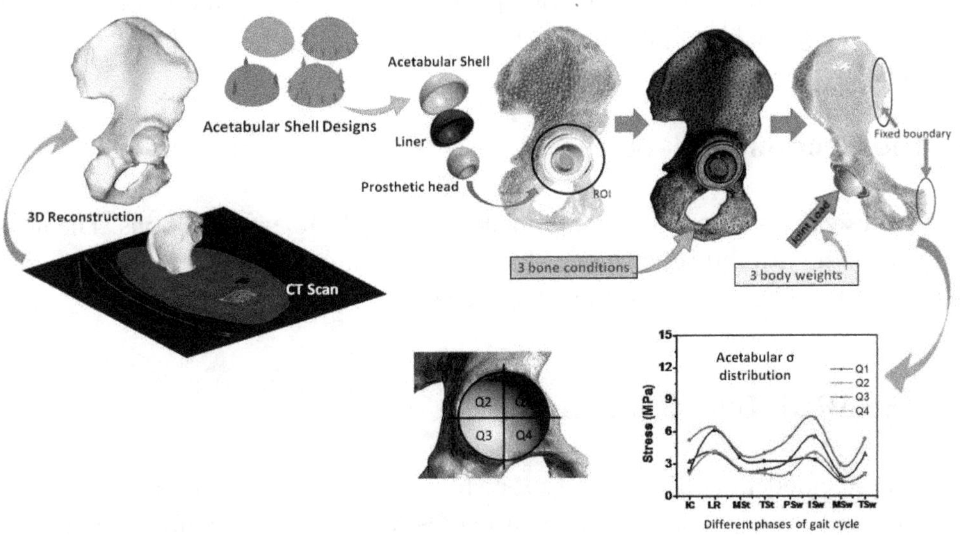

This study involves FE modelling of the entire hemipelvis considering the case of THR. It includes HC−UHMWPE and HDPE−20%HA−20%Al_2O_3 as liner materials and both the above mentioned finned and spiked shell designs. It also includes a new design of combination of these both. The subject parameters included bone condition and bodyweight. Physiologically relevant load cases of a gait cycle were considered.

Methods

The modelling and FE analysis in this study is similar as the previous Case Studies. Cement-less implant has been simulated. The implanted model consisted of the operated hemi-pelvis, acetabular shell, liner and the prosthetic femoral head. The natural model consisted of the natural hemi-pelvis, natural femoral head and articular cartilage. The empirical equations (similar to Eq. 12 and 15) used here to calculate the density and Elastic Modulus from HU are as follows (Anderson and Madigan, 2013; Ghosh and Gupta, 2014; Ghosh et al., 2013b; Ghosh et al., 2015; Taddei et al., 2004):

For pelvis,

$$\rho = 1.028 + (0.000769 \times HU)$$

$$E = 2349 \times \rho^{2.15} \tag{16}$$

For femoral head,

$$\rho = 0.2389 + (0.0008531 \times HU)$$

$$E = 6850 \times \rho^{1.49}$$

21 muscles were considered, whose magnitude was based on body weight (Dalstra and Huiskes, 1995; Mukherjee and Gupta, 2016). 8 significant phases of a complete gait cycle were simulated based on these muscle loadings and also joint load (Dalstra and Huiskes, 1995). Three bone conditions and three body weights were considered.

Solutions and Recommendations

The finned shell design was found to be recommendable for all subject conditions. The new combined design also produced significantly good biomechanical environment either for the stronger or the weaker bone condition. But its effect was insignificant for normal (medium) bone condition.

FUTURE RESEARCH DIRECTIONS

The theory of bone remodelling should be further improved in future that may include the effect of electrical stimulus and consider the loading scenario of activities other than just walking so that the occupation of the person can also be included as a parameter of the remodelling algorithm. Moreover, apart from remodelling of only bone, other remodelling aspects such as angiogenesis, neurogenesis, etc., as a result of the altered mechanical stimulus, should also be modelled in FEM. As a simple extension of the study of acetabular shell design (presented here as Case Study 3) asymmetry in the finned structure may be considered in terms of density and shape/size of the fins in the postero-superior region (the area that is mainly subjected to compressive load during weight bearing) in comparison to the other locations.

CONCLUSION

FEM is an essential tool in investigations in the field of biomechanics. In case of THR, a internally hollow femoral stem is recommended for better biomechanical outcome. While the external shape of the stem may be customized according to the femoral geometry, the shape and orientation of the internal hollowness may be customized according to the density of the existing bone and body weight. Considering the acetabular components, finned structure of the shell has been recommended. In case of dental implants, an increasing gradient of porosity from core towards periphery has been recommended to achieve desired peripheral bone biomechanics. For consecutive multiple dental loss, an implant supported dental bridge configuration is more desirable as it ensures invasiveness without a compromise in biomechanical outcome.

ACKNOWLEDGMENT

These research works were supported by the Department of Biotechnology (DBT), Govt. of India [Grant Numbers: BT/PR13944/MED/32/147/2010; DBTO0455] and Department of Science and Technology, Centre for Mathematical Biology [Grant No. DSTO1303]. The work, presented as Case Study 1, was carried out at Indian Institute of Engineering Science & Technology, Shibpur (IIESTS), Howrah – 711103, West Bengal, India, under the supervision of Prof. Amit Roy Chowdhury and Prof. Santanu Majumder, Department of Aerospace Engineering & Applied Mechanics. The research works, presented as Case Studies 2 and 3 were carried out at Indian Institute of Science (IISc), Bangalore – 560012, Karnataka, India, under the supervision of Prof. Bikramjit Basu, Materials Research Centre.

REFERENCES

Anderson, D. E., & Madigan, M. L. (2013). Effects of age-related differences in femoral loading and bone mineral density on strains in the proximal femur during controlled walking. *Journal of Applied Biomechanics*, *29*(5), 505–516. doi:10.1123/jab.29.5.505 PMID:23185080

Aversa, R., Petrescu, F.I., Petrescu, R.V., & Apicella, A. (2016). *Biofidel FEA modeling of customized hybrid biological hip joint design Part II: Flexible stem trabecular prostheses*. Academic Press.

Baleani, M., Fognani, R., & Toni, A. (2001). Initial stability of a cementless acetabular cup design: Experimental investigation on the effect of adding fins to the rim of the cup. *Artificial Organs*, *25*(8), 664–669. doi:10.1046/j.1525-1594.2001.025008664.x PMID:11531719

Bodhak, S., Nath, S., & Basu, B. (2009). Friction and Wear Properties of Novel HDPE—HAp—Al2O3 Biocomposites against Alumina Counterface. *Journal of Biomaterials Applications*, *23*(5), 407–433. doi:10.1177/0885328208090012 PMID:18667457

Boobalan, V., & Shankar, S. (2013). Investigation on Various Proximal Femoral Stem Shapes for Human Hip Prosthesis Using Finite Element Concepts. *Trends in Biomaterials & Artificial Organs*, 27.

Chatterjee, S., Ghosh, U.B., & Majumder, S., RoyChowdhury, A., & Pal, S. (2013). Customization of Design for Hip Implants. *Indian Journal of Biomechanics*, *4*, 58–62.

Chatterjee, S., Kobylinski, S., & Basu, B. (2018). Finite Element Analysis to Probe the Influence of Acetabular Shell Design, Liner Material, and Subject Parameters on Biomechanical Response in Periprosthetic Bone. *Journal of Biomechanical Engineering*, *140*(10), 101014. doi:10.1115/1.4040249 PMID:30029239

Chatterjee, S., Roy, S., Majumder, S., & RoyChowdhury, A. (2020). Biomechanical Analysis to Probe Role of Bone Condition and Subject Weight in Stiffness Customization of Femoral Stem for Improved Periprosthetic Biomechanical Response. *Journal of Biomechanical Engineering*, *142*(10), 101002. doi:10.1115/1.4046973 PMID:32320044

Chatterjee, S., Sarkar, S., Kalidindi, S. R., & Basu, B. (2019). Periprosthetic biomechanical response towards dental implants, with functional gradation, for single/multiple dental loss. *Journal of the Mechanical Behavior of Biomedical Materials*, *94*, 249–258. doi:10.1016/j.jmbbm.2019.03.001 PMID:30928669

Chen, X.-Y., Zhang, C.-Y., Nie, E.-M., & Zhang, M. (2012). Treatment planning of implants when 3 mandibular posterior teeth are missing: A 3-dimensional finite element analysis. *Implant Dentistry*, *21*(4), 340–343. doi:10.1097/ID.0b013e31825cbc67 PMID:22814561

Cilingir, A. C. (2010). Finite element analysis of the contact mechanics of ceramic-on-ceramic hip resurfacing prostheses. *Journal of Bionics Engineering*, *7*(3), 244–253. doi:10.1016/S1672-6529(10)60247-8

Cipriano, C. A., Issack, P. S., Beksaç, B., Della Valle, A. G., Sculco, T. P., & Salvati, E. A. (2008). Metallosis after metal-on-polyethylene total hip arthroplasty. *American Journal of Orthopedics (Belle Mead, N.J.)*, *37*, E18–E25. PMID:18401490

Dalstra, M., & Huiskes, R. (1995). Load transfer across the pelvic bone. *Journal of Biomechanics*, *28*(6), 715–724. doi:10.1016/0021-9290(94)00125-N PMID:7601870

Driscoll, M. (2019). The Impact of the Finite Element Method on Medical Device Design. *Journal of Medical and Biological Engineering*, *39*(2), 171–172. doi:10.100740846-018-0428-4

Essner, A., Sutton, K., & Wang, A. (2005). Hip simulator wear comparison of metal-on-metal, ceramic-on-ceramic and crosslinked UHMWPE bearings. *Wear*, *259*(7-12), 992–995. doi:10.1016/j.wear.2005.02.104

Fernandes, P., Folgado, J., & Ruben, R. (2004). Shape optimization of a cementless hip stem for a minimum of interface stress and displacement. *Computer Methods in Biomechanics and Biomedical Engineering*, *7*(1), 51–61. doi:10.1080/10255840410001661637 PMID:14965880

Fernandez-Redondo, V., Gomez-Centeno, P., & Toribio, J. (1998). Chronic urticaria from a dental bridge. *Contact Dermatitis*, *38*(3), 178–179. doi:10.1111/j.1600-0536.1998.tb05696.x PMID:9536421

Fisher, J. (2011). Bioengineering reasons for the failure of metal-on-metal hip prostheses. *The Journal of Bone and Joint Surgery. British Volume*, *93*(8), 1001–1004. doi:10.1302/0301-620X.93B8.26936 PMID:21768619

Ghosh, R., & Gupta, S. (2014). Bone remodelling around cementless composite acetabular components: The effects of implant geometry and implant–bone interfacial conditions. *Journal of the Mechanical Behavior of Biomedical Materials, 32*, 257–269. doi:10.1016/j.jmbbm.2014.01.010 PMID:24508712

Ghosh, R., Gupta, S., Dickinson, A., & Browne, M. (2013a). Experimental validation of numerically predicted strain and micromotion in intact and implanted composite hemi-pelvises. *Proceedings of the Institution of Mechanical Engineers. Part H, Journal of Engineering in Medicine, 227*(2), 162–174. doi:10.1177/0954411912461238 PMID:23513987

Ghosh, R., Mukherjee, K., & Gupta, S. (2013b). Bone remodelling around uncemented metallic and ceramic acetabular components. *Proceedings of the Institution of Mechanical Engineers. Part H, Journal of Engineering in Medicine, 227*(5), 490–502. doi:10.1177/0954411913478703 PMID:23637259

Ghosh, R., Pal, B., Ghosh, D., & Gupta, S. (2015). Finite element analysis of a hemi-pelvis: The effect of inclusion of cartilage layer on acetabular stresses and strain. *Computer Methods in Biomechanics and Biomedical Engineering, 18*(7), 697–710. doi:10.1080/10255842.2013.843674 PMID:24156480

Harris, W. H. (2012). Edge loading has a paradoxical effect on wear in metal-on-polyethylene total hip arthroplasties. *Clinical Orthopaedics and Related Research, 470*(11), 3077–3082. doi:10.100711999-012-2330-7 PMID:22644421

Hua, X., Wang, L., Al-Hajjar, M., Jin, Z., Wilcox, R. K., & Fisher, J. (2014). Experimental validation of finite element modelling of a modular metal-on-polyethylene total hip replacement. *Proceedings of the Institution of Mechanical Engineers. Part H, Journal of Engineering in Medicine, 228*(7), 682–692. doi:10.1177/0954411914541830 PMID:24963036

Kärrholm, J., Anderber, C., Snorrason, F., Thanner, J., Langeland, N., Malchau, H., & Herberts, P. (2002). Evaluation of a femoral stem with reduced stiffness: A randomized study with use of radiostereometry and bone densitometry. *JBJS, 84*(9), 1651–1658. doi:10.2106/00004623-200209000-00020 PMID:12208924

Kraaij, G., Zadpoor, A.A., Tuijthof, G.J., Dankelman, J., Nelissen, R.G., & Valstar, E.R. (2014). Mechanical properties of human bone–implant interface tissue in aseptically loose hip implants. *Journal of the Mechanical Behavior of Biomedical Materials, 38*, 59-68.

Langton, D., Jameson, S., Joyce, T., Gandhi, J., Sidaginamale, R., Mereddy, P., Lord, J., & Nargol, A. (2011). Accelerating failure rate of the ASR total hip replacement. *The Journal of Bone and Joint Surgery, 93*(8), 1011–1016. doi:10.1302/0301-620X.93B8.26040 PMID:21768621

Li, Q., Ichim, I., Loughran, J., Li, W., Swain, M., & Kieser, J. (n.d.). Numerical simulation of crack formation in all ceramic dental bridge. Key Engineering Materials.

Li, W., Swain, M., Li, Q., Ironside, J., & Steven, G. (2004a). Fibre reinforced composite dental bridge. Part II: Numerical investigation. *Biomaterials, 25*(20), 4995–5001. doi:10.1016/j.biomaterials.2004.01.011 PMID:15109861

Li, W., Swain, M., Li, Q., Ironside, J., & Steven, G. (2004b). Fibre reinforced composite dental bridge.: Part I: experimental investigation. *Biomaterials, 25*(20), 4987–4993. doi:10.1016/j.biomaterials.2004.01.010 PMID:15109860

Li, W., Swain, M. V., Li, Q., & Steven, G. P. (2005). Towards automated 3D finite element modeling of direct fiber reinforced composite dental bridge. *Journal of Biomedical Materials Research. Part B, Applied Biomaterials*, *74*(1), 520–528. doi:10.1002/jbm.b.30233 PMID:15912531

Lin, C.-L., Wang, J.-C., Ramp, L. C., & Liu, P.-R. (2008). Biomechanical response of implant systems placed in the maxillary posterior region under various conditions of angulation, bone density, and loading. *The International Journal of Oral & Maxillofacial Implants*, 23. PMID:18416413

Lin, D., Li, Q., Li, W., & Swain, M. (2009). Dental implant induced bone remodeling and associated algorithms. *Journal of the Mechanical Behavior of Biomedical Materials*, *2*(5), 410–432. doi:10.1016/j.jmbbm.2008.11.007 PMID:19627848

Majumder, S., Roychowdhury, A., & Pal, S. (2007). Simulation of hip fracture in sideways fall using a 3D finite element model of pelvis–femur–soft tissue complex with simplified representation of whole body. *Medical Engineering & Physics*, *29*(10), 1167–1178. doi:10.1016/j.medengphy.2006.11.001 PMID:17270483

Martin, R. B. (1972). The effects of geometric feedback in the development of osteoporosis. *Journal of Biomechanics*, *5*(5), 447–455. doi:10.1016/0021-9290(72)90003-6 PMID:4667271

Martin, R. B. (1983). Porosity and specific surface of bone. *Critical Reviews in Biomedical Engineering*, *10*, 179–222. PMID:6368124

Massoud, S. N., Hunter, J. B., Holdsworth, B. J., Wallace, W. A., & Juliusson, R. (1997). Early femoral loosening in one design of cemented hip replacement. *The Journal of Bone and Joint Surgery. British Volume*, *79*(4), 603–608. doi:10.1302/0301-620X.79B4.0790603 PMID:9250746

Mattheck, C., Vorberg, U., & Kranz, C. (1990). Effects of hollow shaft endoprosthesis on stress distribution in cortical bone. *Biomedizinische Technik. Biomedical Engineering*, *35*, 316–319. doi:10.1515/bmte.1990.35.12.316 PMID:2078647

Mukherjee, K., & Gupta, S. (2016). The effects of musculoskeletal loading regimes on numerical evaluations of acetabular component. *Proceedings of the Institution of Mechanical Engineers. Part H, Journal of Engineering in Medicine*, *230*(10), 918–929. doi:10.1177/0954411916661368 PMID:27475907

Muratoglu, O. (2001). A novel method of crosslinking UHMWPE to improve wear, reduce oxidation and retain mechanical properties. *The Journal of Bone and Joint Surgery. British Volume*, *83*, 447.

Muratoglu, O. K., Bragdon, C. R., O'Connor, D., Perinchief, R. S., Estok, D. M. II, Jasty, M., & Harris, W. H. (2001). Larger diameter femoral heads used in conjunction with a highly cross-linked ultra–high molecular weight polyethylene: A new concept. *The Journal of Arthroplasty*, *16*(8), 24–30. doi:10.1054/arth.2001.28376 PMID:11742447

Muratoglu, O. K., Bragdon, C. R., O'Connor, D. O., Jasty, M., Harris, W. H., Gul, R., & McGarry, F. (1999). Unified wear model for highly crosslinked ultra-high molecular weight polyethylenes (UHMWPE). *Biomaterials*, *20*(16), 1463–1470. doi:10.1016/S0142-9612(99)00039-3 PMID:10458559

Natali, A.N., Carniel, E.L., & Pavan, P.G. (2010). Modelling of mandible bone properties in the numerical analysis of oral implant biomechanics. *Computer Methods and Programs in Biomedicine, 100*, 158-165.

Nath, S., Bodhak, S., & Basu, B. (2007). Tribological investigation of novel HDPE-HAp-Al2O3 hybrid biocomposites against steel under dry and simulated body fluid condition. *Journal of Biomedical Materials Research. Part A, 83*(1), 191–208. doi:10.1002/jbm.a.31203 PMID:17397040

Nath, S., Bodhak, S., & Basu, B. (2009). HDPE-Al2O3-HAp composites for biomedical applications: Processing and characterizations. *Journal of Biomedical Materials Research. Part B, Applied Biomaterials, 88*(1), 1–11. doi:10.1002/jbm.b.31050 PMID:18338785

Norman, T. L., Shultz, T., Noble, G., Gruen, T., & Blaha, J. (2013). Bone creep and short and long term subsidence after cemented stem total hip arthroplasty (THA). *Journal of Biomechanics, 46*(5), 949–955. doi:10.1016/j.jbiomech.2012.12.010 PMID:23357700

Perona, P. G., Lawrence, J., Paprosky, W. G., Patwardhan, A. G., & Sartori, M. (1992). Acetabular micromotion as a measure of initial implant stability in primary hip arthroplasty: An in vitro comparison of different methods of intial acetabular component fixation. *The Journal of Arthroplasty, 7*(4), 537–547. doi:10.1016/S0883-5403(06)80076-8 PMID:1479374

Pietrabissa, R., Contro, R., Quaglini, V., Soncini, M., Gionso, L., & Simion, M. (2000). Experimental and computational approach for the evaluation of the biomechanical effects of dental bridge misfit. *Journal of Biomechanics, 33*(11), 1489–1495. doi:10.1016/S0021-9290(00)00089-0 PMID:10940408

Quinn, G., Studart, A.R., Hebert, C., VerHoef, J., & Arola, D. (2010). Fatigue of zirconia and dental bridge geometry: Design implications. *Dental Materials, 26*, 1133-1136.

Rieger, J. S., Jaeger, S., Schuld, C., Kretzer, J. P., & Bitsch, R. G. (2013). A vibrational technique for diagnosing loosened total hip endoprostheses: An experimental sawbone study. *Medical Engineering & Physics, 35*(3), 329–337. doi:10.1016/j.medengphy.2012.05.007 PMID:22673003

Roy, S., Das, M., Chakraborty, P., Biswas, J.K., Chatterjee, S., Khutia, N., & Saha, S., & RoyChowdhury, A. (2017). Optimal selection of dental implant for different bone conditions based on the mechanical response. *Acta of Bioengineering and Biomechanics*, 19. PMID:28869633

Ruben, R. B., Fernandes, P. R., & Folgado, J. (2012). On the optimal shape of hip implants. *Journal of Biomechanics, 45*(2), 239–246. doi:10.1016/j.jbiomech.2011.10.038 PMID:22115063

Ruben, R. B., Folgado, J., & Fernandes, P. R. (2007). Three-dimensional shape optimization of hip prostheses using a multicriteria formulation. *Structural and Multidisciplinary Optimization, 34*(3), 261–275. doi:10.100700158-006-0072-4

Saha, N., Dubey, A. K., & Basu, B. (2012). Cellular proliferation, cellular viability, and biocompatibility of HA-ZnO composites. *Journal of Biomedical Materials Research. Part B, Applied Biomaterials, 100*(1), 256–264. doi:10.1002/jbm.b.31948 PMID:22102555

Saha, N., Keskinbora, K., Suvaci, E., & Basu, B. (2010). Sintering, microstructure, mechanical, and antimicrobial properties of HAp-ZnO biocomposites. *Journal of Biomedical Materials Research. Part B, Applied Biomaterials, 95*(2), 430–440. doi:10.1002/jbm.b.31734 PMID:20878929

Schmidt, J., & Hackenbroch, M. H. (1994). The Cenos hollow stem in total hip arthroplasty: First experiences in a prospective study. *Archives of Orthopaedic and Trauma Surgery*, *113*(3), 117–120. doi:10.1007/BF00441616 PMID:8054230

Shi, L., & Fok, A.S. (2009). Structural optimization of the fibre-reinforced composite substructure in a three-unit dental bridge. *Dental Materials, 25*, 791-801.

Sjodin, B. (2016). What's The Difference Between FEM, FDM, and FVM? In S. Mraz, R. Begg, & M. McBurnett (Eds.), *Machine Design*. Endeavor Business Media, LLC.

Slonaker, M., & Goswami, T. (2004). Review of wear mechanisms in hip implants: Paper II–ceramics IG004712. *Materials & Design*, *25*(5), 395–405. doi:10.1016/j.matdes.2003.11.011

Stops, A., Wilcox, R., & Jin, Z. (2012). Computational modelling of the natural hip: A review of finite element and multibody simulations. *Computer Methods in Biomechanics and Biomedical Engineering*, *15*(9), 963–979. doi:10.1080/10255842.2011.567983 PMID:21574077

Taddei, F., Pancanti, A., & Viceconti, M. (2004). An improved method for the automatic mapping of computed tomography numbers onto finite element models. *Medical Engineering & Physics*, *26*(1), 61–69. doi:10.1016/S1350-4533(03)00138-3 PMID:14644599

ten Broeke, R. H., Tarala, M., Arts, J. J., Janssen, D. W., Verdonschot, N., & Geesink, R. G. (2014). Improving peri-prosthetic bone adaptation around cementless hip stems: A clinical and finite element study. *Medical Engineering & Physics*, *36*(3), 345–353. doi:10.1016/j.medengphy.2013.12.006 PMID:24378381

Toniollo, M. B., Macedo, A. P., Palhares, D., Calefi, P. L., Sorgini, D. B., & Mattos, M. G. C. d. (2012). Morse taper implants at different bone levels: A finite element analysis of stress distribution. *Brazilian Journal of Oral Sciences*, *11*, 440–444.

Tripathi, G., & Basu, B. (2014). In vitro osteogenic cell proliferation, mineralization, and in vivo osseointegration of injection molded high-density polyethylene-based hybrid composites in rabbit animal model. *Journal of Biomaterials Applications*, *29*(1), 142–157. doi:10.1177/0885328214520805 PMID:24452882

Tripathi, G., Dubey, A. K., & Basu, B. (2012). Evaluation of physico-mechanical properties and in vitro biocompatibility of compression molded HDPE based biocomposites with HA/Al2O3 ceramic fillers and titanate coupling agents. *Journal of Applied Polymer Science*, *124*(4), 3051–3063. doi:10.1002/app.35339

Tripathi, G., Gough, J. E., Dinda, A., & Basu, B. (2013). In vitro cytotoxicity and in vivo osseointergration properties of compression-molded HDPE-HA-Al2O3 hybrid biocomposites. *Journal of Biomedical Materials Research. Part A*, *101*(6), 1539–1549. doi:10.1002/jbm.a.34452 PMID:23065866

Turkyilmaz, I., Tözüm, T., & Tumer, C. (2007). Bone density assessments of oral implant sites using computerized tomography. *Journal of Oral Rehabilitation*, *34*(4), 267–272. doi:10.1111/j.1365-2842.2006.01689.x PMID:17371564

Virulsri, C., Tangpornprasert, P., & Romtrairat, P. (2015). Femoral hip prosthesis design for Thais using multi-objective shape optimization. *Materials & Design*, *68*, 1–7. doi:10.1016/j.matdes.2014.11.027

Wagner, P., Olsson, H., Ranstam, J., Robertsson, O., Zheng, M. H., & Lidgren, L. (2012). Metal-on-metal joint bearings and hematopoetic malignancy: A review. *Acta Orthopaedica*, *83*(6), 553–558. doi:10.3109/17453674.2012.747055 PMID:23140092

Warashina, H., Sakano, S., Kitamura, S., Yamauchi, K.-I., Yamaguchi, J., Ishiguro, N., & Hasegawa, Y. (2003). Biological reaction to alumina, zirconia, titanium and polyethylene particles implanted onto murine calvaria. *Biomaterials*, *24*(21), 3655–3661. doi:10.1016/S0142-9612(03)00120-0 PMID:12818536

Weinans, H., Huiskes, R., Van Rietbergen, B., Sumner, D., Turner, T., & Galante, J. (1993). Adaptive bone remodeling around bonded noncemented total hip arthroplasty: A comparison between animal experiments and computer simulation. *Journal of Orthopaedic Research*, *11*(4), 500–513. doi:10.1002/jor.1100110405 PMID:8340823

Yamako, G., Chosa, E., Totoribe, K., Hanada, S., Masahashi, N., Yamada, N., & Itoi, E. (2014a). In-vitro biomechanical evaluation of stress shielding and initial stability of a low-modulus hip stem made of β type Ti-33.6 Nb-4Sn alloy. *Medical Engineering & Physics*, *36*(12), 1665–1671. doi:10.1016/j.medengphy.2014.09.002 PMID:25282098

Yamako, G., Chosa, E., Zhao, X., Totoribe, K., Watanabe, S., Sakamoto, T., & Nakane, N. (2014b). Load-transfer analysis after insertion of cementless anatomical femoral stem using pre-and post-operative CT images based patient-specific finite element analysis. *Medical Engineering & Physics*, *36*(6), 694–700. doi:10.1016/j.medengphy.2014.02.018 PMID:24629623

Yang, C.-T., Wei, H.-W., Kao, H.-C., & Cheng, C.-K. (2009). Design and test of hip stem for medullary re-vascularization. *Medical Engineering & Physics*, *31*(8), 994–1001. doi:10.1016/j.medengphy.2009.06.001 PMID:19581119

Zhang, W., Titze, M., Cappi, B., Wirtz, D., Telle, R., & Fischer, H. (2010). Improved mechanical long-term reliability of hip resurfacing prostheses by using silicon nitride. *Journal of Materials Science. Materials in Medicine*, *21*(11), 3049–3057. doi:10.100710856-010-4144-z PMID:20725769

ADDITIONAL READING

Bonet, J., Gil, A. J., & Wood, R. D. (2012). *Worked examples in nonlinear continuum mechanics for finite element analysis*. doi:10.1017/CBO9781139088046

Borst, R., Crisfield, M. A., Remmers, J. J. C., & Verhoosel, C. V. (2012). *Non-linear finite element analysis of solids and structures* (2nd ed.). Wiley. doi:10.1002/9781118375938

Erdemir, A., Guess, T. M., Halloran, J., Tadepalli, S. C., & Morrison, T. M. (2012). Considerations for reporting finite element analysis studies in biomechanics. *Journal of Biomechanics*, *45*(4), 625–633. doi:10.1016/j.jbiomech.2011.11.038 PMID:22236526

Geng, J., Tan, K. B. C., & Liu, G. (2001). Application of finite element analysis in implant dentistry: A review of the literature. *The Journal of Prosthetic Dentistry*, *85*(6), 585–598. doi:10.1067/mpr.2001.115251 PMID:11404759

Koutromanos, I. (2018). *Fundamentals of Finite Element Analysis: Linear Finite Element Analysis*. Wiley.

Kraaij, G., Zadpoor, A. A., Tuijthof, G. J. M., Dankelman, J., Nelissen, R. G. H. H., & Valstar, E. R. (2014). Mechanical properties of human bone–implant interface tissue in aseptically loose hip implants. *Journal of the Mechanical Behavior of Biomedical Materials*, *38*, 59–68. doi:10.1016/j.jmbbm.2014.06.010 PMID:25023868

Oñate, E. (2009). *Structural Analysis with the Finite Element Method. Linear Statics* (Vol. 1&2). Springer. doi:10.1007/978-1-4020-8733-2

KEY TERMS AND DEFINITIONS

Bone Remodelling: It is a continuous process over time by which the old bone tissue is replaced by new ones. Additionally, through this process, the bone re-adjusts its structure and density in a location wise manner if there is any alteration in the biomechanical environment which may be a pathology, altered activities, etc.

Element (in FEM): It is a building block of the volume of an object and is of a regular geometry which is any one of the list of geometries that a FE software package contains.

Implant: A medical device surgically inserted within the body. Their uses include replacement of a joint or support to a fractured bone (orthopaedic), replacement of lost tooth/teeth (dental), maintenance of arterial lumen (stent), reconstruction of breast or buttocks (cosmetic), etc.

Mesh (in FEM): The process of virtual splitting of a volume into multiple blocks of regular shape (elements) in a FE model.

Node (in FEM): The points through which the various elements connect to each other and transfers force, displacement, etc. in a meshed FE model.

Stress-Shielding: Reduction of physiological stress from the bone at a particular location, often due to implantation, resulting in depletion of bone tissue in that location, as governed by Wolff's law.

Total Hip Replacement: Surgical process of replacing the articular surfaces of the hip joint by the acetabular and femoral implant components.

APPENDIX: ABBREVIATIONS

AOI : Area of interest
APDL : Ansys parametric design language
BRA : Bone remodelling algorithm
CAD : Computer aided design
CT : Computed tomography
DICOM : Digital imaging and communications in medicine
FDM : Finite difference method
FEM : Finite element method
FVM : Finite volume method
HC–UHMWPE : Highly cross linked UHMWPE
HDPE–HA–Al_2O_3 : High density polyethylene composite with hydroxyapatite and alumina
HU : Hounsfield Units
IS : Implant screw.
ISDB : Implant supported dental bridge
LS : Locking screw
MRI : Magnetic resonance imaging
PDE : Partial differential equations
STP : Standard pressure and temperature
THR : Total hip replacement
UHMWPE : Ultra high molecular weight polyethylene

Chapter 3
Design for Additive Manufacturing in Medical Devices

Shammodip Roy
Stryker Corporation, USA

ABSTRACT

The chapter will cover the fundamentals of product design principles for additive manufacturing in medical devices, with a key focus on orthopedic implants and related devices. The chapter will describe commonly used additive manufacturing processes for medical devices and orthopedics, key design considerations, and how they have opened up possibilities in healthcare. The chapter will detail the fundamentals of design principles, which are expanding the boundaries of rapid, meaningful innovation and positively impacting the innovation cycles of a wide range of industries. It will describe how topology optimization, with the help of computing power, is providing design engineers with the tools to accurately understand structure-functional relationships of designs and in the process re-imagine the biomedical designs of tomorrow.

INTRODUCTION

Additive manufacturing is a general term for technologies which, based on a geometrical representation, creates physical objects by successive addition of material (ISO/ASTM 52900:2015). These technologies are used across applications in various industries such as healthcare, aerospace, architecture, automobiles, archaeology, consumer goods, toys, food and entertainment.

All definitions for terms and nomenclature associated with additive manufacturing technology used herein are according to ISO/ASTM 52900:2015.

Applications of additive manufacturing, also referred to as "3D printing", span across multiple applications in both metallic materials and plastics. Common additive manufacturing techniques used for metallic materials include Laser Rapid Manufacture (LRM) or Direct Metal Laser Sintering (DMLS) and Electron beam Melting (EBM). Commonly used additive techniques for plastic materials are Fused

DOI: 10.4018/978-1-7998-9078-2.ch003

Deposition Modeling (FDM), Material Jetting (MJ), Stereo-lithography (SLA), Multi Jet Fusion (MJF) and Selective Laser Sintering (SLS). Most additive manufacturing methods make use of one or more of the below principles:

1. Vat photopolymerization: This is an additive manufacturing process in which a liquid photopolymer is selectively cured by light-activated polymerization in a vat[1].
2. Material extrusion: This is an additive manufacturing process in which a material is selectively extruded through a nozzle or orifice[1].
3. Material jetting: This is an additive manufacturing process in which droplets of a material are selectively deposited[1].
4. Binder jetting: This is an additive manufacturing process in which a binder liquid is deposited on layers of powdered materials, selectively joined together, and then followed by a densification process (Mostafaei et al., 2021).
5. Powder bed fusion1: This is an additive manufacturing process in which thermal energy is used to selectively fuse regions of a powder bed.

Depending on the materials that each of these techniques can build, they are used for different types of applications across both product prototyping and large-scale manufacturing. For example, while FDM is widely used to prototype complex geometries out of relatively low-cost plastic materials within a short amount of time, SLS is more commonly used to manufacture more structurally sound parts with better surface finish over a longer build time. Figure 1 describes some of the common additive manufacturing technologies, their underlying principles and their material compatibility for different types of applications.

Figure 1. Additive manufacturing technologies
Source: 3D Hubs, Knowledge Database

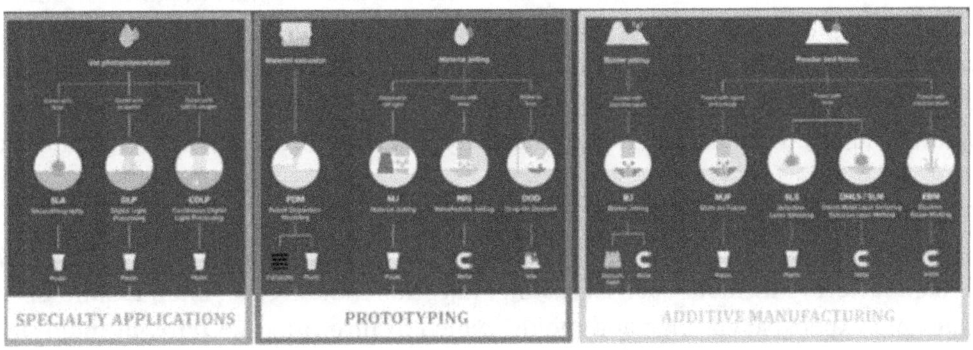

In healthcare industry and research, both plastic and metal additive technologies are being rapidly adopted globally across multiple applications. The orthopedic implant industry has seen an explosion of additively manufactured prostheses over the last two decades, and currently is one of the highest commercial users of large-scale additive manufacturing.

Additively manufactured spinal cages are currently used to treat patients undergoing spinal fusions, 3D printed hip stems and acetabular cups are used to treat patients undergoing total and partial hip arthroplasty procedures, and additively manufactured femoral and tibial implants are being increasingly used to treat patients undergoing total and partial knee arthroplasty procedures.

There are several technical and financial advantages of shifting from traditional manufacturing techniques such as milling, forging, casting etc. to additive manufacturing, when it comes to fabricating orthopedic implants. Some key advantages are described below:

Patient-Specific Applications

Human anatomy is structurally complex, hence difficult and expensive to accurately fabricate using traditional manufacturing techniques. Additive technologies, when used in conjunction with 3D segmentation and modeling tools, are able to fabricate anatomical models of interest as well as patient-specific implant models from medical scans of patients. Examples include 3D printing of mandibular surgical models for planning of oral surgery, 3D printing of custom pelvic plate for cancer patients with severe pelvic bone loss, 3D printing of custom dental implants and 3D printing of patient-specific knee cutting guides for knee arthroplasty surgery.

Light-Weight Structures

Additive manufacturing allows the use of specific lattice structures, with pre-determined porosity, to be integrated within build volumes of medical devices, which can reduce overall material weight while maintaining the required strength of the part. Such modulation of micro-structure within a part is not possible with traditional manufacturing techniques. Different lattice structures such as diamond and cylindrical cell patterns are commonly used in additive designs. Topology optimization is another technique used to light-weight part designs, such that material is added where needed and removed from where it is not needed, depending on the actual function of the part. A topology optimized part is often a geometrically more complex version of its parent version that lends itself to additive manufacturing techniques while being lighter, cheaper and often stronger than the parent part. Light-weighting has several advantages such as lowering stress shielding effect in implants, reducing material and manufacturing cost, and improving surface properties of the device (surface porosity can allow for bone on-growth). Example is a 3D printed porous hip stem for hip arthroplasty surgery that allows for cement less fixation into bone.

Accelerate Product Development Cycles

Traditional manufacturing methods such as casting, injection molding etc. often require up-front capital investment into expensive castings and molds, which in turn often go through multiple iterations depending on the evolution of the part design. Additive manufacturing does not require such iterative capital investments, and hence is quicker and cheaper to iterate a design through. This greatly reduces the overall time and cost of product development cycles.

Part Consolidation

Two or more parts of a multi-part device assembly can be combined and consolidated into a single part build, using additive manufacturing techniques, in the process reducing the number of components and eliminating the need for subsequent assembly process(es). This is common in automobile and aerospace industries, where multiple parts are built concurrently in the "as assembled" state, resulting in significant cost and time savings. Surgical instruments often involve multiple component assemblies, and hence offer potential opportunities for part consolidation in the future.

DESIGN FOR ADDITIVE MANUFACTURING

Given the promise of additive manufacturing today, especially in healthcare and orthopedics, it is important to understand that designing for additive manufacturing technologies requires a different approach than traditional manufacturing methods. To a design engineer, additive manufacturing promises a substantially greater design freedom than traditional manufacturing techniques such as milling, casting, forging, molding, etc. There is a paradigm shift from "design what we can make" for traditional manufacturing technologies to "make what we can design" for additive manufacturing techniques (Zeidler, 2020).

Figure 2 shows a staircase model illustrating different levels of adoption for additive manufacturing[5], at the tip of which is Design for Additive manufacturing (DfAM), which when optimized can deliver new product designs that are truly customized for performance and deliver use benefits over their lifetimes.

Figure 2. Staircase model showing different models of adoption of additive manufacturing[5]
Source: Additive manufacturing staircase, 2022.

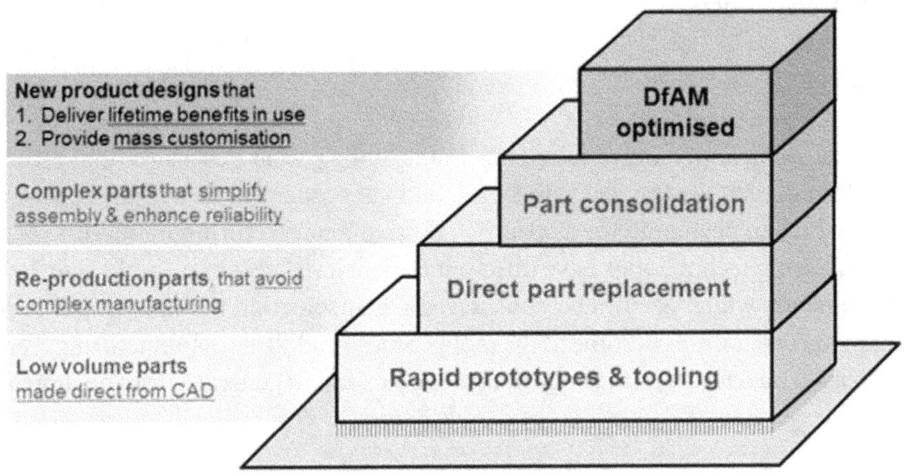

When starting to design for additive manufacturing, it is critical to select the most suitable additive manufacturing process for the given application, out of the many available 3D printing techniques and materials. Below are the considerations of the end application to keep in mind, when selecting an optimal additive manufacturing process:

Form

It is important to understand the required print size, surface finish, visual and tactile feel. This will be dependent on what this part will be used for – For example, is it a quick prototype to communicate general shape? Or, is it intended to communicate final look and feel of the part to an end user?

Fit

The required accuracy and precision of the detailed features on the printed part, and subsequent need for post-processing (milling, sanding, etc) are important aspects to consider, when choosing the additive manufacturing process. This could be dependent on what this part is intended to be used with - Is it intended to mate with a machined part with tight tolerances? Or, is it intended to mate with another 3D printed part? Is the interface a tight clearance fit or a loose fit?

Function

The required structural strength and material properties of the final part need to be understood with regards to the limitations of the different additive manufacturing techniques as well as extent of required support structures. This is dependent on what this part will ultimately be used for – For example, is it intended to be subjected to a mechanical test against pre-determined criteria, to gain confidence on structural integrity of the design?

Cost and Time

Depending on the size, complexity and required attributes of the part as described above, the preparation time, build time, clean up and post processing times, and subsequently the overall cost of the part will need to be considered when designing for a specific additive manufacturing method.

Additive manufacturing technologies have different dimensional accuracy, build size, layer thickness and support requirements, which need to be considered when selecting the most suitable process for a given design. Below are capabilities of some of the more common additive manufacturing technologies for both metallic and plastic materials that are used in orthopedics and in general across healthcare industries.

FDM

1. Dimensional Accuracy: For desktop FDM printers, the dimensional accuracy is within ± 0.5%, with maximum accuracy within ± 0.5 mm. For industrial FDM printers, the dimensional accuracy is within ± 0.15%, with minimum accuracy within ± 0.2 mm.
2. Layer Height: Layer height for FDM ranges between 50 – 400 µm, with typical layer height of 200 µm.
3. Build Size: Typical build size for desktop FDM printers is 200 x 200 x 200 mm, and upto 900 x 600 x 900 mm for industrial FDM printers.
4. Support Structures: Design for FDM does not always need support structures. Moreover, dissolvable support structure materials are available.

SLA

1. Dimensional Accuracy: For desktop SLA printers, the dimensional accuracy is within ± 0.5%, with maximum accuracy within ± 0.10 mm. For industrial SLA printers, the dimensional accuracy is within ± 0.15%, with minimum accuracy within ± 0.05 mm. Overall, SLA offers a greater dimensional accuracy as compared to FDM for a given design.
2. Layer Height: Layer height for SLA ranges between 25 – 100 μm, with typical layer height of 50 μm. Hence, SLA offers a greater z-resolution than FDM for a given design.
3. Build Size: Typical build size for desktop SLA printers is 145 x 145 x 175 mm, and upto 1500 x 750 x 500 mm for industrial SLA printers. Whereas desktop SLA printers have smaller build volumes overall as compared to FDM, industrial SLA printers can build taller components as compared to FDM, provided those designs can be oriented accordingly.
4. Support Structures: Design for SLA always needs support structures, unlike FDM.

Material Jetting

1. Dimensional Accuracy: The dimensional accuracy is within ± 0.1%, with maximum accuracy within ± 0.05 mm. So material jetting offers a greater dimensional accuracy as compared to both FDM and SLA for a given design.
2. Layer Height: Layer height for material jetting ranges between 16 – 30 μm, with typical layer height of 16 μm. Hence, material jetting offers a greater z-resolution than both FDM and SLA for a given design.
3. Build Size: Typical build size for material jetting printers is 380 x 250 x 200 mm, and upto 1000 x 800 x 500 mm. Hence, material jetting printers can build designs bigger than that of FDM printers but still smaller than those for SLA printers.
4. Support Structures: Design for material jetting always needs support structures, unlike FDM. However, unlike SLA, support structure materials are always dissolvable.

Binder Jetting

1. Dimensional Accuracy: The dimensional accuracy is within ± 0.2 mm. So binder jetting offers a comparable dimensional accuracy to FDM, but inferior to SLA and material jetting for a given design.
2. Layer Height: Layer height for binder jetting is typically 100 μm. Hence, binder jetting offers a superior z-resolution to FDM, but inferior to SLA and material jetting for a given design.
3. Build Size: Typical build size for binder jetting printers is 400 x 250 x 250 mm, and upto 1800 x 1000 x 700 mm. Hence, binder jetting printers can build designs bigger than that of FDM, SLA and material jetting printers.
4. Support Structures: An added advantage of design for binder jetting is that does not need support structures, unlike FDM, SLA and material jetting.

SLS/MJF

1. Dimensional Accuracy: For SLS and MJF printers, the dimensional accuracy is within ± 0.3%, with maximum accuracy within ± 0.3 mm. So both these plastic 3D printing processes offer a dimensional accuracy comparable to FDM and binder jetting, but inferior to SLA and material jetting for a given design.
2. Layer Height: Layer height for SLS and MJF printers ranges between 70 – 120 μm, with typical layer height of 100 μm. Hence, SLS and MJF offer a z-resolution that is comparable to binder jetting, superior to FDM, and inferior to SLA and material jetting for a given design.
3. Build Size: Typical build size for SLS and MJF printers is 300 x 300 x 300 mm, and upto 750 x 550 x 550 mm. Hence, SLS and MJF printers can build designs comparable to FDM, but smaller than that of SLA, material jetting and binder jetting printers.
4. Support Structures: An advantage of design for SLS and MJF printers is that it does not need support structures, unlike FDM, SLA and material jetting.

DMLS

1. Dimensional Accuracy: For DMLS printers, the dimensional accuracy is within ± 0.1mm. So this metallic 3D printing process offers a dimensional accuracy comparable to that of SLA and material jetting, while being superior to FDM, binder jetting, SLS and MJF for a given design.
2. Layer Height: Layer height for DMLS printers ranges between 30 - 50 μm. Hence, DMLS offers a z-resolution that is comparable to SLA, slightly inferior to material jetting, while being superior to FDM, binder jetting, SLS and MJF processes for a given design.
3. Build Size: Typical build size for DMLS printers is 250 x 150 x 150 mm, and upto 500 x 280 x 360 mm. Hence, DMLS printers can build smaller designs than other additive manufacturing processes.
4. Support Structures: A disadvantage of design for DMLS printers is that it always needs support structures.

In addition to considering the end application of a given design as well as the capabilities of the different additive manufacturing processes, a design engineer also needs to consider specific design features and their impact on manufacturability using the different additive processes, when selecting the most optimal method. Below are some key considerations of design features.

Build Envelope

The very first step is to determine the overall size of the part and its build envelope corresponding to an additive manufacturing process. The build envelope for a given design is limited by the maximum build volume of the machine and powder supply. When designing the build envelope for a given design, it is important to orient the part such that its outer boundaries do not violate the build envelope along any dimension, including along the z-direction (direction of build) in the fully built state.

Whenever possible, multiple parts should be closely spaced on the build plate of the machine, such that they do not violate the build envelope as well as each other in the fully built state. This would ensure efficient usage of the build plate for a single run of the machine.

Resolution

Multiple factors contribute to the resolution of a 3D printed part, such as layer thickness, minimum wall thickness and STL model mesh size.

The primary factor affecting part resolution and surface finish of a 3D printed part is the layer thickness of the additive manufacturing process used to fabricate it. Figure 3 shows the effect of different layer thicknesses on surface finish of a 3D printed part.

Figure 3. Macro view of FDM prints of varying layer thickness (50, 200 & 300 μm layer height from left to right) at the same scale2

As described earlier, it is important to weight the pros and cons of the different additive processes to select the most suitable layer thickness for the end application of the printed part. Reducing the layer thickness proportionately increases the build time and vice versa, so it is important to implement the largest layer thickness that will still produce a part of sufficient resolution for the end application.

The minimum wall thickness of the part design determines the smallest size feature that needs to be resolved in the printed part. It is important to understand the minimum wall thicknesses that different additive manufacturing processes can support, in order to choose a suitable method for the given design.

Table 1 specifies the minimum recommended wall thickness for some common additive manufacturing processes.

Table 1. Minimum recommended wall thickness for additive manufacturing processes

Additive manufacturing process	Minimum wall thickness (mm)
FDM	0.8
SLA	0.5
Material jetting	1.0
Binder jetting	1.0
SLS / MJF	0.7
DMLS	0.4

DMLS offers the greatest resolution for metallic designs, whereas SLA provides the best part resolution for plastic components.

The inherent 3D resolution of a part is driven by the size of the triangular mesh that is used to define the part surface geometry within the "standard tessellated language" (STL) format of the digital CAD file. A given part design needs to be first converted from its native CAD format to the STL format, while specifying the triangular surface mesh size. Each of these triangular facet elements is enclosed within three vertices and its orientation is defined by a unit normal vector. The smaller the size of the triangular mesh, the higher is the part resolution and vice versa.

Orientation

Orientation of a given part on the build plate needs to be designed such that the part is well supported throughout the additive build process while minimizing support structures. Improper orientation of a part will trigger the need for additional supports, which will increase the need for post-processing as well as affect final surface finish of the part.

In general, orienting a part such that it has inclined surfaces at an angle greater than 45^0 with respect to the horizontal ensures that the structure is self-supporting, whereas a part orientation where an inclined surface is at an angle lesser than 45^0 requires external support structure to be added to the inclined surface.

The surface finish of a fabricated part is greatly dependent on part orientation on the build plate. In order to obtain a good surface finish, it is important to orient a part such that most of its surfaces or features are at an angle within the range of 60^0 - 140^0 with respect to the horizontal. Surfaces inclined closer to the horizontal (either at an angle $< 60^0$ or $> 140^0$) are likely to have a rougher surface finish.

Supports

When designing for additive manufacturing, it is important to keep in mind the intended build orientation, resulting locations of support structures and accordingly specify critical features and dimensions.

As mentioned above, supports are needed to provide stability to surfaces that are inclined close to the build plate during the build process. In addition, supports are also required to prevent warpage in a part with large overhangs, that sustain large residual stresses during the build.

Part locations, where support structures are required, generally come out rough when support structures are removed. Hence, significant post-processing steps, in the form of machining, sanding, blasting etc, need to be done on those surfaces to bring those features within specifications. When designing a part, up-front understanding of this process will allow a design engineer to accordingly avoid those surfaces when specifying inspection datums or critical dimensional tolerances, so as to reduce the need for extensive post-processing and limit finished part cost.

Overhangs

Large overhanging features are unstable and likely to sink or warp during an additive build process. Hence, it is important to design overhanging features for stability during an additive build.

Horizontal overhangs greater than 1 mm length should be avoided, if possible. Wherever possible, they should be replaced with inclined surfaces that are self-supporting, convex or concave surfaces.

If a part design has internal undercut features, it is very important to convert the horizontal overhang features (figure 4 left image) into angular or curved, self-supporting geometry (figure 4 right image), to ensure that support structures are not trapped within the part.

Figure 4. Design for additive manufacturing - overhang features

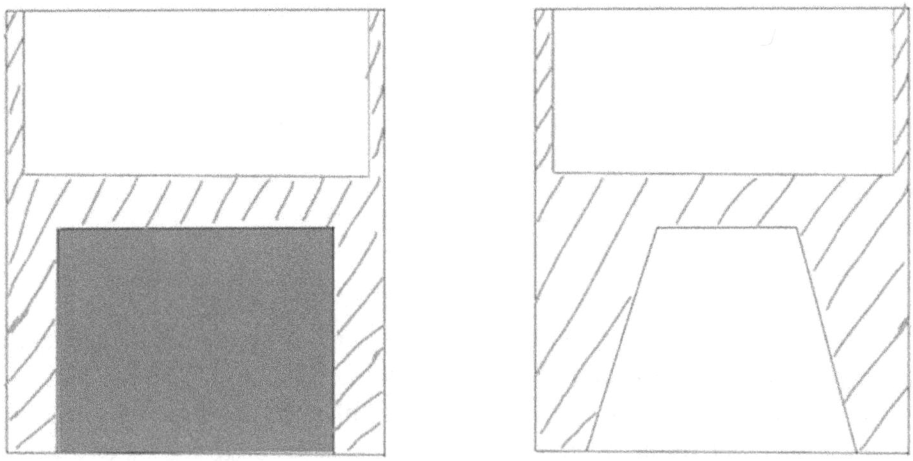

Holes

Hole features within a design need to be oriented either axially perpendicular to the build plate, or inclined such that the angle between the hole axis and build plate is within the range $60^0 - 90^0$. This will ensure that that the form error of the cylindrical hole feature and support structures are both minimized.

If hole features are oriented parallel to the build plate, support structures formed within the hole will be difficult to be removed, especially in the case of blind holes. In the case of unavoidable hole features that are oriented parallel to the build plate, consider designing the hole shape to be non-circular in order to compensate for distortion in the final shape, or build to size and accept some distortion. Alternatively, as shown in figure 5, consider designing for a "tear-drop" or "diamond-shape" hole, leaving perimeter material stock around the hole feature, for subsequent machining of the stock material to create circular hole feature in finished part.

Material Removal

When additively manufacturing a hollow part, escape holes need to be included within the design to allow for removal of the unused and trapped material or powder from within the hollow part.

The location of the escape hole should be designed, wherever possible, such that it is aided by gravity in the trapped powder flowing out of it.

The size of the escape hole is dependent on the material being used as well as the design of the part. In general, a minimum 3 mm diameter is recommended for escape holes.

Figure 5. Design for additive manufacturing - hole features

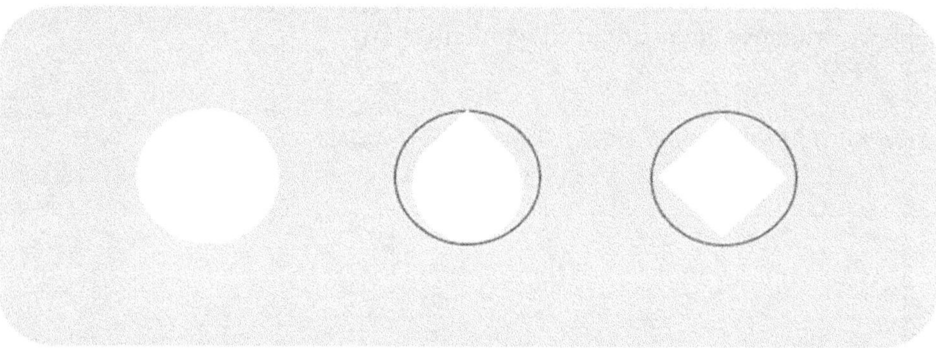

Windows

Similar to hole features, use self-supporting design elements such as inclines, radii and arches to improve dimensional accuracy and surface finish of the window feature.

Window edges might be likely to sag towards the build plate in the direction of gravity, so design in more dimensional tolerance in the direction of sagging as opposed to the other direction. This will allow any mating relationships of the additively manufactured window feature to be preserved at worst case material condition.

Threads

Thread features are, in general, difficult to build to required form and size specification as well as required surface finish at the end of an additively manufactured process. Subsequent machining of the thread feature is generally required to meet required specifications.

External threads are easier to fabricate by additively manufactured processes than internal threads, mainly due to the relative ease of access to the helical thread turns for an external thread feature. Internal threads, if part of a design, need to be accordingly modified to preserve extra stock material that can be subsequently machined away to create the final thread form. Care should be taken to ensure that any support structures and powder are removed after the additive process, prior to machining of internal thread features.

Cavities

As described earlier, cavities or internal features need to be designed such that they are self-supporting, such that a subsequent difficult, expensive and time-consuming support removal process is not required.

In the case of designs where self-supporting internal cavities are not possible, powder escape features, such as through holes, channels and slots, need to be implemented in the design, to allow for removal of excess powder and proper cleaning of the printed part.

Any porous surfaces should be made externally accessible such that a power wash, cleaning or blasting process can be subsequently applied to remove the excess powder from within the pores of the additively manufactured part.

Labelling

One of the advantages of additively manufacturing a design is the ability to integrate part labelling within the part fabrication process. Part identifying information such as part number, lot number etc. can be applied onto or cut into a surface of the design, which will then be printed as a part feature. Additional part marking steps such as laser marking, embossing, engraving etc. are not needed, as is the case of traditional manufacturing processes.

Distortion

For additive manufacturing processes, temperature fluctuations constantly occur at the micro scale during the build process, which lead to residual stresses that can cause local distortions in part geometry over the course of a build. Such distortions show up in the final printed part as feature deformation and warpage.

It is important to address this build-up of residual stresses throughout the build process for a given design, and accordingly add reinforcing structures such as flanges, connecting ribs, thicker walls etc around areas of the design that are more susceptible to distortion.

Figure 6 shows an example design on the left that has thin wall sections which are likely to distort during the build process due to residual stresses, resulting in form error in the finished part. A way to address this would be to add a connecting rib at the bottom of the design (right image), which strengthens the wall sections and stiffens them against residual stresses, thus creating a part free of distortion at those features.

Figure 6. Design for additive manufacturing – preventing distortion

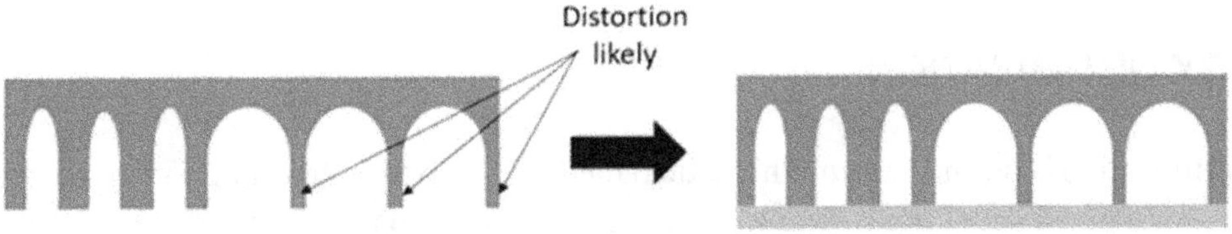

Aspect Ratio

Aspect ratio is another important consideration when additively building tall structures upon a build plate. Aspect ratio is defined as the ratio between the height (above build plate) of a feature and its largest cross-sectional dimension.

For a given cross-sectional size, as part height increases, the likelihood of instability and distortion of an unsupported feature also increases. Hence, it is important to consider strengthening tall features within a design accordingly, and as much as possible, avoid features with very high aspect ratios. As a general rule of thumb, it is recommended to maintain a design aspect ratio for minimally supported features to less than 10, and not to exceed an aspect of 20.

Porous Surfaces

Additively manufactured porous surfaces cannot be subsequently machined. It is important for a design engineer to delineate porous and solid additively manufactured regions within a part design, and allow for sufficient machined stock within a base solid region over which a porous region is built.

If a porous region is adjacent to a load-bearing region in a design, consider designing in a solid rim region bordering the intended load-bearing region that separates the porous region from it, since porous structures are significantly weaker sections of a part for any given material.

For any subsequent machining of part features after the additive build process, in order to create the final part, it is important to identify solid regions within the part, and avoid porous features, as holding faces for fixturing as well as for proper datum dimensioning for purposes of final part inspection.

Weight Reduction

Part weight or volume is a key factor in the overall time and cost to additively manufacture a part. Any opportunity to reduce part volume and weight can improve part yield, reduce cost and potentially also reduce part distortion.

It is important to identify and understand the functional requirements of a design, in order to streamline a part geometry for optimal material volume/weight while still ensuring that the part design meets all its performance requirements.

One technique that allows a design engineer to achieve this is called Topology Optimization. Topology Optimization is a computational method to iteratively achieve an optimized topology of a part, such that its volume is minimized, while ensuring that a pre-determined set of structural and strength requirements are met. The following sections describe Topology Optimization in more detail.

TOPOLOGY OPTIMIZATION

Topology optimization is defined as a mathematical method that optimizes material layout within a given design space, for a given set of loads, boundary conditions and constraints, with the aim of maximizing the performance of the system[6].

In medical device design, topology optimization is used to create an optimal material distribution within a given implant or device design using the finite element method (FEM). Other common areas of applications of topology optimization are aerospace, civil engineering and automobile engineering. For a given design of a medical device or orthopedic implant, topology optimization can be used to explore innovative design concepts early on in the development cycle, and in the process identify an optimized geometry and material distribution of the design that can be additively manufactured.

The objectives for topology optimization on a given design may include one or more factors such as overall stiffness, maximum principal stress, displacement, temperature etc, depending on the functional requirements of the device. The constraints within which topology optimization is performed could be mass, volume, specific features to include or exclude, among others. For example, in a given design loading areas and fastener interfaces can be excluded from topology optimization and hence retained as part of the output design. The goal is always to maximize the performance of a design for a given set of loads, boundary conditions and constraints. The input design is almost always a discretized (meshed) CAD model that is used to create a finite element model using the above parameters, which in turn is iteratively solved to generate an output faceted model. The output model often has a different number of mesh elements than the input model and requires additional modeling changes to make it manufacturable by one or more of the additive processes.

Some common computer-aided engineering (CAE) tools used to perform topology optimization are ANSYS Workbench (version 18 or higher), Autodesk NetFabb and SolidThinking Inspire, among others. The sequential steps of a topology optimization workflow for a given design are described below, when implemented using ANSYS Workbench. Other topology optimization tools use a very similar workflow.

Generate Input Model

A given design CAD model is brought into SpaceClaim environment for model editing. Any feature detail such part labelling and other cosmetic features, not required to be within the scope of topology optimization, may be removed here. In addition, sharp corners of the native CAD model should be rounded to prevent mathematical discontinuities within the simulated solution. Any part of a feature likely to be under a load or boundary condition can be delineated as a separate surface at this stage, to be able to select that specific area for applying the load or boundary condition later.

Discretize Model

The SpaceClaim model is then brought into ANSYS Workbench for performing topology optimization. The first step is to generate a mesh spanning the entire model. It is important to make sure that the mesh size is small enough to be able to generate at least 3 mesh elements across the smallest feature detail or thinnest cross-section of the design. Specific areas of a component may need to have a finer mesh than other areas, depending on the geometry. When a multi-component assembly is being analyzed, different mesh sizes may need to be implemented for different components, depending on their size and geometric complexity.

Apply Material Properties, Loads, Boundary Conditions and Constraints to Model

Depending on the functional requirements that are being analyzed as part of the topology optimization exercise, apply the different material properties, loads and boundary conditions on the discretized model that will simulate intended use of the medical device. In addition, identify any areas of the design that would need to be excluded from optimization scope, or in other words, areas of the design where material should be conserved. These are known as "exclusion constraints". In general, loading areas and supporting boundary conditions are applied as exclusion constraints, and hence regions of the design that remain conserved between the input and output models. Another common constraint that is applied is a minimum threshold mass that the output model should not violate, when compared to the input model mass. For example, a 30% mass constraint would imply that the analysis can continue to optimize until

70% of the mass of the input model has been removed, but no further. After applying all loads, boundary conditions and constraints, care should be taken to ensure that the model is neither under-constrained nor over-constrained.

Set Objectives for Topology Optimization

Depending on the end application of the design and manufacturability of the design, objectives should be set for the analysis. Commonly used objectives used in structural analyses include specifying a maximum principal stress criterion, a minimum tensile yield safter factor, a maximum displacement value, among others. These criteria are boundary limits for one or more of the variables that the solver is calculating at each iteration, after modifying the material distribution of the design space. As soon as the analysis identifies that one or more of these criteria are reached at a given iteration, it will stop iterating further and generate the faceted model from the last iteration as the final output. In some cases, one or ore more structural parameters of interest might be mutually opposing in nature (for example, stiffness and displacement), and care should be taken to apply the worst-case criterion as the objective that is applicable to the end application of the design.

Run Topology Optimization

Once the finite element model is ready, the topology optimization analysis can be run. The FEM solver calculates physical parameters of interest, dependent on the material properties, loads and boundary conditions of the model, at each of the mesh elements. After each iteration, mesh elements within the optimization design space, that have parameter values farthest away from the pre-defined objective criteria, are removed, such that defined constraints are not violated. The above process continues to repeat itself until one or more of the objective criteria is met, and the final faceted model is generated as output of the analysis (Diegel et al, 2019).

Refine Design According to Optimized Model

The output faceted model generally has rough edges and an uneven topology. Hence it is typically overlaid onto the original CAD model in a CAD environment (Creo, SolidWorks etc), and non-overlapping material is removed from the parametric CAD model. An alternative method is to re-model the design based on the boundaries of the topology optimized faceted model. Once the final parametric CAD model is generated in this way, this optimized model represents the best material distribution of the original design, given the functional requirements of that device (Leary, 2019).

Validate Optimized Model and Prepare For 3D Printing

It is always a best practice to validate the optimized CAD model by running a finite element analysis (FEA) on it that is identical in materials, loads and boundary conditions to the topology optimization model. The resulting parameters of interest (example, maximum principal stress, stiffness, displacement etc) should be near identical in magnitude and relative location to that of the topology optimization model. This validates the topology optimization analysis. The optimized CAD model can then be converted into STL format, and is ready for additive manufacturing (Mirzendehdel & Suresh, 2017).

Figure 7 shows an example of design evolution of an additively manufactured aluminum alloy bracket, using topology optimization. This reduced the number of parts from 4 to 1, eliminated 44 rivets, thus reducing the overall weight by 35% and increasing the part stiffness by 40%. This illustrates the power of topology optimization, as an integral part of design for additive manufacturing, in improving a design and significantly reducing its cost (Bendsoe & Sigmund, 2011).

Figure 7. Evolution of multi-part bracket to a single part (Image: Airbus Defense and Space)
Source: Hornick, 2015

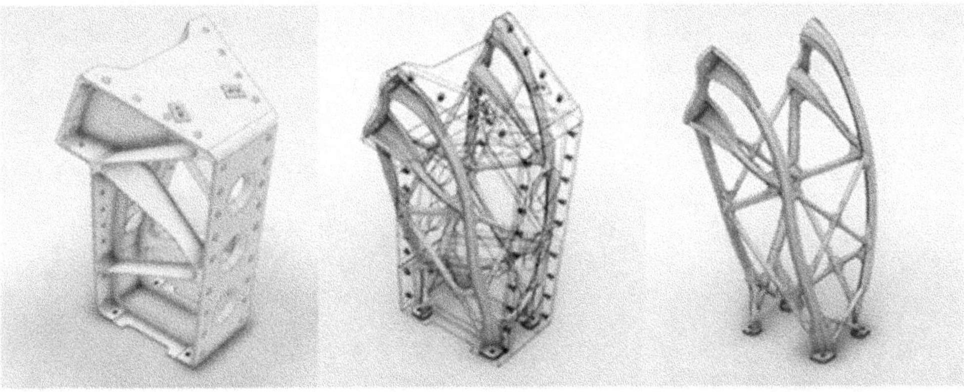

In summary, design for additive manufacturing has enormous potential in design and innovation across hi-tech industries such as aerospace, automobiles and particularly medical devices. Topology optimization is an integral part of design for additive manufacturing, and is being increasingly used early on in development cycles in the orthopedic implant industry, and continues to hold tremendous promise in re-imagining medical device designs of tomorrow.

REFERENCES

Additive manufacturing staircase. (2022). https://resources.renishaw.com/en/details/additive-manufacturing-staircase--78019

Diegel, O., Nordin, A., & Motte, D. (2019). *A Practical Guide to Design for Additive Manufacturing* (1st ed.). Springer.

Hornick, J. (2015). *3D printing will rock the world.* Paseo Publishing.

Hubs. *(2017)* https://www.hubs.com/knowledge-base/

ISO/ASTM 52900:2015 Additive manufacturing — General principles — Terminology. https://www.iso.org/obp/ui/#iso:std:iso-astm:52900:ed-2:v1:en

Leary, M. (2019). *Design for Additive Manufacturing* (1st ed.). Elsevier.

Mirzendehdel, A. M., & Suresh, K. (2017). *A Hands-on Introduction to Topology Optimization* (1st ed.). Createspace Independent Pub.

Mostafaei, A., Elliott, A. M., Barnes, J. E., Li, F., Tan, W., Cramer, C. L., Nandwana, P., & Chmielus, M. (2021). Binder jet 3D printing—Process parameters, materials, properties, modeling, and challenges. *Progress in Materials Science, 119*, 100707.

Sigmund, O., & Bendsoe, M. P. (2011). *Topology Optimization* (2nd ed.). Springer.

Zeidler, S. (2020). *Additive Manufacturing in Orthopedics.* https://www.sme.org/technologies/articles/2020/may/additive-manufacturing-in-orthopedics/

Chapter 4
Gait Analysis:
An Effective Tool to Measure Human Performance

Monisha Gowri S.
National Institute of Technology, Rourkela, India

Ravi Kant Avvari
https://orcid.org/0000-0001-5586-746X
National Institute of Technology, Rourkela, India

Mirza Khalid Baig
National Institute of Technology, Rourkela, India

Thirugnanam Arunachalam
National Institute of Technology, Rourkela, India

ABSTRACT

The impact of the musculoskeletal system on human locomotion is acquired using force platforms, motion analysis systems, and EMG markers. The kinetic and kinematic parameters can be obtained using the force platform, EMG markers, and motion analysis system. The videos are captured using the motion analysis system and analyzed using suitable software. If the typical gait pattern is irregular, it indicates a disorder or abnormality. An abnormal gait cycle or pattern may rise due to joint pain, muscle strain, deformities of bone, weakness, and other impairments in limbs. The gait analysis is used to study various gait abnormalities. The obtained pattern of abnormal subjects is clinically correlated for the assessment of gait disorders. Gait analysis has a wide range of applications like sports for enduring athletes' performance and injury prevention, in rehabilitation, post-surgery analysis, design of orthotics and shoes, and biomechanics studies of astronauts.

DOI: 10.4018/978-1-7998-9078-2.ch004

INTRODUCTION

Biomechanics deals with the study of living organisms, particularly their structure and function, using the principle of mechanics. Human locomotion is the act of walking, running, jumping that includes coordinated joint movement through muscle forces that involve motor control and the musculoskeletal system (Nutt et al., 1993). The biomechanics that deals with human locomotion are termed gait analysis. It is the method to assess human movement and measure the Range of Motion (ROM) during locomotion.

The gait analysis is the quantitative assessment measuring motion with or without force. The kinetic parameters are Ground Reaction Force (GRF), Centre of Mass (CoM), Centre of Pressure (CoP), whereas the kinematic parameters are velocity, acceleration, and angle. The techniques involved in analyzing gait analysis are motion capture, inertial system, EMG (Electromyography) and force platforms. The importance of gait analysis is assessing gait disorder due to bone deformities and motor control impairment. The clinical approach for analyzing gait disorders is concerned with evaluating the body balance and its symmetry of motion pattern. The gait analysis has vital parameters that help to assess the movements. Various computational techniques are used to measure the kinetics and kinematics parameters during gait analysis.

This chapter discusses the various phases of a gait cycle, gait parameters, gait abnormalities and computational techniques available for measuring gait. The chapter also briefs about different applications of gait.

GAIT CYCLE

Gait is a cyclic process, where the same set of movements repeatedly occurs to enable the forward progression of the body. This set of coordinated movement aim to move the body with support and transfer of weight of the body. Thus, a gait cycle is a period between the heel strike of a foot which has support function to subsequent heel strike of the same foot while other foot advances for the next support position.

Phases of Gait Cycle

The gait cycle is divided into the following events:

- Heel strike
- Foot flat
- Heel-rise
- Opposite initial contact
- Toe-off
- Feet adjacent
- Tibia vertical
- Heel strike

The events of the gait cycle of a leg are categorized into two periods: stance and swing (Perry, 1992). Figure 1 represents the different phases and sub-phases of the gait cycle. The stance phase is when the reference foot is grounded, and the swing phase is when the reference foot is no longer touching the ground.

Figure 1. Schematic diagram of different phases of the gait cycle.

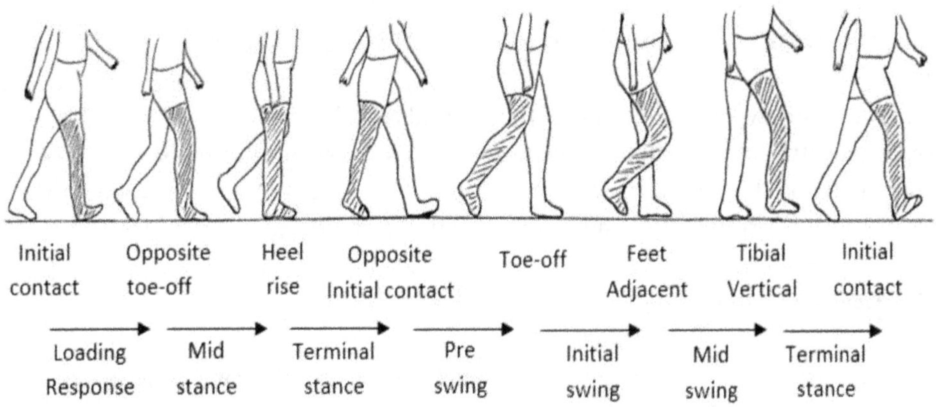

The gait cycle has a period called double support, where both the limbs are in contact with the ground, which occurs twice in the stance period. The single support period in the stance phase occurs when only one leg is in contact with the ground. The stance period has the supportive role and the swing period has a forward progression role. The block diagram in Figure 2 represents the division of the gait cycle.

Figure 2. Block diagram of the gait cycle showing various sub-phases.

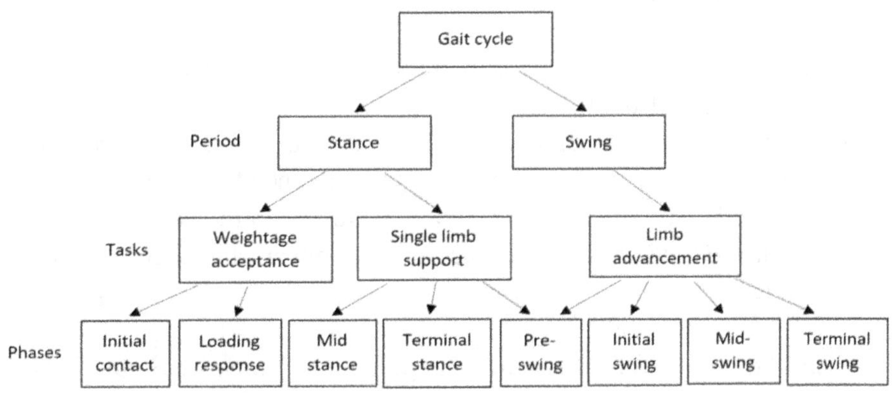

These phases perform three tasks to move the body in forward progression:

- Weight acceptance
- Single limb support
- Limb advancement

Weight Acceptance

The weight acceptance task aims to absorb shock limb stabilization and preserve progression of the body

- Initial contact occurs when the feet strike the ground, also known as heel strike. Repeated progress occurs over heel strike to foot flat in the stance limb for forward progression of CoG.
- Loading response is the phase where the weightage of acceptance occurs where the knee flexes slightly for shock absorption as the heel strikes on the ground and stabilize the single-limb support.

Single Limb Support

The single-limb support task is responsible for weight-bearing stability and progression of the body.

- Mid-stance of single-limb support is the first phase where the advancement of the swing limb occurs and moves ahead of the stance limb as the weight is transferred to the forefoot of the stance limb.
- The terminal stance of single-limb support is the last phase which starts with heel rise of stance limb and proceeds until the other (swing) foot strikes the ground.

Limb Advancement

The limb advancement task helps in foot clearance from the ground, position the limb for the swing phase, and executes limb advancement.

- Pre-swing is the final double support period in preparing the supportive limb for the swing period and lifting off the stance foot from the ground.
- Initial swing is the first phase of the swing period. During this phase, the knee flexes to the maximum. The initial swing starts with the toe-off with the reference foot (stance limb).
- Mid-swing is the second phase of the swing period, ending with the vertical tibia.
- The terminal swing is the final phase with the complete limb advancement and positions the limb for the next gait cycle.

A gait cycle consists of two double and two single supports, as explained in Figure 3. The approximate proportion of double and single support is 1:4. Overall, the stance period is 60%, and the swing period is 40% of the gait cycle (Kharb et al., 2011).

Table 1 explains the various sub-phases of a gait with hip, knee, ankle movement during locomotion.

Centre of Gravity

The centre of gravity (CoG) is the point where the body's whole weight is concentrated and relates to the body's posture. The CoG is the vertical entity on which the balance and stability of any object depend. It is crucial to understand and visualize the line of gravity acting downward through CoG while determining the balance or stability of the individual. A person is stable when the line of gravity is within the walking base. An increased or decreased walking base can cause abnormal gait patterns. The greater displacement of CoG from the midpoint is due to the larger Base of Support (BOS).

Figure 3. The time dimension of a gait cycle

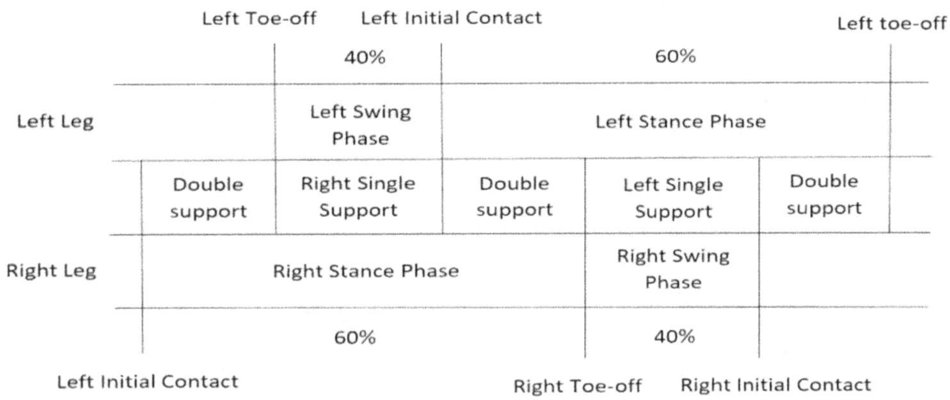

Table 1. Kinematics of lower extremities during gait

Phases	Hip	Knee	Ankle
Initial Contact(IC)	30° flexion by contraction of rectus femoris muscle.	5° flexion by hamstring and extension by quadriceps.	Plantar flexion by tibialis anterior muscle.
Loading response	Adductor magnus and gluteus maximus muscles contraction.	Flexion of 15° to 20° occurs.	10° to 15° plantar flexion is increased.
Mid-stance	10° extension by contraction of gluteus medius muscle.	Maximum flexion and then extension occurs.	Supinated and dorsiflexed by contraction of triceps surae.
Terminal stance	10° to 13° goes to hyperextension and then flexes.	Flexed from 0° to 5°.	Supinated and Plantar flexed.
Pre-swing	Less extended.	35° to 40° flexed.	Plantar flex 20°.
Initial swing	Extend to 10° and flexed by iliopsoas muscle contraction.	Flexes of 40° to 60° occurs.	From 20° plantar flexion to dorsiflexion.
Mid-swing	It is flexed to 30° by adductors.	It flexes 60° and sartorius muscle contraction of 30°.	Tibialis anterior muscle dorsiflexed.
Terminal swing	Flexes 25°to 30°.	Locked extension.	Neutral position.

The CoG is the point that lies in front of (approximately 2 cm for a typical adult human) the second sacral vertebra. However, the anatomical position of the CoG varies with the different postures of the body. The CoG moves forward in a sinusoidal curve vertical or lateral during walking or running. The vertical trajectory of CoG viewed in the sagittal plane traces the sinusoidal curve with two higher and lower curves in a gait cycle. During loading response and pre-swing, the CoG is lowest, while in the mid-stance and mid-swing phase, the CoG is highest. The lower curves are governed by pelvic rotation, knee, ankle, and foot interaction, whereas the higher curves are governed by pelvic tilt and knee flexion in the stance period. Figure 4 illustrates the vertical displacement of CoG at different phases in a gait cycle. The lateral displacement of CoG occurs predominantly during the mid-stance. In contrast, the lateral displacement of the pelvis limits the lateral displacement of CoG.

The CoG plays an essential role in analyzing the asymmetric gait. Since the kinetic and kinematic data are used in analyzing the gait pattern, the CoG needs to be optimized for the regular pattern (Tesio & Rota, 2019).

Figure 4. The Centre of Gravity (CoG) vertical displacement at different gait phases.

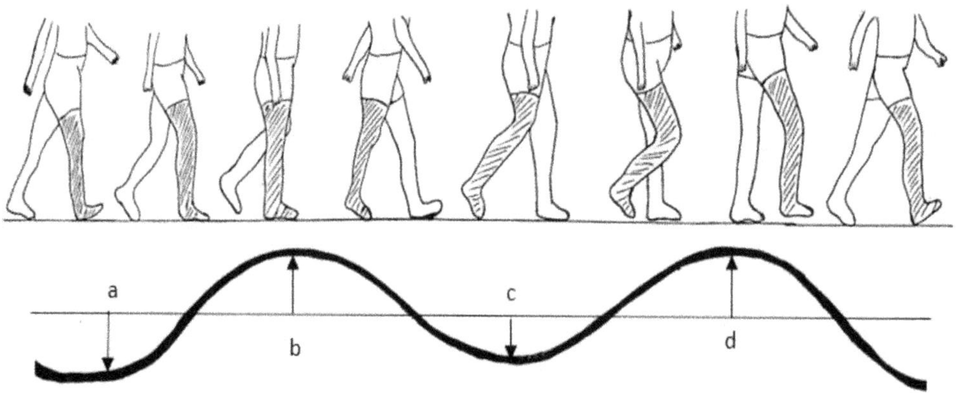

Determinants of Gait

Six determinants are considered to minimize the energy expenditure during a gait cycle. These are

- Pelvic rotation
- Pelvic tilt
- Stance knee flexion
- Foot mechanisms
- Knee mechanisms
- Hip adduction

Pelvic Rotation

The pelvic rotates in the transverse plane during the swing phase's forward progression, preventing the vertical excursion. The maximum pelvic rotation occurs just before the initial contact of the limb with 4° rotation at each side of the plane. The pelvic rotation reduces the angle of hip flexion and extension. During pelvic rotation, the stride length is longer than the pelvic flexion of the advancing limb (Saunders et al., 1953).

Pelvic Tilt

Pelvic tilt occurs in the frontal plane, which is slightly tilted over the swinging limb. It prevents the horizontal displacement of CoG by the maximum tilt of pelvic occurring at the mid-swing phase with a 5° drop at the swing limb. The pelvic tilt reduces the height of the apex of the CoG

Knee Flexion

The knee flexion shortens the limb in the middle of the stance phase. Knee flexion reduces the vertical height of the CoG.

Ankle and Foot Mechanism

The foot or ankle is dorsiflexed in the early stance period when the knee is extended. The CoG is lowered in the vertical displacement. As the knee is flexed, the foot or ankle is plantarflexed at the late stage of the stance period while the ankle lengthens the limb at heel strike. Both smoothen the CoG curve and reduce the lowering of CoG, thereby minimizing the energy expenditure.

Lateral Displacement of the Pelvis

The lateral oscillation of CoG occurs by horizontal movement of the pelvis to provide balance. The lateral displacement of the pelvis reduces the displacement of the CoG in the horizontal plane. In contrast, the other displacements reduce the CoG in the vertical plane. These determinants should function together, exhibiting a smooth CoG curve to reduce energy expenditure. Since determinants determine the displacement of the CoG, it helps study the kinetics and kinematics of the gait (Banaszkiewicz & Kader, 2014). Figure 5 illustrates the six determinants of gait showing minimum displacement of CoG.

Figure 5. Schematic showing six determinants of gait during A) pelvic rotation B) pelvic tilt C) knee flexion D) ankle mechanism E) foot mechanism F) lateral displacement of the pelvis

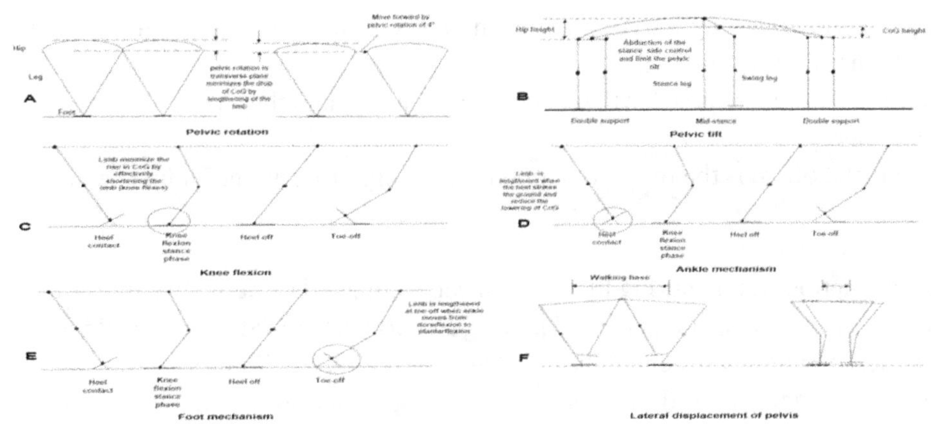

Ground Reaction Force

The ground reaction force is the external vertical force exerted on the body. The vertical anteroposterior and mediolateral GRF act between the foot and the ground as the body progresses forward. The subject's weight corresponds to the GRF during the anatomical position. As the body moves, the GRF increases with the increase in acceleration of the body.

The GRF can be calculated as:

$$GRF_i = m(a_i + g) \quad (1)$$

where,

m = mass of the body.

a = acceleration of the body.

g = acceleration due to gravity.

i = axis (x, y, z)

The centre of pressure (CoP) is the position of the vertical projection of the GRF vector acting downward at the area of contact with the ground. The CoP is considered a reflection of the body's neuromuscular responses to imbalances of the CoG (Winter, 1995). The difference between the CoG and CoP is proportional to the horizontal acceleration of CoM.

Energy Expenditure

The body experiences energy expenditure during locomotion. Expenditure of kinetic energy and generation of potential energy is required for gait as the body moves forward (Winter et al., 1976). The energy used by an individual during walking can be divided into three types:

- Energy is consumed during the contraction and relaxation of muscles as they accelerate and deaccelerate the limb for movement.
- The heart's activity increases due to muscular movements and thus also improves breathing, which consumes energy.
- The basal metabolism is the minimum amount of energy that an individual can consume at rest is irreducible.

Oxygen consumption occurs during locomotion can estimate the energy expenditure of the gait. The energy exchange occurs between kinetic and potential energy produced during walking, and the energy transfer occurs between limb segments. The head, arm and trunk (HAT) was studied through the sagittal plane, whereas other planes showed a negligible energy consumption. The linear motion's kinetic energy was considered neglecting angular motion's kinetic energy (Waters & Mulroy, 1999). At the stance period, the CoG of HAT elevates vertically, producing kinetic energy for the forward progression. The elevation reaches maximum converting kinetic energy to potential energy during the mid-stance phase.

The energy can be expressed as energy/unit time (oxygen consumption) or energy/unit distance (oxygen cost). The energy/unit time includes the basal metabolism and the energy expenditure of the heart and the breathing. Oxygen consumption is the amount of O_2 consumption per minute. The oxygen cost is the amount of O_2 consumption per unit distance travelled during walking (ml/kg/meter). The rate of O_2 consumption is the power required per minute (ml/kg/minute). The O_2 cost is the energy consumed to perform walking. The O_2 cost is the rate of O_2 divided by the walking speed. It is reported that the O_2 cost is higher in children and decreases in adults (Harris & Wertsch, 1994). The energy expenditure associated with gait abnormalities is higher, as the patient experience more stress in the muscles for locomotion. Thus, the gait analysis measures the energy expenditure to decrease the stress on the muscles and the joints (Chambers & Sutherland, 2002).

GAIT PARAMETERS

It is essential to know the gait parameters to understand the gait analysis and characterize the gait pattern. The main parameters of gait analysis are spatial-temporal parameters. The other parameters include kinetic, kinematic, anthropometry and dynamic EMG based parameters.

The Spatio-temporal parameters are basic gait analysis parameters, including cadence, step length, stride length, velocity, and speed. Few parameters are explained below.

- Step length is the point of the heel strike of one foot to the heel strike of another foot. Two steps make one stride, also known as a gait cycle, which is defined from the initial point of contact of the reference foot to the initial point of contact of the same reference foot.
- The sum of the two-step length is the stride length.
- The number of steps per unit time (steps/min) is known as cadence. The average cadence for a normal healthy subject is between 90 to 120 steps per minute.
- Cycle time (s) = 120/cadence (average cadence for the healthy subject)
- Speed (m/s) = (stride length*cadence)/120 or stride length/cycle time
- Step time is the interval between one foot's initial contact to another foot's contact.

The parameters involved in kinetics are GRF, joint reaction forces, moments and plantar pressure. Kinematic parameters involved in measuring the motion of body segment without considering the forces are angles of the different joints (ankle, knee, hip), acceleration, velocity and angular motion of the joint segments. The dynamic EMG records the electrical activity of muscles by recording and measuring the voltage potential generated by the muscle's electrochemical activities. The electrical activity of the muscle is one of the critical parameters used for gait analysis. Anthropometric parameters involved in measuring body dimension are BMI (body mass index), height, weight and limb length.

GAIT DISORDERS

An abnormal gait is the irregular pattern of the gait cycle due to neurological impairment (loss of motor control and coordination), orthopaedic problem (skeletal deformities and its degeneration) and other medical illness (heart failure, obesity). The gait pattern is altered and may vary with age, gender and physical health. The gait disorders are classified according to neurologic, orthopaedic, and other dysfunctions. The soft tissue impairment causes passive mobility, preventing standard postures and ROM at the joints required for walking and running. Contractures are one form of deformity experienced on a subject when positioned in a single posture for a longer period of time. The fibrous connective tissue undergoes a structural change due to prolonged inactivity of soft tissue, which stiffens, thereby causing a reduced movement around that location.

Other deformities that could affect the gait are abnormal joint contours, which again affect the ROM, or congenital disorders like a club foot (talipes equinovarus). The club foot is an inward position of the foot instead of being neutral and unable to rest the foot on the ground completely. Club foot is congenital where the children experiencing this are unable to land their foot flat.

The next factor that could influence the gait pattern is muscular problems. If the muscles are weak, insufficient strength in the muscle occurs during locomotion for proper coordination. The other cause of muscle weakness is atrophy, thinning or loss of muscle function. Disuse causes thinning and weakening muscles, whereas muscle cramping occurs when overexcited. Spasticity is another muscular problem where the muscle gets overexcited for prolonged contraction. It causes neurological problems like cerebral palsy, which affects the brain.

Pain is another cause of pathological gait, which occurs due to excessive tissue tension. Joint pain could be another cause for a person with impairment to walk. Pain can lead to deformities like contractures (Simon, 1993).

The various gait disorders are classified as follows:

Musculoskeletal Disorder

Antalgic Gait

A person suffering from unilateral antalgic gait disorder will have a shorter stance period. The body's weight is loaded on the affected limb to avoid pain, causing a limp. The person lifts his foot and lowers it in a fixed ankle position to reduce the load on the affected limb. The lower extremities abnormalities such as knee osteoarthritis, sprain in ankle and pain in the hip cause antalgic gait. Walking aids are used to decrease the pain in the affected limb.

Coxalgic Gait

The coxalgic gait is characterized by shifting the upper torso of the painful hip and is caused by the decrease in the hip abductor. The antalgic gait may constitute a coxalgic gait pattern that minimizes the affected limb's weight-bearing load. A gluteus medius lurch, abductor lurch, or the Duchenne sign is the torso's shifting during the single-limb stance period. The shifting of the upper torso toward the affected hip reduces the moment arm of the body's gravity. This reduces the abductor forces to maintain the statc equilibrium, which leads to a decrease in joint reaction force.

Trendelenburg Gait

The Trendelenburg gait is the tilt of the pelvis that occurs with a similar appearance to the coxalgic gait. It is characterized by a single stance period of the affected hip by the drop of the contralateral hemipelvis. The upper torso is shifted to the affected limb, similar to the coxalgic gait. The Trendelenburg gait is caused by a decrease in abductor movement and is common in polio patients. As in coxalgic gait, the abductor force decreases but cannot maintain pelvis level, dropping the contralateral hemipelvis.

Muscular and Myelopathy Disorder

Steppage Gait

The steppage gait is the foot drop where an individual has a weakness in ankle dorsiflexion due to lifting the limb higher than the usual so that the foot does not drag on the ground. As the foot is dropped, the toe of the affecting limb makes contact with the ground before the foot's heel hits the ground. This gait disorder is also known as slap gait, as it makes a thud sound when the toe hits the ground. The patients having steppage gait is unable to stand on their heel.

Waddling Gait

Waddling gait exhibits weakness in hip and thigh muscles which causes instability of the pelvis. The person will have difficulty in standing from a sitting position. Elevation of the hip occurs with the swing limb which leads to a waddling appearance. It characterizes muscular dystrophy. The patients bend the trunk towards the stance period to prevent hip elevation during the swing phase.

Myelopathic Gait

Myelopathic gait is cervical spondylotic myelopathy causing disturbance in gait and balance problems. The hypertrophy of ligaments and degeneration of osteophytes cause the spinal canal to narrow and compress the cervical spinal cord.

Neurological Disorder

Cautious Gait

The cautious gait is the excessive degree of changes in an individual's gait pattern. The patients have a broad walking base, slow gait, stooped posture. Phobic gait is one of the cautious gait disorders where the patients have an extreme fear of falling. Walking aids like crutches or holding the other person's hand can be used to improve cautious gait.

Spastic Gait

The spastic gait involves both the lower and upper extremities causing spasticity. It is characterized by dragging the stiffened limb in semi-circular motion on the affected side (circumduction). The spastic gait is categorized as spastic paraparetic and spastic hemiparetic gait. The spastic paraparetic gait is characterized by bending of legs in an adducted position near the hip, bending or extension of knee and feet in plantar flexion postion. The spastic hemiparetic gait is characterized by inward bending of arms, pronated forearm and flexed fingers. The lateral movement of the leg is the characteristic of hemiparetic, also known as Wernicke-Mann gait.

Ataxic Gait

Ataxic gait is the lack of coordination of movement causing unsteady walk with a wide base. The patients suffering from ataxic gait have varying step lengths of irregular movement. The patients with cerebellar dysfunction experience ataxic gait. Titubation occurs when the patient's body oscillates back and forth or side to side.

Parkinsonian Gait

Parkinson gait is characterized by impaired postural stability. These patients have bradykinesia causing small steps with difficulty in initiating the steps. They have higher step frequency rather than higher step length while walking faster. Further impairment increases as the disease progresses, resulting in a slow gait with a broader walking base and short step length.

TECHNIQUES OF GAIT ANALYSIS

Various equipment, accessories, and software used for measuring the gait pattern are discussed so that the reader will oversee various types of tools/software for analyzing the comprehensive gait pattern. In earlier days, only Spatio-temporal parameters were measured with less sophisticated equipment, resulting in inappropriate gait evaluation (Davis, 1988). The latest advancement in technology helps measure additional data like joint angle, GRF and gait energetics (Baker, 2007). The main parameters to be measured are categorized as kinematic and kinetic data. The kinematic data characterize the motion through the geometrical description of the human segments in 2D or 3D. The motion analysis system comprises infra-red cameras and reflecto-markers are used to obtain the kinematic parameters such as displacement, velocity and acceleration. An inertial measuring unit (IMU) sensor is used as a wearable device, a combination of accelerometer, gyroscope, and magnetometer for kinematic measurements. The videos are captured using the motion analysis system and analyzed using a suitable track management software like Qualisys Track Management (QTM™) and Visual 3D software. The limitations of kinematic data are mechanical artefacts and correct placements of the reflecto markers.

The kinetic data characterize the forces and moments. The kinetic parameters such as GRF, muscle forces, and foot pressure are measured using force platforms and pressure mats. The limitations of kinetic data are frictional forces and visco-elastic properties of soft tissues causing a possible error in calculating muscle forces concerning movement. Before carrying out the experiments, calibration of the gait equipment/tools is essential. Figure 6 explains the gait analysis techniques used for measuring the gait profile.

Figure 6.Various techniques for measuring gait profile

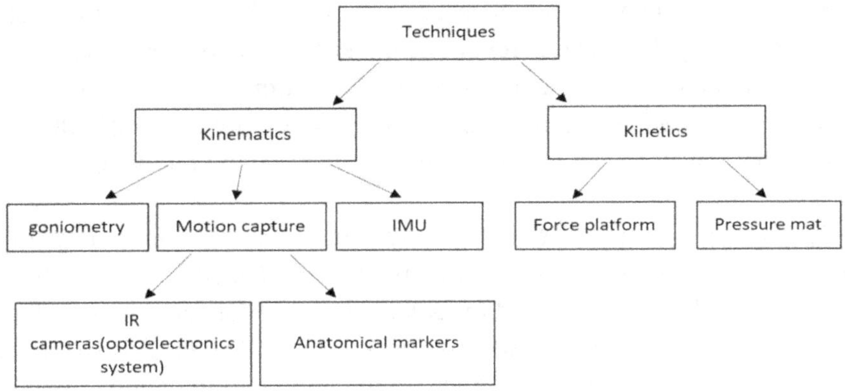

Kinematics

The kinematics deals with studying human motion without considering the forces involved in the movement. The individual body's portrayal is streamlined to a progression of a series of rigid bodies or sections moving in 3D (6 degrees of freedom). The clinical implications of the information recorded are the movement of anatomical joints across the various planes (sagittal, transverse and frontal) at the foot, lower leg, knee and hip level. Cadence, step length, walking speed, stride length, stride time, duration of single-limb support and double limb support is additionally part of kinematic data. It also characterizes the joint angles, pelvic tilt, extension/flexion of the hip, extension/flexion of the knee and plantarflexion/ dorsiflexion of the ankle. The most commonly used tools for measuring the kinematic data are motion capture unit, goniometry and IMU sensors.

Goniometers

The goniometers are potentiometers used to measure the movement of joints as angle measurement. The latest goniometer used is the electro goniometer, a rotary type potentiometer. These goniometers use a variable resistor attached to the limb segment, which is fixed by cuffs around the limb and the joint to be measured. The resistance of the potentiometer varies when the movement occurs by producing an electrical output. The other goniometers used for gait measurement are flexible strain gauges and polarized light goniometers. The limitation of using these goniometers is that the angle can be measured only for one plane at a time and requires accurate calibration for the joint axis.

Motion Analysis

The motion capture analysis has advancement from simple to sophisticated equipment with more accuracy, multimodal, camera-based, and cartesian optoelectronic anthropometer(Hachinski, 2008). Previously, the markers system was used for motion analysis and placed according to the planes of the body. There are three landmarkers with regard to the planes: sagittal plane landmarkers, coronal plane landmarkers, transverse rotation markers. The sagittal landmarkers of the hip is the greater trochanter since the centre of the hip position is difficult to locate. The knee joint is more prone to direct placement of markers in the sagittal plane. The ankle is between the centre of the joint and the lateral malleolus. The landmarkers for the ankle is placed on the lateral malleolus. The sagittal landmarkers are the posterior-anterior and superior iliac spines at the pelvis. The coronal landmarkers of the pelvis is placed at two anterior superior iliac spines. The middle of the patella is the coronal landmarkers for the knee. For the ankle, the middle of the distal tibia is the coronal landmarkers. The arc of rotation is difficult to find the location and capture. Thus, the transverse rotation markers are placed at the middle of the limb's segment anteriorly or lateral side of the limb.

Currently, optoelectronic technology is used for measuring the motion analysis of the human gait. The optoelectronic system uses marker points that convert the reflected or emitted light signals into electrical signals from the markers and are used for gait analysis model construction. The light signals are emitted by active markers where small LEDs attach to the body in the region of interest, whereas the lights are reflected by the passive markers, which are IR (infra-red) light and are again captured by the cameras. For an accurate calculation, the markers are placed close to the joint's centre of rotation.

Before developing these systems, therapists and doctors view the patient's walking pattern to evaluate the clinical gait analysis without a basic visual system. First, they need to identify the deviation of the gait pattern from the normal pattern with a detailed and systemic view. Secondly, the identified abnormalities should be described as a term of pathology. As it is not a direct result, it is the combination of the patient's recompense and the effect of pathology.

Data is collected as "frames", which varies between 50 and 1 kHz depending on *the system used for acquisition. Apart from calibrating, the error may occur due to misplacement of markers at anatomical landmarks and the soft tissue, skin, muscle artefact. Thus, there should be knowledge of displacement of tissue and muscle during the evaluation of gait to better evaluate the movements (Baker, 2006).

Inertial Measuring Unit (IMU)

The IMU is the sensor embedded with an accelerometer, gyroscope, and magnetometer that measures the acceleration, angular rate, and magnetic field direction. It is based on micro-electromechanical systems (MEMS), which are low cost, compact and used for real-time tracking in the orientation of the objects where it is used. Placing the sensor at each limb segment helps determine the limb segment's orientation independently. This sensor is used as 6 DOF OR 9 DOF depending upon the application. The accelerometer uses Newton's second law of motion, where the net force acting on the body is proportional to the acceleration of the body. The joint angle can be measured by integrating the angular velocity obtained by integrating the acceleration from the sensor, which is expressed as,

$$\theta = \iint v = \iint a \tag{2}$$

θ = joint angle

v = angular velocity

a = angular acceleration

The transducers like piezoelectric and piezoresistive sensors are used for accelerometers as IMU sensors (Muro-de-la-Herran et al., 2014). The gyroscope principle is based on the Coriolis effect, which gives the angular velocity, segment inclination, joint angle of the body segment where it is placed. The magnetometer is used to increase the reliability of the combination with accelerometer and gyroscope and make the global reference frame possible (Surer & Kose, 2011).

The IMU sensors give the joint's measured angle and angular rotation for gait analysis. It is used to diagnose abnormal gait, fall detector, and segment orientation. It is placed on the limb segment such as the thigh and shank to determine the joint's angle and segment orientation through acceleration data and angular rotation (Seel et al., 2014).

The joint angle measurement through IMU is used with 3D optical markers, which can be used for precise joint angles. VICON™ optical motion system and Xsens™ IMU sensor are used to estimate the joint's angle, resulting in a novel method from real-time measurement (Seel et al., 2014).

Kinetics

The kinetic parameters are governed by Newton's second law and Euler's law of motion. A state of equilibrium exists at each joint such that the moments balance the externally applied forces. The product of body weight (in percentage) and leg length gives the moment. Once the moments, angular velocities and joint angles are determined, the power at various joints is calculated. The GRF has a load of vertical patterns with normal cadence as an 'M' shaped curve (Harris & Wertsch, 1994). Figure 7 illustrate a schematic diagram of the 'M' shaped curve of GRF.

Figure 7. The 'M' shaped curve of the vertical GRF

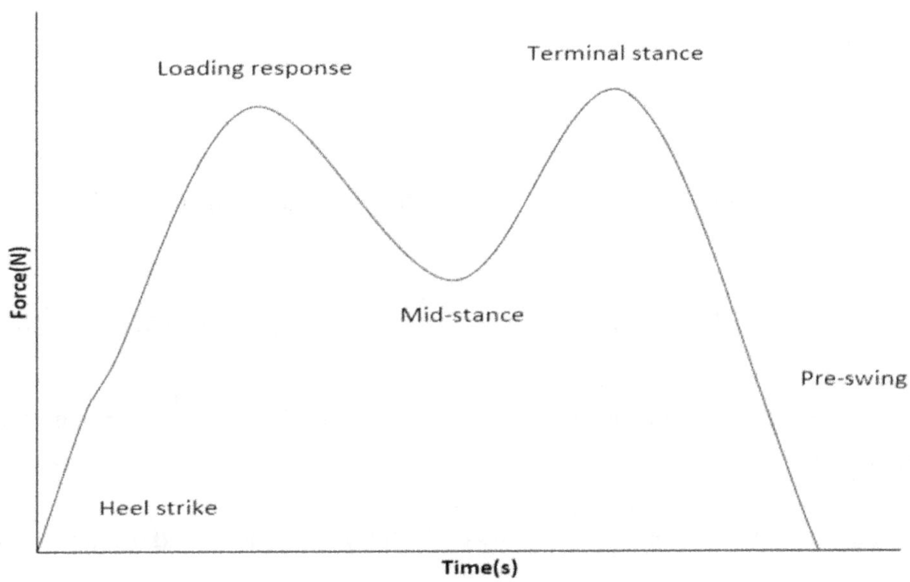

$$\sum F_x = m * acc_x \tag{3}$$

$$\sum F_y = m * acc_y \tag{4}$$

$$\sum F_z = m * acc_z \tag{5}$$

where,

$\sum F_x, \sum F_y, \sum F_z$ = summation of joint reaction forces in x, y, z components.

m = mass of the segment.

acc_x, acc_y, acc_z = x, y, z components of linear accelerations of the centre of mass of the limb segment.

The joint moments are calculated using Euler's law of motion

$$\sum M_x = I_{xx} a_x + \left(I_{zz} - I_{yy}\right) \omega_y \omega_z \tag{6}$$

$$\sum M_y = I_{yy} a_y + \left(I_{xx} - I_{zz}\right) \omega_z \omega_x \tag{7}$$

$$\sum M_z = I_{zz} a_z + \left(I_{yy} - I_{xx}\right) \omega_x \omega_y \tag{8}$$

where,

$\sum M_x, \sum M_y, \sum M_z$ = summation of x,y,z components of joint moments applied to the limb segment

I_{xx}, I_{yy}, I_{zz} = mass moments of inertia of the limb segment along x,y,z direction respectively.

a_x, a_y, a_z = the angular accelerations of the centre of mass of the limb segment around the x, y, and z directions, respectively.

$\omega_x, \omega_y, \omega_z$ = angular velocities of the centre of mass of the limb segment around the x, y, and z directions, respectively.

Power is calculated as

$$P_x = M_x \omega_x \quad (9)$$

$$P_y = M_y \omega_y \quad (10)$$

$$P_z = M_z \omega_z \quad (11)$$

Total power: $\sum P = P_x + P_y + P_z$ (12)

Force Platform

The force platform is used for measuring kinetic data. The force platform/plates use the sensors on the platform, which calculated the forces and moment with a flat surface as a tracking path. The force plate has a piezoelectric or force transducer typically. This force platform is insole with embedded sensors or pressure mat, which helps establish pressures or force profiles. The output of the force platform gives the components of the GRF vector as vertical, anteroposterior, mediolateral and the position of CoP of the body.

Electromyography

It measures the electrical activity of muscles by recording the muscle action potential generated in the muscle. The instrumentation unit for recording EMG signals is the surface electrode, amplification and conditioning of signal, transmission of signal, data display, and storage unit. The surface and fine-wired electrodes are commonly used in gait analysis. The surface electrodes are non-invasive and are placed on the skin using a conductive gel to minimize the electrical impedance. The limitation with the surface electrode is that they cannot record a specific part of the muscles. In order to overcome the limitation of the surface electrodes, a fine-wired electrode is used, which invasively measures the electrical activity of an individual muscle. It consists of a pair of fine wire inserting into the muscle through a hypodermic needle. The amplitudes of EMG signal from surface and fine-wire are small, and without amplification, it does not allow direct recording. A differential amplifier is used for signal conditioning. A high common-mode rejection ratio (CMRR) eliminates electrical noise by these electrodes. The EMG signals recorded by fine-wired and surface electrodes have a different known frequency range of spectral characteristics. The EMG signals from the surface electrode range from 10 Hz to 350 Hz, whereas the signals from fine-wired electrodes have a frequency range of 10 Hz to 1000 Hz (Abu-Faraj et al., 2015).

The EMG measured from muscle indicates the muscular effort level used during an activity. The difference in amplitude of the EMG represents the measure of muscle tension. When the amplitude increases, the gait speed also increases with comfortable movement. The timing of EMG is important in gait assessment as it can relate to the gait cycle phases. In pathological gait, EMG can be used to evaluate neuromuscular disorders deformities of the hip and ankle (Abu-Faraj et al., 2015).

Other Techniques

The other techniques used in gait analysis are imaging systems like ultrasound and MRI used to measure anthropometry data (height, BMI, waist to hip ratio) by analyzing the dimension of the body from the image obtained and modelling the geometric structure in 3D view (Ganley & Powers, 2004). The magnetic system using ferromagnetic markers are used and analyze the body locomotion by movement pattern of the markers. The acoustic system that uses ultrasound pulse determines the position of the movements. Machine learning is recently used in gait analysis for diagnosis and rehabilitation (Prakash et al., 2018).

The combined kinematic/kinetic techniques are used with multiple regression techniques for asymmetry gait (Jena et al., 2021).

APPLICATIONS OF GAIT ANALYSIS

Gait analysis is used in many applications like sports for enduring performance and injury prevention for athletes, rehabilitation for patients with movement impairment, and design orthotics/shoes for differently-abled subjects. In sports, the loading behaviour of foot and neuromuscular activities are assessed for many games. An athlete's energy expenditure limb segment acceleration is evaluated by gait analysis. The gait parameters are assessed during the exercise and training of the athlete, which helps to understand the athlete's body motion and improvement of performance. In rehabilitation, the gait pattern needs to be restored. A robot assistive system is used to drive the gait performance, increase the load-bearing, improve EMG signal. The lower and upper extremities amputation is evaluated by gait pattern analysis. The gait analysis helps in studying the body's motion by modelling the body segment and analyzing the movement by simulation. The clinical gait analysis is used to diagnose the disorders associated with locomotion.

Gait Analysis in Sports

In sports, gait analysis is essential to improve the athlete's performance and prevent injury. High-speed and high-resolution camera systems are used to evaluate the body posture during sports activities. In the triple jump, gait is used to study and analyze the neuromuscular activities of the athlete to obtain maximum horizontal speed during the jump with a higher degree of impact force during take-off. By analyzing the neuromuscular response, proper training and exercise can be suggested to improve the athlete's performance (Perttunen & Heinonen, 2000). The kinematic data of soccer players is measured using wearable and vision sensors as a gait analysis system. Recurrent Neural Network (RNN) is used for classifying and recognizing soccer gait in real-time. The study revealed the foot kinematics data as the most reliable in classifying and recognizing soccer gait patterns (Akhtaruzzaman et al., 2016).

The typical injury that occurs in runner athletes is patellofemoral pain. A recent study determined the altering step time and patellofemoral joint loads during running. The study revealed that an increase in step rate and reduction in stride length decreases the patellofemoral load and influential factors in biomechanics (Heiderscheit, 2014). Videotaped observation gait analysis is used to assess athletes related to the injury or pain. It measures the kinetic and kinematic parameters through the retro-reflective tape, digital camera and treadmill (Whittle, 1985). It is also important to evaluate an athlete's strength, stability, flexibility, mobility for improved performance in static and dynamic conditions.

Gait Analysis in Rehabilitation

Rehabilitation is vital in a patient's life with injury. It provides recovery of the injured body segment to the maximum degree of recovery and improves the quality of the patient's life. Restoring the gait with a natural pattern is one of the main factors in rehabilitation. In evaluating ROM and subjective gait (parameters), a physiotherapist assesses the gait analysis of an injured patient (Keegan, 2011). A disorder like cerebral palsy, Parkinson are examined through gait analysis and proper rehabilitation treatment is given for their recovery. A robot assistive system drives gait performance, increases load-bearing, and improves the EMG signal. The robotic system reduces the burden of the therapist by controlling the forces and positions with accuracy and reliability. During the recovery period, the force between the affected limb of the patient and robot should not be greater, leading to secondary injury and limiting rehabilitation.

Other Applications

The upper and lower extremities amputation is evaluated by obtaining and analyzing the gait pattern. The upper extremities include the shoulder, spine and arm, whereas the lower extremities include the hip, knee and ankle. These extremities are impaired or injured due to neuromuscular disorder or trauma. Gait analysis determines the pattern of the subject with disorder or trauma and suggests a suitable rehabilitation. The prosthesis design is aided by determining the gait pattern of the subject. The problem of fitting the prosthesis and its gait deviation can be studied for proper fit and performance of the prosthesis. The gait analysis is also used to study a high-heeled shoe's rollover characteristics over an inclined surface. The study showed that walking surfaces have large radii of curvature at the early stage and late stage of the stance period in an inclined plane (Mishra et al., 2019).

In total joint arthroplasty, the replacement for hip, knee, or ankle evaluation is needed to analyze the longevity and biomechanics properties of the implants. Thus, gait analysis is important in studying the joint during replacements. The total knee arthroplasty showed a difference in stability and forces at the knee joint affect longevity and patient satisfaction.

Orthotic devices maintain the gap between the heel and the ground for proper ankle motion. The study determines the proper ground-level walking using an orthotic polypropylene material device, which reduces the plantar foot pressure and maintains a proper ROM. The finite element analysis shows a decrease in the plantar foot pressure with an orthotic device (Jena et al., 2020).

SUMMARY AND CONCLUSION

Human gait is a bipedal movement that involves the movement of the hip, knee and ankle. The computational measurement of kinetics, kinematics, musculoskeletal activity, foot pressure, energy expenditure as gait parameters can be assessed by gait analysis. The gait analysis assesses the performance of athletes and in rehabilitation for restoring the differently-abled patients. With advancements in technologies, efficient, portable, non-invasive techniques for monitoring, measuring, diagnosing the gait pattern minimizes testing time with improved clinical evaluation. Design and development of low-cost devices with multi-sensing measurements are essential for enhanced usability in gait analysis. Improved computational algorithms are designed for signal processing and provide significant benefits in real-time applications for analyzing and interpreting the data. The evolution of technology in the near future can replace 3D motion capture with imaging systems like 3D MRI and fluoroscopy. A more integrated system needs to be developed to measure the gait parameters that are impossible to measure directly from the body, like net joint moment and intersegmental forces. As the future scope, advanced modelling of the musculoskeletal system and its simulation can be used to study the motion analysis of the body. Therefore, gait analysis could be a potential diagnosing tool for analyzing human motion in rehabilitation.

REFERENCES

Abu-Faraj, Z. O., Harris, G. F., Smith, P. A., & Hassani, S. (2015). Human gait and clinical movement analysis. Wiley Encyclopedia of Electrical and Electronics Engineering, 1-34. doi:10.1002/047134608X.W6606.pub2

Akhtaruzzaman, M., Shafie, A. A., & Khan, M. R. (2016). Gait Analysis: Systems, Technologies, and Importance. *Journal of Mechanics in Medicine and Biology*, *16*(7), 1630003. doi:10.1142/S0219519416300039

Baker, R. (2006). Gait analysis methods in rehabilitation. *Journal of Neuroengineering and Rehabilitation*, *3*(1), 1–10. doi:10.1186/1743-0003-3-4 PMID:16512912

Baker, R. (2007). The history of gait analysis before the advent of modern computers. *Gait & Posture*, *26*(3), 331–342. doi:10.1016/j.gaitpost.2006.10.014 PMID:17306979

Banaszkiewicz, P. A., & Kader, D. F. (2014). Classic papers in orthopaedics. *Classic Papers in Orthopaedics*, *10*(15), 1–624.

Chambers, H. G., & Sutherland, D. H. (2002). A practical guide to gait analysis. *The Journal of the American Academy of Orthopaedic Surgeons*, *10*(3), 222–231. doi:10.5435/00124635-200205000-00009 PMID:12041944

Davis, R. B. (1988). Clinical Gait Analysis. *IEEE Engineering in Medicine and Biology Magazine*, *7*(3), 35–40. doi:10.1109/51.7933 PMID:18244073

Ganley, K. J., & Powers, C. M. (2004). Determination of lower extremity anthropometric parameters using dual energy X-ray absorptiometry: The influence on net joint moments during gait. *Clinical Biomechanics (Bristol, Avon)*, *19*(1), 50–56. doi:10.1016/j.clinbiomech.2003.08.002 PMID:14659930

Hachinski, V. (2008). A historical review of gait analysis. *Neurosciences*, *13*(4), 460. PMID:21063384

Harris, G. F., & Wertsch, J. J. (1994). Procedures for gait analysis. *Archives of Physical Medicine and Rehabilitation*, *75*(2), 216–225. doi:10.1016/0003-9993(94)90399-9 PMID:8311681

Heiderscheit, B. C. (2014). Increasing running step rate reduces patellofemoral joint forces. *Medicine and Science in Sports and Exercise*, 557–564. PMID:23917470

Jena, S., Arunachalam, T., & Panda, S. K. (2020). Experimental and numerical investigation of a polypropylene orthotic device for assistance in level ground walking. *Proceedings of the Institution of Mechanical Engineers. Part H, Journal of Engineering in Medicine, 234*(4), 356–369. doi:10.1177/0954411919894091 PMID:31854229

Jena, S., Sakhare, G. M., Panda, S. K., & Thirugnanam, A. (2021). Implementation of Multiple Regression Technique for Detection of Gait Asymmetry Using Experimental Gait Data. *Journal of Medical and Biological Engineering, 41*(1), 1–10. doi:10.100740846-020-00533-8

Keegan, K. (2011). Gait analysis. *Clinical Veterinary Advisor: The Horse*, (February), 736–740.

Kharb, A., Saini, V., Jain, Y., & Dhiman, S. (2011). A review of gait cycle and its parameters. *International Journal of Computational Engineering & Management, 13*, 78–83.

Mishra, E., Jena, S., Bhoi, C., Arunachalam, T., & Panda, S. K. (2019). Effect of high heel gait on hip and knee-ankle-foot rollover characteristics while walking over inclined surfaces— A pilot study. *The Foot, 40*, 8–13. doi:10.1016/j.foot.2019.03.004 PMID:30981083

Muro-de-la-Herran, A., García-Zapirain, B., & Méndez-Zorrilla, A. (2014). Gait analysis methods: An overview of wearable and non-wearable systems, highlighting clinical applications. *Sensors (Switzerland), 14*(2), 3362–3394. doi:10.3390140203362 PMID:24556672

Nutt, J. G., Marsden, C. D., & Thompson, P. D. (1993). Human walking and higher-level gait disorders, particularly in the elderly. *Neurology, 43*(2), 268–279. doi:10.1212/WNL.43.2.268 PMID:8437689

Perry, J. (1992). *Gait analysis normal and pathological function*. SLACK Incorporated.

Perttunen, J., & Heinonen, A. (2000). Biomechanical loading in the triple jump. *Journal of Sports Sciences, 18*(5), 363–370. doi:10.1080/026404100402421 PMID:10855682

Prakash, C., Kumar, R., & Mittal, N. (2018). Recent developments in human gait research: Parameters, approaches, applications, machine learning techniques, datasets and challenges. *Artificial Intelligence Review, 49*(1), 1–40. doi:10.100710462-016-9514-6

Saunders, J. B., Inman, V. T., & Eberhart, H. D. (1953). Six Determinants of Gait. *Journal of Bone and Joint Surgery, 35*(3), 543–558. doi:10.2106/00004623-195335030-00003

Seel, T., Raisch, J., & Schauer, T. (2014). IMU-Based Joint Angle Measurement for Gait Analysis. *Sensors (Basel), 14*(4), 6891–6909. doi:10.3390140406891 PMID:24743160

Simon, S. R. (1993). Gait Analysis, Normal and Pathological Function. *The Journal of Bone & Joint Surgery, 75*(3), 476–477. doi:10.2106/00004623-199303000-00027

Surer, E., & Kose, A. (2011). Method and technologies for gait anlaysis. In A. A. Salah & T. Gevers (Eds.), *Computer Analysis of Human Behavior* (pp. 105–123). Springer. doi:10.1007/978-0-85729-994-9_5

Tesio, L., & Rota, V. (2019). The Motion of Body Center of Mass During Walking: A Review Oriented to Clinical Applications. *Frontiers in Neurology*, *10*, 1–22. doi:10.3389/fneur.2019.00999 PMID:31616361

Waters, R. L., & Mulroy, S. (1999, July). (199 9). The energy expenditure of normal and pathologic gait. *Gait & Posture*, *9*(3), 207–231. doi:10.1016/S0966-6362(99)00009-0 PMID:10575082

Whittle, M. (1985). Applications of gait analysis. In Gait Analysis (pp. 177–193). doi:10.1016/B978-075068883-3.50010-6

Winter, D. A., Quanbury, A. O., & Reimer, G. D. (1976). Analysis of instantaneous energy of normal gait. *Journal of Biomechanics*, *9*(4), 253–257. doi:10.1016/0021-9290(76)90011-7 PMID:1262360

APPENDIX: LIST OF ABBREVIATIONS

BMI: Body Mass Index
BOS: Base of Support
CoG: Centre of Gravity
CoM: Centre of Mass
CoP: Centre of Pressure
EMG: Electromyography
GRF: Ground Reaction Force
HAT: Head, Arm and Trunk
ROM: Range of Motion
RNN: Recurrent Neural Network

Chapter 5
Biomechanical Analysis on Vancouver Periprosthetic Fracture in Femur Using the Finite Element Modeling

Raja Dhason
SRM Institute of Science and Technology, India

Sandipan Roy
SRM Institute of Science and Technology, India

Shubhabrata Datta
SRM Institute of Science and Technology, India

ABSTRACT

After total hip arthroplasty (THA), high demands occur in the femoral fixation, especially for elderly patients due to periprosthetic fractures. Fractures happened in the femur after THA was classified based on Vancouver periprosthetic femoral fracture. Choosing the fixation for these types of specific fracture types are challenging due to pattern and orientation of the fracture in the THA prosthesis. Several researchers have reported the clinical, experimental, and computational studies about the failure of fixation methods, and they highlighted the remedies and scope in the further studies. Most of the authors recommended the computational studies having the advantage to predict the inner behaviour of the bone because invivo/invitro study availability of the specimen are limited and it needs ethical clearance. The current study focussed on computational studies in periprosthetic fractures and aims to discuss the three dimensional model creation, mesh generation, material properties, boundary conditions/loading, limitation, opinions, and future thoughts in implant design.

DOI: 10.4018/978-1-7998-9078-2.ch005

INTRODUCTION

Degenerative disease often affects the elderly patient's that may causes injury in the joint structures and cartilage tissue. Total hip arthroplasty (THA) is a surgical procedure that has been used for pain relief and to restore the physiological function from various hip diseases. Total hip arthroplasty (THA) adopted in two well established methods for implant fixation namely cemented and uncemented (Press fit). The signals received to the patients for THA in advanced level of osteoarthritis, fracture in the femur neck, developmental dysplasia of the hip, advanced level of osteonecrosis of the femoral head and spondylitis of the hip joint. Almost 90% of the patients achieved physiological motion and complete pain relief after THA (Beckenbaugh, & Ilstrup, 1978), (Siopack et al. 1995). To assess the treatment options and to study the biomechanical behaviour of the fixation post-surgery. More than one decade, research was adopted on periprosthetic fracture in femur using in vitro and in silico methodologies. In vitro studies were conducted in the laboratories, current studies higher number of cadaveric femora specimen (healthy/ osteoporotic) were used for biomechanical studies. Limitations of in vitro studies is availability of the fresh cadaver in wide range of sample in a short period (Konstantinidis et al., 2010), (Lehmann et al., 2010), (Lenz et al. 2013), (Lenz et al., 2014). To overcome this problem, synthetic femur was used to test experimentally to mimic healthy, injured, repaired and healed conditions (Ebrahimi et al., 2012). Over the last decades, rapid rise in the use of in silicon studies to analyse the phenomena of inner structure behaviour that cannot be clarified by experimental methods. Finite Element models of the intact femur bone and femur fixation has been developed based on the medical image to evaluate the surgical interventions and also to investigate the risk of fracture in the post operative condition in femur (Periprosthetic Femoral Fracture). In this chapter the FE modelling procedure, material, meshing, loads, and detailed biomechanical behaviour that can provide the intrinsic parameters like stress and strain is discussed.

METHODOLOGY

CT image is the basic requirement to develop the surface model of the femur bone to perform the biomechanical analysis. The surface model derived from CT images in dicom format will be represents three-dimensional anatomical shape of the femur. These models are created by implementing no. of arrays of segmentation and region tool to define the smooth surface model without any noise. Mesh was generated in the different components of the femur bone and fixation assembly to define material properties, loading conditions to get the strain, stress distribution of the whole structure of the femur by studying the relation between the displacement and force on every element.

GEOMETRICAL MODELLING OF FEMUR AND ITS FIXATIONS

The geometrical information of the femur was extracted from subject specific femur to construct the three-dimensional model using three techniques like CT scan, Synthetic bone scanning, dry bone scanning and simplified bone model. Initial stage, X-ray was performed on the subject to ensure not affected by disease. Then CT scanning was performed on the subject at the lower extremity region to get the details of the femur. The slice-by-slice details of the CT image was taken at the minimum slice for the healthy person. After scanning the total length of the femur was compared with the literature (Wang et al., 2016).

Due to ethical constrain and limited availability of the volunteers, researchers moved to developing the synthetic femur for experimentation and computational studies. Synthetic femur was scanned using computed tomography (CT) at a slice thickness of 0.5mm and received in the DICOM format (Gee et al., 2021), (Chen et al, 2012), (Duboy et al., 2011). In other case, a dry bone of the femur without any surface damage was carefully selected and it was scanned using portable coordinate measuring machine to generate the surface model of the dry femur bone. In the above all the cases, CT scanned data was imported in to Mimics software (The Materialise Group, Leuven, Belgium) to generate the surface model. The surface model having gap in the surfaces to be rectified by smoothening the surface to remove the noise using Geomagic (3D Systems, US). The bone plate, stem and screws are modelled using modelling software based on information specified in the manufacturer's catalogue and literature (Moazen et al, 2013), (Saha et al., 2011) (Samiezadeh et al., 2015). In some studies, the model was simplified as a standard geometrical model for further studies (Kim et al., 2011), (Leonidou et al., 2015).

FINITE ELEMENT MODEL OF FEMUR

The development of tetrahedral or hexahedral finite element mesh in the surface model is based on geometrical parameters generated from subject specific models. After meshing convergence test was caried out by increasing the number of elements or changing the size of the element. Parameters like stress and displacement (< 1% to 5%) are kept for solution convergence provide marginal change to select the size or number of elements for the model. The bones and implants material properties were assumed to be isotropic and linearly elastic, with an appropriate range of young's modulus and Poisson ratio. Material properties of cortical bone, E = 12.4 – 20 GPa, υ = 0.26 – 0.30; Trabecular bone, E = 0.155 GPa, υ = 0.30; Cancellous bone, E = 0.1 – 0.104 GPa, υ = 0.30 – 0.35; Bone plate, E = 96 – 113.8 GPa, υ = 0.30; Stem and screws, E = 200 – 210 GPa, υ = 0.30; PMMA cement mantle, E = 2 – 2.45 GPa, υ = 0.18 – 0.30 were set to fixed values. The material properties were derived from the previous studies [9 -17].

BOUNDARY CONDITIONS

The boundary conditions of the finite element model were arresting and applying the pressure on the surface of the model. Loads acting on femur bone at an adduction angle of range between 7° to 15° depend on the aims of the analysis. The distal portion of the femur bone was rigidly fixed and the transferred load from the upper extremity to the proximal region of the femur was applied. An axial load of 500 N to 3000 N was applied on the ball tip by considering a person weight to 3 – 4 times of the body weight. The contact surface between bone bone–screw, bone–cement, and plate–screw, were set to bond (Duboy et al., 2011).

RESULT AND DISCUSSIONS

Stress and strain distributions were obtained from the biomechanical analysis by applying different loading conditions were discussed in the following section. The different type of fixations was used to study the stability of the fixation after periprosthetic fracture. Leonidou et al. (2015) used multidirectional locking plate fixation for intact and osteoporosis conditions. Stress and strain were lower by using this internal fixation and also stronger and more stable compare to other group of devices (Wang et al., 2016). Similarly, Gee at al used semi rigid plates to rise the healing rate by maintaining the maximum stresses in cortical bone in compare with metal plate (Gee et al., 2021). Additionally, proximal wire was added with locking plate obtained better fixation stability in the treatment of Vancouver type B1 periprosthetic fracture (Chen et al., 2012). In most of the studies, maximum amount of load was observed by metal implants that lead to 'stress shielding' even after fracture healing (Ebrahimi et al., 2012). To avoid stress shielding, the composite plate was proposed for the alternate replacement of metal implant to increase the compressive force and relative movement in the fracture site for better healing (Kim et al., 2011), (Dhason et al., 2020).

Some of the limitations were adopted during simulation, muscle, ligament and soft tissue effects are not considered in most of the studies. Weight bearing applied by the patient after post-surgery almost two times higher than the recommended by surgeons. Similarly, for bone isotropic material properties were assumed to simplify the problem, but in real bones the characteristic might be nonlinear, anisotropic, and viscoelastic properties in the different region of the bone. Additionally, relative motion between the bone and implants, bone and bone cement, implant and bone cement are not considered in most of the studies for approximation. In future studies, effects of realistic boundary condition, realistic material properties of bone and feasibility of using composite bone in-vivo performance to be studied (Wang et al., 2016), (Gee et al., 2021), (Dhason et al., 2020).

REFERENCES

Beckenbaugh, R. D., & Ilstrup, D. M. (1978). Total hip arthroplasty. *The Journal of Bone and Joint Surgery*, *60*(3), 306–313. doi:10.2106/00004623-197860030-00005 PMID:649633

Chen, D. W., Lin, C. L., Hu, C. C., Wu, J. W., & Lee, M. S. (2012). Finite element analysis of different repair methods of Vancouver B1 periprosthetic fractures after total hip arthroplasty. *Injury*, *43*(7), 1061–1065. doi:10.1016/j.injury.2012.01.015 PMID:22336128

Dhason, R., Roy, S., & Datta, S. (2020). A biomechanical study on the laminate stacking sequence in composite bone plates for vancouver femur B1 fracture fixation. *Computer Methods and Programs in Biomedicine*, *196*, 105680. Advance online publication. doi:10.1016/j.cmpb.2020.105680 PMID:32763643

Dubov, A., Kim, S. Y. R., Shah, S., Schemitsch, E. H., Zdero, R., & Bougherara, H. (2011). The biomechanics of plate repair of periprosthetic femur fractures near the tip of a total hip implant: The effect of cable-screw position. *Proceedings of the Institution of Mechanical Engineers. Part H, Journal of Engineering in Medicine*, *225*(9), 857–865. doi:10.1177/0954411911410642 PMID:22070023

Ebrahimi, H., Rabinovich, M., Vuleta, V., Zalcman, D., Shah, S., Dubov, A., Roy, K., Siddiqui, F. S., Schemitsch, E. H., Bougherara, H., & Zdero, R. (2012). Biomechanical properties of an intact, injured, repaired, and healed femur: An experimental and computational study. *Journal of the Mechanical Behavior of Biomedical Materials*, *16*(1), 121–135. doi:10.1016/j.jmbbm.2012.09.005 PMID:23182385

Gee, A., Bougherara, H., Schemitsch, E. H., & Zdero, R. (2021). Biomechanical design using in-vitro finite element modeling of distal femur fracture plates made from semi-rigid materials versus traditional metals for post-operative toe-touch weight-bearing. *Medical Engineering & Physics*, *87*, 95–103. doi:10.1016/j.medengphy.2020.11.015 PMID:33461680

Kim, J. H., Kim, S. H., & Chang, S. H. (2011). Estimation of the movement of the inter-fragmentary gap of a fractured human femur in the presence of a composite bone plate. *Journal of Composite Materials*, *45*(14), 1491–1498. doi:10.1177/0021998310383730

Konstantinidis, L., Hauschild, O., Beckmann, N. A., Hirschmüller, A., Südkamp, N. P., & Helwig, P. (2010). Treatment of periprosthetic femoral fractures with two different minimal invasive angle-stable plates: Biomechanical comparison studies on cadaveric bones. *Injury*, *41*(12), 1256–1261. doi:10.1016/j.injury.2010.05.007 PMID:21288467

Lehmann, W., Rupprecht, M., Hellmers, N., Sellenschloh, K., Briem, D., Püschel, K., Amling, M., Morlock, M., & Rueger, J. M. (2010). Biomechanical evaluation of peri- and interprosthetic fractures of the femur. *The Journal of Trauma Injury Infection and Critical Care*, *68*(6), 1459–1463. doi:10.1097/TA.0b013e3181bb8d89 PMID:20093986

Lenz, M., Perren, S. M., Gueorguiev, B., Höntzsch, D., & Windolf, M. (2013). Mechanical behavior of fixation components for periprosthetic fracture surgery. *Clinical Biomechanics (Bristol, Avon)*, *28*(9–10), 988–993. doi:10.1016/j.clinbiomech.2013.09.005 PMID:24080369

Lenz, M., Perren, S. M., Gueorguiev, B., Richards, R. G., Hofmann, G. O., Fernandez Dell'Oca, A., Höntzsch, D., & Windolf, M. (2014). A biomechanical study on proximal plate fixation techniques in periprosthetic femur fractures. *Injury*, *45*(Suppl. 1), S71–S75. Advance online publication. doi:10.1016/j.injury.2013.10.027 PMID:24252576

Leonidou, A., Moazen, M., Lepetsos, P., Graham, S. M., Macheras, G. A., & Tsiridis, E. (2015). The biomechanical effect of bone quality and fracture topography on locking plate fixation in periprosthetic femoral fractures. *Injury*, *46*(2), 213–217. doi:10.1016/j.injury.2014.10.060 PMID:25467710

Moazen, M., Mak, J. H., Etchels, L. W., Jin, Z., Wilcox, R. K., Jones, A. C., & Tsiridis, E. (2013). The effect of fracture stability on the performance of locking plate fixation in periprosthetic femoral fractures. *The Journal of Arthroplasty*, *28*(9), 1589–1595. doi:10.1016/j.arth.2013.03.022 PMID:23642449

Samiezadeh, S., Tavakkoli Avval, P., Fawaz, Z., & Bougherara, H. (2015). On optimization of a composite bone plate using the selective stress shielding approach. *Journal of the Mechanical Behavior of Biomedical Materials*, *42*, 138–153. doi:10.1016/j.jmbbm.2014.11.015 PMID:25482217

Shah, S., Kim, S. Y. R., Dubov, A., Schemitsch, E. H., Bougherara, H., & Zdero, R. (2011). The biomechanics of plate fixation of periprosthetic femoral fractures near the tip of a total hip implant: Cables, screws, or both? *Proceedings of the Institution of Mechanical Engineers. Part H, Journal of Engineering in Medicine*, *225*(9), 845–856. doi:10.1177/0954411911413060 PMID:22070022

Siopack, J. S., Jergesen, H. E., & Francisco, S. (1995). Conferences and Reviews Total Hip Arthroplasty. *The Western Journal of Medicine*, *162*(2), 243–249. PMID:7725707

Wang, G., Wang, D., Mao, J., Lin, Y., Yin, Z., Wang, B., He, Y., & Sun, S. (2016). Three dimensional finite-element analysis of treating Vancouver B1 periprosthetic femoral fractures with three kinds of internal fixation. *International Journal of Clinical and Experimental Medicine*, *9*(4), 7557–7564.

Chapter 6
Recent Advancements in Design and Material Perspective of the Implants in the Biomechanics of the Cervical Spine

Pechimuthu Susai Manickam
https://orcid.org/0000-0001-7507-5726
SRM Institute of Science and Technology, India

Balamurugan S.
SRM Institute of Science and Technology, India

Sandipan Roy
SRM Institute of Science and Technology, India

ABSTRACT

Understanding biomechanics helps us to understand the problems involved in the cervical spine. The most common problems in the cervical spine are cervical spondylosis, herniated disc, degenerative disc disease, and spinal stenosis. The surgeries to treat these problems are anterior cervical discectomy and fusion surgery, cervical spinal fusion, laminectomy, laminoplasty, and artificial disc replacements. The post-surgery effects of the implants and the surgical studies can be simulated using numerical simulation. In the numerical study, by varying the design of the implants and by varying the material models, the simulation can be conducted and the best comes can be used further for in vitro study. In the in vitro study, the local stress and strains cannot be calculated but the finite element study provides a clear in sight in the stress and strain distribution across the full structure. In summary, the presurgical clinical evaluation can be conducted using the numerical study and the in vitro experiments to understand the biomechanics of cervical spine with implants.

DOI: 10.4018/978-1-7998-9078-2.ch006

INTRODUCTION

Anatomy and Bio Mechanics of Cervical Spine

The human beings spine known are a complex arrangement that provides support, control, and transmission of the trunk and extremities also safeguard the spinal cord (Oxland, 2015) (Oxland, 2016). The cervical spine segment is an intricate arrangement of bones, spinal segments, joints, nerves, muscles and connective tissues. The anatomy of the bony spine is separated into the cervical vertebrae - seven, the thoracic vertebrae - twelve, the lumbar vertebrae - five, and the ossified sacral - five and coccygeal segments – four respectively (Shapiro, & Risbud, 2014). The human spinal column is a mechanical wonder that provided both rigidity and flexibility. The arrangement of the spinal stake allows sufficient motion in the sequence from the head, trunk, and pelvis; offers safeguard of the spinal cord; handovers force due to weight and bending moments from the superior body to pelvis; provides a suspension and assists as a pivot for the head (Galbusera, 2018). From C1 to S1, the vertebral bodies are formed by the articulating parts of vertebrae, which are separated by, the posterior facet joints and intervertebral discs (IVD's) (Galbusera, 2018) (Sharabi, et al., 2018). The IVD's tend to be statically load supporting joints, while the role of the facet is dynamically sliding and gliding joints (Malandrino, 2018), (Wade, 2018). A curved design of the column has amplified resistance to compression forces. The cervical spine is the topmost spine of the human spine and consists of seven vertebrae (Oxland, 2015). The important role of the cervical spine acting as the mobile bearing of the human head and safeguard the spinal cord. The cervical spine anatomy divided into lower, middle and upper cervical spine. The kinematic parameters which are involved in the spine are the elastic zone and the neutral zone. The region which restrict the motion is called as the neutral zone and the elastic zone will have a stiffer and generates a non linear resistance towards the applied motions. The IAR is the point which allows the cervical spine to bend and rotate and it is formed as a natural point.

FUNCTIONAL ANATOMY OF VERTEBRAE, INTERVERTEBRAL DISC, LIGAMENTS AND MUSCLES

The vertebrae is the small bone that stacks on top of one another. It supports the body weight, safeguard the spinal cord and attach the ligaments in the spine to allow flexibility and movement. The frontal portion of a vertebra is termed the vertebral body. The latter portion is named the neural arch or vertebral. The vertebral is further separated into the posterior elements and the pedicles.The following elements are the transverse processes, Interior articulated process, spinous process, articular processes, and laminae (Yoganandan et al., 2000). Facet joints are attached directly to the neural arch and transfer a substantial amount of compressive load to the lower segments. The intervertebral discs are made up of annulus (a tough, collagen fibers elastic ring in outer) surrounding a soft gel center (nucleus). The intervertebral discs damping the spine during loading and bending actions. The ligaments are fibrous elastic bands or connective tissues in the sheet form of connecting two or more bones, cartilages, or structures together. It provides constancy to a joint during movement and rest. In the cervical spine, the muscles perform the role for holding up posture, rotation of the neck, and ranging the neck backward.

MECHANICAL PROPERTIES OF CERVICAL SPINE

The mechanical properties of the vertebral trabecular bone were the discussed here. The lading modes were tensile, compressive and shear. In the tensile load the ultimate strength is 1.33 MPa to 3.53 MPa (Kopperdahl et al., 2002). The compressive load the ultimate strength is 0.038 MPa to 10.6 MPa (Kopperdahl, & Keaveny, 1998). The yield strength during the compression load is 0.1 MPa to 24 MPa.The ultimate strain in the compressive load varies from 0.96% to 2.30%. and during the load is 1.09% to 2.51%. The yield strain in the tensile load varies from 0.71% to 0.88%. The compressive load varies from the 0.65% to 0.87%. The load transfer mechanism in the spine is achieved by the intervertebral disc the mechanism and the pattern are the main factors to be understood. It transmits the load from the upper segment to the lower segment and maintains the stability of the spine. In the intervertebral disc the annulus is the outermost component which is made of fibers and the nucleus is made of collagen. The circumferential tension was applied and the failure was occurred at 2.7 MPa. and the failure strain is 0.3 In the tensile behavior the maximum strain rate is 4 percentage per seconds (Newell et al. (2017).

MAJOR ISSUES IN HUMAN CERVICAL SPINE

The major problems in the human cervical spine are disc degeneration, muscle disorder, joint dysfunction, arthritis, vertebral fracture. Muscle disorder is the superior back pain is caused by irritation of muscle, tension, or tightness. Joint Dysfunction is the pain caused by the attachment ribs in the spine at respective level of the thoracic spine. Herniated discs - In the spine, disease due to degenerative disc causes pain. Arthritis is caused because of swelling due to arthritis in the spine can cause pressure to the nerve, tenderness, and restricted range of motion. Disorders in the facet joint of the spine can effect from osteoarthritis. Vertebral fractures in terms of compression fractures due to osteoporosis are a major cause of spine pain in the elder age.

FINITE ELEMENT STUDY IN CERVICAL SPINE

The finite element study of the cervical spine will provide valuable information for clinician in understanding the treatment strategies. In the FE models the model generation is a tedious process to mimic the real behavior of the soft tissues. The material models are considered to define the behaviour of the soft tissues. Generally, in the FE models the bones are considered as the rigid and solids and the intervertebral disc is considered as the both linear and non-linear material models (Arab et al., 2020), (Biswas et al., 2019), (Bhattacharya et al., 2019). The ligaments of the cervical spine is considered as from the force deflection curves which are calculated from the experimental results. In some models the linear material properties also considered. The loading conditions are defined as both static and dynamic models. In the static models generally the range of motion and the stress distribution across the vertebrae, intervertebral disc and the ligaments are calculated. The loading conditions applied in the static modes are compressive preload 50 N is considered it replicated the head weight (Manickam et al, 2021), (Manickam, & Roy,a, 2021), (Manickam, & Roy,b 2021). The moments applied in the top of the endplate varies from ±2 Nm for all the physiological motion. In the dynamic models used to study the injury prediction can be simulated in that the compression load will be varied 200 N to 1200 N. All the finite element simulations were validated against the experimental data.

IMPLANTS IN THE CERVICAL SPINE

The different implants are cages, screws, rods and plates. In the anterior and the posterior interbody fusion are done using the cages. The main purpose of the cages is it restores the height of the intervertebral disc after the implantation. For the disc degeneration disease and the interbody fusion cage can be the alternate solution. Cages acts as a spine stabilizer which is implanted between the vertebrae to maintains the intervertebral height and the lordosis curve (Warburton et al., 2000). The grafts were provided to fuse with the adjacent vertebrae. The major problem in the cages are subsidence, implant loosening and migration of the cage. The shape of the cage plays major role in making the fusion with adjacent segments. The screws are used to hold the vertebrae together with the plates and rods. The rods are used to improve the stability to the spinal implant. The spinal rods are named as Harrington rods which adds stability for the spinal structure. The plates used in the spinal stabilization and it prevent the pullout of the implants. Pedicle screws are used in all types of spinal surgery for deformity correction, tumor decompression and stabilization, fracture stabilization, and fusion surgery for degenerative disease (Gupta et al., 2021), (Biswa et al., 2018), (Manickam et al., 2022), (Roy et al., 2014). The S-type dynamic cage is the dynamic implant which provides motion after the anterior cervical discectomy and fusion surgery (Manickam et al., 2021), (Manickam & Roy, a,b, 2021).

MATERIALS USED FOR THE IMPLANT IN THE CERVICAL SPINE

The cages are manufactured using the Titanium, PEEK, ceramic and acrylic materials. The new upcoming materials which are newly introduced are bioactive glass, Silicon Nitride, Apatite-Wollastonite, poly(ε-caprolactone) +HA (biodegradable). The screws are made up of titanium doped with hydroxyapaptite, calcium phosphate and tantalum and the new upcoming carbonated apatite (Warburton et al., 2000). The rods used in the spinal implant are Titanium, cobalt chromium alloys, PEEK, stainless steel and nitinol. The new upcoming materials are titanium molybdenum, biodegradable materials. The plates are made up of titanium and the new materials are biodegradable materials.

CONCLUSION

The basic biomechanics and the anatomical structure of the cervical spine was discussed. The major issues related to the cervical spine was elaborately discussed and the finite element study acts as tool to analyze the structure. The implants used for the cervical spine and the materials used for the development of the implant were discussed here.

REFERENCES

Arab, A. Z., Merdji, A., Benaissa, A., Roy, S., Bouiadjra, B. A., Layadi, K., Ouddane, A., & Mukdadi, O. M. (2020). Finite-Element analysis of a lateral femoro-tibial impact on the total knee arthroplasty. *Computer Methods and Programs in Biomedicine*, *1*(192).

Bhattacharya, S., Roy, S., Rana, M., Banerjee, S., Karmakar, S. K., & Biswas, J. K. (2019). Biomechanical performance of a modified design of dynamic cervical implant compared to conventional ball and socket design of an artificial intervertebral disc implant: A finite element study. *Journal of Mechanics in Medicine and Biology*, *19*(04), 1950017. doi:10.1142/S0219519419500179

Biswas, J. K., Roy, S., Majumder, S., Karmakar, S. K., Saha, S., & Roychowdhury, A. (2018). Artificial Intervertebral Disc Replacement to Provide Dynamic Stability to the Lumbar Spine: A Finite Element Study. *Journal of Long-Term Effects of Medical Implants*, *28*(2), 101–109. doi:10.1615/JLongTermEffMedImplants.2018025397 PMID:30317959

Biswas, J. K., Roy, S., Pradhan, R., Rana, M., & Majumdar, S. (2019). Effects of cervical disc replacement and anterior fusion for different bone conditions: a finite element study. *International Journal for Multiscale Computational Engineering, 17*(4). doi:10.1615/IntJMultCompEng.2019030212

Galbusera, F. (2018). The spine: its evolution, function, and shape. In *Biomechanics of the Spine* (pp. 3–9). Academic Press. doi:10.1016/B978-0-12-812851-0.00001-X

Gupta, Y., Iyer, R., Dommeti, V. K., Nutu, E., Rana, M., Merdji, A., & Roy, S. (2021). Design of dental implant using design of experiment and topology optimization: A finite element analysis study. *Proceedings of the Institution of Mechanical Engineers. Part H, Journal of Engineering in Medicine*, *235*(2), 157–166. doi:10.1177/0954411920967146 PMID:33094686

Kopperdahl, D. L., & Keaveny, T. M. (1998). Yield strain behavior of trabecular bone. *Journal of Biomechanics*, *31*(7), 601–608. doi:10.1016/S0021-9290(98)00057-8 PMID:9796682

Kopperdahl, D. L., Morgan, E. F., & Keaveny, T. M. (2002). Quantitative computed tomography estimates of the mechanical properties of human vertebral trabecular bone. *Journal of Orthopaedic Research*, *20*(4), 801–805. doi:10.1016/S0736-0266(01)00185-1 PMID:12168670

Malandrino, A. (2018). Intervertebral disc. In *Biomechanics of the Spine* (pp. 89–103). Academic Press.

Manickam, P. S., Ghosh, G., & Roy, S. (2022). Optimization of Bone Graft Shapes of S-Type Cervical Cage through Genetic Algorithm. *International Journal for Multiscale Computational Engineering*, *20*(1), 55–68. doi:10.1615/IntJMultCompEng.2021039717

Manickam, P. S., & Roy, S. (2021). The biomechanical effects of S-type dynamic cage using Ti and PEEK for ACDF surgery on cervical spine varying loads. *The International Journal of Artificial Organs*, *44*(10), 748–755. doi:10.1177/03913988211039525 PMID:34387526

Manickam, P. S., & Roy, S. (2021). The biomechanical study of cervical spine: A Finite Element Analysis. *The International Journal of Artificial Organs*. Advance online publication. doi:10.1177/0391398821995495 PMID:33645324

Manickam, P. S., Roy, S., & Shetty, G. M. (2021). Biomechanical evaluation of a novel S-type, dynamic zero-profile cage design for anterior cervical discectomy and fusion with variations in bone graft shape: A finite element analysis. *World Neurosurgery, 154*, e199–e214. doi:10.1016/j.wneu.2021.07.013 PMID:34246827

Newell, N., Little, J. P., Christou, A., Adams, M. A., Adam, C. J., & Masouros, S. D. (2017). Biomechanics of the human intervertebral disc: A review of testing techniques and results. *Journal of the Mechanical Behavior of Biomedical Materials, 69*, 420–434. doi:10.1016/j.jmbbm.2017.01.037 PMID:28262607

Oxland, T. R. (2015). A history of spine biomechanics. *Der Unfallchirurg, 118*(1), 80–92. doi:10.100700113-015-0087-7 PMID:26526281

Oxland, T. R. (2016). Fundamental biomechanics of the spine—What we have learned in the past 25 years and future directions. *Journal of Biomechanics, 49*(6), 817–832. doi:10.1016/j.jbiomech.2015.10.035 PMID:26706717

Roy, S., Panda, D., Khutia, N., & Chowdhury, A. R. (2014). Pore geometry optimization of titanium (Ti6Al4V) alloy, for its application in the fabrication of customized hip implants. *International Journal of Biomaterials, 2014*, 1–12. doi:10.1155/2014/313975 PMID:25400663

Shapiro, I. M., & Risbud, M. V. (2014). Introduction to the structure, function, and comparative anatomy of the vertebrae and the intervertebral disc. In *The intervertebral disc* (pp. 3–15). Springer. doi:10.1007/978-3-7091-1535-0_1

Sharabi, M., Wilke, H. J., & Haj-Ali, R. (2018). The Vertebral Bone. In *Biomechanics of the Spine* (pp. 71–87). Academic Press.

Wade, K. (2018). Vertebral Endplates. In *Biomechanics of the Spine* (pp. 125–140). Academic Press.

Warburton, A., Girdler, S. J., Mikhail, C. M., Ahn, A., & Cho, S. K. (2020). Biomaterials in spinal implants: A review. *Neurospine, 17*(1), 101–110. doi:10.14245/ns.1938296.148 PMID:31694360

Yoganandan, N., Kumaresan, S., & Pintar, F. A. (2000). Geometric and mechanical properties of human cervical spine ligaments. *Journal of Biomechanical Engineering, 122*(6), 623–629. doi:10.1115/1.1322034 PMID:11192384

Chapter 7
Injury Prevention and Improving the Performance of Athletes

Tharani Kumaran
National Institute of Technology, Rourkela, India

Mirza Khalid Baig
National Institute of Technology, Rourkela, India

Ravi Kant Avvari
https://orcid.org/0000-0001-5586-746X
National Institute of Technology, Rourkela, India

Thirugnanam Arunachalam
National Institute of Technology, Rourkela, India

ABSTRACT

Sports biomechanics helps one to understand the sport movements and helps increase the performance of individuals and prevent injury. Understanding the mechanics involved in every sport is crucial for kinesiologists help patients to recover from traumatic or overuse injuries. Each sport activity has different phases, and in each phase, the kinetics and kinematics are studied for better understanding. Computational technologies help to understand sports movements and forces developed during motion and provide proper guidance that can be followed to avoid injuries and enhance performance. The aspects of biomechanical analysis of sports, injuries and their causes in sports, various techniques for recording and analysis of sports movements, and lastly, performance improvement and preventing injury using the analysis are discussed in this chapter.

INTRODUCTION

Major thirst for studying sports biomechanics is to understand movements in each sports activity and help the athletes to enhance their performance and prevent injuries resulting from the wrong movement (Knudson, 2007). The primary reason for the damage is the lack of coaching on proper movement

DOI: 10.4018/978-1-7998-9078-2.ch007

techniques right from their juvenile stage of practising and playing (Gearity & Murray, 2011). Sports biomechanics research can assist athletes and coaches in training and educating individuals to improve performance while reducing the risk of injuries (Gee, 2010). Sports biomechanics is associated with both kinematics and kinetics. Kinematics helps to understand the movements involved in a sports activity. Generally, movements are termed based on planes and the axis they occur. Movements in the sagittal plane are flexion and extension, movement in the frontal plane is adduction and abduction, movement in the horizontal plane is internal rotation and external rotation. Some sport includes projectile motion, and this is based on optimum projection angle. On knowing this projection angle, flight time and range are calculated, which helps to improve the performance (Jeklin, 2016). Kinetics is the study of the action of movements caused by force. Translational and rotational kinetics are also involved in the sports activity. This chapter discusses sports surfaces, injuries in performing sports, and techniques used to measure the performance. The biomechanical aspects of different sports activities will also be discussed here.

TYPES OF SPORTS SURFACES

There are different types of sports surfaces. The performance of sports surfaces depends on the properties and combination of materials used to build the sports surface. Sports surfaces evaluation should be done priorly because their mechanical property might affect the performance of a sports person (Katkat et al., 2009). Sports techniques can be improved on making a perfect surface for a particular sport. They are natural and synthetic surfaces for indoor and outdoor games. If the surface property is changed, it is necessary to change the technique, as each surface has its characteristics; this might help reduce the injuries caused by sports surfaces (Dixon et al., 2015). The sports surface also helps in improving the performance of the individual. Sports surfaces behaviour is based on surface characteristics such as friction, traction, compliance, resilience and hardness. Sports surfaces are classified as natural and artificial surfaces.

Natural Surfaces

The natural surface is made up of grass and sand. The soil texture determines the mechanical strength of the sand-based sports turf (Guisasola et al., 2010). It is necessary to provide the required amount of recovery time after every use, which would lead to an increased lifetime of grass than expected. Also, when surfaces are used frequently, wear damage of surface is increased (Stiles et al., 2009).

Artificial Surfaces

The artificial surface is prepared using various synthetic materials based on the requirement of the sport. It is made up of polymeric material like PVC, polyurethane, maple hardwood, vulcanised rubbers for indoor sports. It is usually made with sand dressed in artificial grass courts and polymeric surfaces for outdoor games for athletic events. Various artificial surfaces are explained in the following section.

Concrete and Asphalt Surfaces

They are usually used for running and marathons. These surfaces have better traction, high resilience and low compliance. These surfaces are a bit riskier and are prone to injury (Dixon et al., 2015).

Sprung Wooden Floors

These surfaces are usually prepared for practising gymnastics. On proper maintenance of these wooden floors, its lifetime can be increased. These surfaces reduce impact shock. They behave elastically for more significant weight, whereas surface manifest point elasticity for small weights (Walker & Subic, 2013).

Cast in-Situ Elastomers

For athletics events, the most preferred surface is cast in-situ elastomers. These surfaces have good durability and are easy to install. The top layer is made with cross-linked polymers such as polyurethane or with natural/synthetic latex, beneath this asphalt layer of about 25 mm, 100 mm of bitumen macadam material, 200 mm basement with crushed compact. As the polymeric surfaces get aged, they start to degrade and wear. The thermal effect on degradation might lead to a loss of surface resilience.

Synthetic Fibre Textile Surfaces

These surfaces have improved dimensional stability. PVC or foam rubber is used for the backing to have more compliance. They are laid loosely and bounded to subfloors by adhesive joints, and each section is connected by overlapping and cutting (Schoukens, 2009).

Synthetic Turf

Synthetic turf surfaces are synthetic fibres or ribbon woven, knitted or tufted into backing fabric to create a mattress like structure. The pile strands are fixed to the backing using a rubber latex binder to get dimensional stability, structural integrity, and better flexibility. They are placed on the concrete over many layers of stone and soil by placing a flexible foam shock pad in between. Shock pads help reduce injury due to falls and provide good resilience and a suitable environment for playing. These surfaces are used in American football, hockey, soccer, tennis and baseball, which use synthetic fibres. Energy absorption, recovery and compliance are affected by temperature (Jastifer et al., 2019).

Generally, the sports surface system is distinguished by sub-base, mid-layer and performance-surface layer. The role of the sub-base and mid-layer is to bear the load and absorb the shock, respectively. Currently, there are many sports surfaces engineered for both indoor and outdoor games based on the requirements of players-surface interaction and ball-surface interaction in different sports. Even though few sports such as football, cricket and golf stick to natural sports surfaces, most sports are getting adapted to synthetic sports surfaces because the maintenance of artificial surfaces is more accessible than the natural surface. Therefore, designing a sports surface that can maintain high performance even under high usage is the main challenge for surface engineers.

INJURIES TO JOINTS AND TISSUES IN SPORTS

Injuries occur when joints or muscles experience more loads than usual when the load applied exceeds the tolerance limit (Twist & Eston, 2005). Sports injury reduces the performance of the individual. Based on the time taken to recover from injuries caused, a sports injury can be classified as minor, moderate, and severe. In minor injury, athletes take one to seven days to recover, and in mild injury, athletes take eight to twenty-one days for their recovery. In contrast, athletes require more than twenty-one days to heal in severe injury. Different types of injury occurrence depend on load characteristics such as load, load rate, the magnitude of load and frequency of load repetition (Edwards, 2018). Injuries are classified as traumatic injury and overuse injury. Traumatic injury is due to external forces, whereas overuse injury occurs due to frequent trauma and does not allow tissues to undergo self-repair (Finch & Cook, 2014; Smoljanovic et al., 2009). Also, sports injuries have an impact due to the usage of sports equipment and footwear. Generally, it gets fractured when the load is applied to a structure beyond its strength. Energy dissipation is associated with the magnitude of the applied load. When energy dissipation is higher, intramolecular bonds in soft tissues and bones are cracked, and more tissues are destroyed and cause more complex fractures (Bahr, 2009). Mainly, injuries occur at the joints and in their surrounding tissues. Usually, joint injuries are a combination of bone injury and soft tissues associated with it, namely ligaments, cartilage and muscle tendons (Getgood et al., 2019). So, soft tissue injury involves the repair mechanism such as inflammations in damaged tissue area, repairing the damaged tissues and degeneration of most damaged tissues (Ashkavand et al., 2013).

Articular Cartilage

Cartilage is classified into three types such as articular cartilage, elastic cartilage and fibrous cartilage. They are categorised based on their elasticity, structure and strength. The intervertebral disc and meniscus are classified as fibrous cartilage; the outer ear is classified as elastic cartilage, which protects the joints by forming extra protective layers are classified as hyaline or articular cartilage (Bhosale & Richardson, 2008). The primary function of articular cartilage is to impart low friction to the respective joints, which enables the joints to accept the weight in various ranges of motion for performing the necessary activity (Sophia Fox et al., 2009). Articular cartilage is liable to deformation; it enlarges the load-bearing area to subsidise the stress concentration. Articular damage occurs due to progressive degeneration of articular cartilage or traumatic incidents. Sometimes, articular cartilage cells can heal based on the spot of injury and range of damage (Buckwalter, 1998). Cartilage injuries occur without any fractures. When they experience too much load (Mithoefer et al., 2009) than their usual bearing capacity, wear and tear of cartilage occur, further progressing into osteoarthritis. On repeated loading conditions, swelling occurs, cartilage gets fractured in case of severe traumatic injury and overuse of joints leads to osteoarthritis (Thelin et al., 2006).

Ligaments

It is known that ligaments are connective tissues that connect bone to bone. The primary function of ligaments is to provide stability to joints, control joint movements, and impart the load at joints (Anderson, 2001). Usually, sprains occur when joints undergo excessive motion, which is not severe, but when ligaments are stretched far from the normal range, this might cause everlasting deformation of ligaments

(Smith, 2005). Ligament failure occurs when the limb experiences twisting and bending loads, and also it is dependent on load rate (Chen et al., 2019). Ligament tear has many effects: losing limb stability, misalignment of joints, loss of mechanoreceptors that sense motions, and irregular contact pressure. Recovery of injury may take several months, but the recovered ligament does not have mechanical property equivalent to the previous ligament (Hauser & Dolan, 2011). Those with less muscle strength or any records with ligament injury are more susceptible to ligament injuries (Murrell et al., 2001).

Tendons

Tendon injury is a common injury. Muscle injuries or sudden strain to muscle might occur due to improper muscle relaxation. Tendons usually connect the bones to the muscle (Lin et al., 2004). Muscle fibres and tendons are damaged during traumatic injuries or overuse of muscles. Tendon injuries are of two types: chronic and acute tendon injuries. Chronic tendon injury is characterised by decreased strength and reduced movements in tendons, whereas acute tendon injury is associated with complete or partial rupture (Nourissat et al., 2015). These injuries occur in rapid muscle contraction.

Joint Injury

The Pelvis and Hip Joint

Even though the pelvis is made of cancellous bones, they have enormous strength. These cancellous bones have an excellent shock-absorbing property which helps in reducing the stress concentration (Tyler et al., 2010). Pelvic girdle injuries are rare in sports, but in some cases, due to compression of high energy loads, this might affect pelvic girdles. When the femur is experiencing torsion, axial compression, shear and bending loads in combination, femoral fracture occurs. The hip joint transmits the force to the femur, which has bending, shear, and compressive components, showing more probability of getting fractured at different sites. Continuous loading in femur bone increases the fracture rate in joggers and long-distance runners (K. Anderson et al., 2001).

The Knee Joint

Knee injuries occur when they experience bending, trauma, or twisting, which is unusual from anatomical design. Other factors for causing knee injury are overuse of the knee, improper guidance to athletes, and high contact sports, which results in a frequent change in direction (Thacker et al., 2003). Knee injuries can be fatal with more complications, sometimes leading to dysfunction, or sometimes reversible with few damages. Soft tissue injuries are also associated with the knee joint, such as meniscus damage, ligamental tear and sprains. The knee joint comprises the patellofemoral joint, tibiofemoral joint, meniscus, posterior cruciate ligament, medial collateral ligament, medial meniscus, lateral meniscus, lateral collateral ligament and the anterior cruciate ligament (Allen et al., 1995). The primary function of the meniscus is to distribute stress, shock-absorbing, acting as a lubricant to the joints and providing stability in the medial-lateral direction and the anteroposterior direction. Cruciate ligament withstands hyperextension of the knee, lateral collateral ligament is resistant to adduction and abduction (Miller & Thompson, 2014). Despite fractures, ligamental injuries are much more common. Usually, ligamental damage occurs when the knee experiences axial load and movement in the outwards direction with external rotation. This loading condition causes ligament tear, medial collateral ligament, medial meniscus, and the anterior cruciate ligament (Weiss & Whatman, 2015). Usually, it is the anterior cruciate ligament that experiences more disruption. Meniscus tear usually happens on a rotating body about the knee bearing weight. Sudden change in direction with more significant acceleration is also responsible for a meniscus tear (Takeda et al., 2011).

The Ankle Joint

Runners often experience ankle-foot injuries, and gymnasts are prone to ankle injuries (Vormittag et al., 2009). Soft tissues surrounding joints are more prone to inflammations and sprains. Plantar flexions of the ankle and inversion load-bearing lead to sprain in the lateral ligaments, and eversions cause sprains in the medial ligament. Forced dorsiflexion causes damage to the tibiofibular ligaments (Bonnel et al., 2010). Damage to the posterior tibial tendon or the peroneal tendon causes tendonitis, which usually occurs in runners—continuous serving movements or jumping leads to rupture of the Achilles tendons (Longo et al., 2009).

In summation, injuries to the human musculoskeletal system by loading effect was discussed in this section. The difference between overuse and traumatic injury in soft tissues and bones were discussed. The impact of injuries in soft tissues and its characteristic were also discussed.

USE OF SPORTS EQUIPMENT AND TECHNIQUES IN INJURY

Sports equipment is generally used to improve performance provide safety to the performer. Though they are helpful to the athlete, this causes some inconvenience and injuries. Sports equipment includes footwear, surfaces, protective equipment and clothing. Slight changes in the comfort of shoes or surfaces might alter the muscle pattern (Nigg et al., 1999; Nurse et al., 2005).

The interface between the footwear and the sports surface is crucial for lower limb injuries. Both compliant and non-compliant surfaces impact injuries for athletes (Mears et al., 2018). Compliant surfaces might lead to fatigue injuries; though non-compliant injuries have improved impact loading conditions, it causes injury to athletes. It is mainly stated that natural surfaces are suitable for running because they have good smoothness, desirable resilience, and proper compliance. Non-compliant and hard surfaces cause lower back pain and injuries in the lower extremity. Synthetic surfaces are stiffer, which causes injuries in tendons and joints. The hard ground has less compliance which increases chances for tendon injury and causes calf muscle inflammations. Usage of inappropriate footwear might increase the incidence of injury. Generally, normal walking foot strike patterns vary for individuals using normal flat footwear and pointed heels (Mishra et al., 2019). So, when it comes to sports activity, it is suggested to use proper footwear that satisfies the sports person's requirements to avoid injuries.

Shoes with poor grip affect the balance and might cause muscle tear. Different shoe soles can be designed to improve performance and comfort (Reinschmidt & Nigg, 2000). Also, how this affects the foot strike pattern is analysed by performing gait analysis (Morio et al., 2009). Tendons stretch more on subjecting to heavy load, which causes Achilles' tendinitis. This occurs on wearing poorly fitted shoes (Schepsis et al., 2002). Protective equipment such as wrist guards knee and ankle braces, prevents or reduce ligament sprains. However, usage of this equipment might increase the chances of injury (Finch & Cook, 2014).

Sports technique is the most critical parameter to be focused on by an athlete for improving performance. Every athlete follows different techniques in their field of sports. Sports like running, javelin throw, long jump, high jump and others have different techniques. Sometimes, a change in technique or practising a wrong technique leads to injuries. A few examples are discussed in this section. The first example briefly discusses the different running techniques and changing running style.

Runners are classified based on the striking of the foot. The types of strikes are Rearfoot strike (RFS), forefoot strike (FFS) and midfoot strike (MFS). Heel strikes the ground first and is followed by contact of toe in RFS, whereas in FFS, ball of the foot strikes the ground first, and in MFS, both the ball of the foot and heel strike the ground simultaneously. Running techniques can be changed, but this change in practice can be helpful or detrimental (L. M. Anderson et al., 2020). Sometimes, it helps in preventing injuries. Usually, coaches prefer to change the running technique from RFS to non-rear foot runners (NRFS) to manage or prevent injuries. On this transition, it is observed that athletes with patellofemoral pain changing to NRFS helps to reduce pains in lower limbs. When runners shift from the FFS technique to RFS, runners are capable of performing all common mechanical characteristics as persistent RFS runners (L. M. Anderson et al., 2020).

According to the study done by Daoud et al. (2012), different running styles influence injury rates. Both intrinsic factors and extrinsic factors are responsible for running injuries. Intrinsic factors include any history of injuries, abnormalities in biomechanical aspect, body mass index and sex. The extrinsic factors that influence the damage are core strength, shoes, flexibility and duration of training. Also, studies say that emerging runners usually adopt landing in FFS or MFS to reduce injuries and improve speed. Factors affecting running injuries are the stress factor caused by the magnitude of joint movements and the rate of joint movements. This continuous stress causes musculoskeletal damage and damages to connective tissues. In this study, 52 experienced and trained athletes were observed over some time. All these athletes were long distant runners and middle-distant runners trained by the same coach. All the details of daily routine training, including total distance covered daily, duration of training, and performance, were recorded for every athlete. The 2D video analysis technique was used to identify the foot strike pattern. It is pretty difficult to assess the athlete's foot strike in every training, but the overall majority of foot strike pattern used by athletes is identified. Based on the analysis, it is reported that when the heel makes initial contact with the ground with a positive plantar angle is categorised as RFS.

Similarly, when the ball of the foot is making initial contact with the ground, it is categorised as FFS and when both heel and ball of the foot makes simultaneous contact in-ground is classified as MFS. Statistical analysis was done between FFS and RFS on various factors such as injury rate, FFS injury rate, traumatic injury rate and RFS injury rate. Based on the obtained results, it is classified that 69% of athletes are habitual FFS runners, 31% of athletes are regular RFS runners, and traditional MFS runners are not present. Many repetitive injuries were observed during the monitoring. It is classified that 25% of athletes faced mild injury, 40% had a moderate injury, and 35% had a severe injury. Among traumatic injuries, 54% had mild injuries, 33% had moderate injuries, and 13% had severe injuries. Common repetitive injuries observed in athletes are iliotibial band syndrome, Achilles tendinopathies, medial tibial stress syndrome and patellofemoral pain syndrome. Foot strike pattern is a significant factor for causing injury. It can be concluded from this study that runners should be familiar with types of running form and injuries associated with it. Also, the athlete should know about reducing the loading impacts. Moreover, it is hypothesised that habitual FFS experience minor injury compared to RFS runners.

Biomechanics uses the principle of kinematics to study human motion. It helps people understand the damages occurring in the body due to heavy loads or any accidental injury. Second example briefs about the occurrence of injury in javelin throwers. According to the study done by Wei & Yalong (2021), injuries that occurred in the last stage for javelin throwers in the elbow is studied, and the reason for the cause of injury and improving performances are analysed. The aerodynamic factor is the most predominant factor that can improve the performance of the javelin throwers. The primary reason for throwing elbow injury is improper training or techniques, leading to poor muscle strength and ligament toughness around the elbow, which fails to adapt to high-speed throwing and strength. The case study, reports that it is tough to get accustomed to the speed of the swinging arm, which results in elbow joint damage and ligament damage, when the antagonistic and coordinating muscle is not balanced. Due to repetitive shots, local muscles and ligaments undergo fatigue, which damages the muscles and ligaments of the throwing elbow joint. Injury prevention can be done by using elbow pads to keep up the stability of joints. Elbow joints have to be activated before training and relaxation to boost blood circulation and nutrition at the joints. To improve athletes' performance, athletes must focus on building up muscle strength ligament toughness. Proper guidance must be given on training new techniques; otherwise, this might lead to misguidance. Muscle fatigue can be reduced by exercising regularly with appropriate guidance. This helps control the movement speed and helps to attain a good recovery effect. Focusing on training and techniques on proper throwing can reduce injury and improve performance.

BIOMECHANICAL ANALYSIS OF SPORTS PERFORMANCE

Performance improvement is a significant factor that has to be focused on in sports biomechanics. Generally, performance is increased by the coordinated movement pattern. Sports movement is also associated with the coordination of joints and the musculoskeletal system. For every movement to occur, bones, joints and soft tissues surrounding the joints are involved. Every sport has different activities and has distinct actions. Particular joints or muscles experience force, pressure, moment, and friction at every action or movement. Coordinated movement pattern in the body is the most critical parameter in sports biomechanics and exercise. Biomechanics or body movements are commenced from the neural signals of the central nervous system (Prochazka & Ellaway, 2012).

Biomechanical analysis of sports can be grouped into three levels: qualitative analysis, quantitative analysis, and semi-quantitative analysis. In qualitative research, observations are based on recorded videos, cine films, or real-time (Bartlett, 1999). This analysis helps us to know whether the technique is done correctly. Qualitative research helps the coach identify the flaws in performance and provide information to improvise the training to improve the sports performance of the individual.

Quantitative analysis is done to study the kinematic of the sport and help to improve the performance. This analysis helps to perform the detailed evaluation of techniques used by sportsperson and help in the betterment of performance. It also helps in the thorough assessment of movements and helps to define performance parameters that individuals can implement to enhance their performance. The semi-quantitative analysis is done when appropriate quantitative analysis does not measure a particular movement. Knowing the biomechanics of every activity is very important for analysing, and this knowledge helps improve sports performance and reduce injury. So, here in this section few examples of sports activities such as ballistic activity, running and javelin throw are discussed.

Analysis of Ballistic Movements

Ballistic sports activity such as hitting, throwing, jumping and kicking skills has three phases: preparation, action, and recovery (Bartlett, 1999). Every step has its different biomechanical characteristics. The long-jump is a ballistic activity. Muscle control and joint coordination are essential activities to monitor performing various movements or actions. In the long jump, take-off and landing decide the performance of the individual. The angle at which the athlete take-off is coordinated with the athlete horizontal take-off speed. Greater the take-off speed, the greater the range (Bridgett & Linthorne, 2007). Biomechanical characteristics of different phases in ballistic activity are generally discussed below.

Preparatory Phase

Biomechanical activity in the preparation phase is mentioned as follows:

- The body prepares to put itself in a favourable position for getting into the action phase.
- It maximises the acceleration path.
- It initiates the sequence of muscle action.
- Here, the agonist's muscles are activated and stretched. The response is decided based on requirements like force and speed, such as phasic or tonic responses. The phasic response requires a faster backswing, whereas the tonic response requires a greater backswing. The preparation phase must be short and slow to control the output of both phasic and tonic responses to achieve reasonable accuracy.
- Muscle length of agonist's muscle is increased to the point at which tension is maximum by using length-tension relationship.
- The elastic energy of the agnostic muscle is stored, and it pays off in the action phase.
- The Golgi tendon organ is activated to interrupt muscle contraction. It relaxes the agonist muscle by activating the antagonist muscle to contract in the action phase.

Action Phase

In this phase, many of the principles of coordinated movement patterns are observed. Muscle action is sequenced according to the movement pattern of each segment at the required time. Muscle sequence is initiated from a large muscle group followed by smaller, faster muscles distal to limb. This sequence increases the speed of movement as segmental movement increases. Actions are commenced with minimum inertia. Also, force in each segment is applied in the direction of motion. At last, redundant degrees of freedom is controlled (Bartlett, 1999).

Recovery Phase

In this phase, movements are decelerated in a controlled fashion by appropriate muscle contraction. Also, a temporary stability position is achieved in this phase.

Running

Running is cyclic activity; running stride is defined as toe-off of one foot to toe-off of the same foot. Running has two supporting and recovery phases (Geiringer, 1995). In the supporting phase, one leg is off the ground, whereas both legs are not in contact with the ground in the recovery phase. This phase is also known as the flight phase. Running is distinguished from walking by the absence of a double support phase. The recovery phase is short for jogging, and as the running speed increases recovery phase gets elongated.

Phase Analysis of Running

Running activity is divided into two-phase phases, such as the supporting phase and the non-supporting phase. Each phase is subdivided into three sub-phases. The Block diagram in Figure 1 explains the various stages of running activity.

Figure 1. Different phases of running

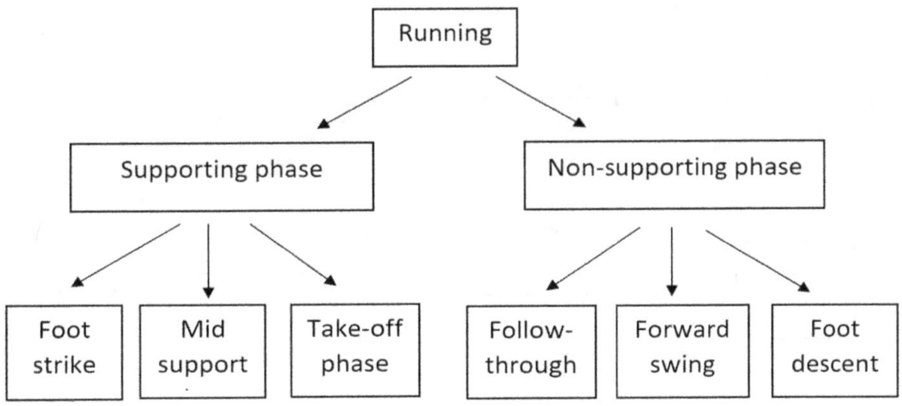

Supporting Phase

The first phase is a foot strike; it is the moment where the foot strikes the ground. In this phase, the primary function is impact absorption. The mid-support phase follows this. To hold up bodyweight, this phase prolongs to maintain forward momentum. This phase is distinguished by shortening limb length about the lowest centre of mass position. The third sub-phase is the take-off phase. This phase involved the extension of the limb and accelerated the body to move forward and lift the body upwards at the same time.

Non-Supporting Phase

The non-supporting phase has three subphases: follow-through, forward swing, and foot descent. In the follow-through phase, hip extension is reduced, followed by the flexion of the thigh and initiates the knee flexion. The forward swing phase prepares the foot descendent and characteristic function to amplify the upward and forward ground reaction thrust. The recovery leg starts to swing forward, harmonising with the opposite leg during take-off. In the last subphase, it tends to arrest the forward motion of the foot and leg. This phase imparts an active landing for an easy and smooth foot strike.

Javelin Throw

The javelin throw is a highly coordinating sport in which all joints in the body need to coordinate. The throwing distance is affected by aerodynamic factors like gravity wind. Three types of throwing movement are classified: sidearm, overarm, and underarm (Bartlett, 2014).

The javelin throw is an example of the overarm throw. Movements involved in this are withdrawal of javelin followed by crossover, then start of delivery stride followed by foot landing and release. The ultimate goal in javelin throw is to increase the throwing distance, generally called "maximisation of throwing phase" (Menzel, 1986). Distinguishable phases of javelin throw are as follows: Approach run, release phase, braking phase and flight of the javelin. The approach run and release phase is subdivided into two sub-phases. The Block diagram in Figure 2 explains the various phases of javelin throw.

Figure 2. Different phases of javelin throw

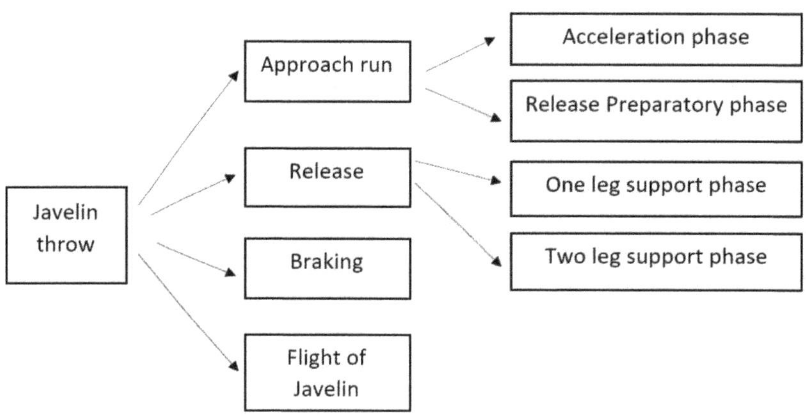

Approach Run

The approach run is divided into the acceleration and release-preparatory phases.

- Acceleration phase

The target of the acceleration phase is to achieve the highest approach velocity. This is to increase the throwing distance of the javelin. This phase is initiated by an approach run and ends with the start of the withdrawal of the javelin. Along with the highest approach velocity, other factors influence biomechanical activity, including stride length, acceleration phase duration, and stride rate.

- Release preparatory phase

The preparatory release phase is initiated with the javelin's withdrawal and terminated with the release of javelin by stepping into the braking phase. This phase aims to attain biomechanical objectives such as achieving the highest velocity and proper body position on the foot's landing on the throwing arm side. Factors affecting biomechanical response are stride length and landing angle.

Release Phase

The release phase is subdivided into one leg support phase and two leg support phases. Two leg support phase is also known as the bracing phase. The release phase in Figure 3 begins with placing the bracing leg on the throwing arm side and terminates when the hand loses its contact with the javelin. This phase mainly focuses on the releasing velocity of the javelin, the angle at which javelin is released. The rate at which javelin is released significantly impacts the throwing distance. The aerodynamic factor affects the angle of release, which α0 denotes. The angle of release in javelin is less than shot put because of the influence of aerodynamic characteristics. Along with releasing angle of the javelin, the angle of attitude (β0) also impacts the javelin's flight. In the javelin throw, the horizontal distance is reduced if the angle of attitude is high; this hurts the athlete's performance.

To have high releasing velocity, impulse transmission from the trunk to javelin should be proper. The athlete achieves the maximum momentum to throw. The momentum and energy are transferred from the lower extreme (hip) to the javelin. However, lower limbs and joints attain maximum velocity before the upper extremities. The knee angle in the bracing leg should be varied between 160 to 180 degrees to reduce the trunk and lower extremities' rate, reducing the momentum. The bracing leg should be placed with full grip to have more excellent impulse transmission. Also, the bracing stride should be more significant; this helps sound impulse transmission (Menzel, 1986).

Figure 3. Release phase of javelin

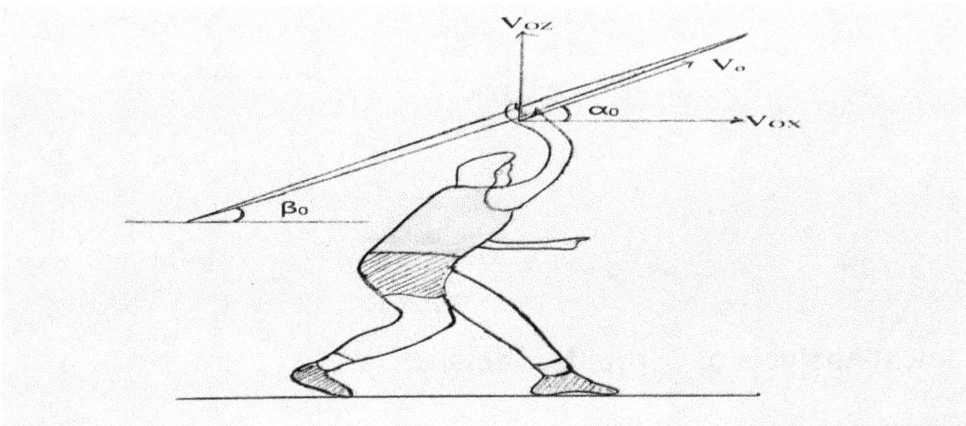

Braking

In the braking phase, the main objective is to reduce the braking distance and the approach velocity. The approach velocity has to be reduced by the athlete to avoid crossing the foul line at the final stage. Two biomechanical factors influencing this phase are the centre of gravity (CG) velocity and body position at the end of the release phase. Even though approach velocity is reduced in the bracing phase, the CG velocity of the athlete remains high at the time of release. Minimising the CG velocity in the release phase helps the athlete reduce the final phase length.

Flight of the Javelin

The flight phase is dealt only with the javelin. It is the release period of a javelin from the athlete's hands to the initial contact with the ground. The aerodynamic factor or force significantly impacts the javelin during its flight phase. This force functions concerning the javelin's centre of pressure (CP) because of the difference in velocity of wind and javelin. Figure 4 represents that CP and CG are located in different positions, creating pitching moment (PM). Turing of the direction of javelin downwards is guided by PM. If the PM is large, the distance between the CP and CG is also significant. The distance between the CP and CG is more remarkable for the short distance covered. Two aerodynamic forces operating on javelin are the lift force and drag force. The force acting against the flight direction is called drag force (D), and the force acting vertically is known as lift force (L). Both aerodynamic forces depend on airflow, flight direction, attitude angle and wind direction.

Figure 4. Aerodynamic factors in-flight phase

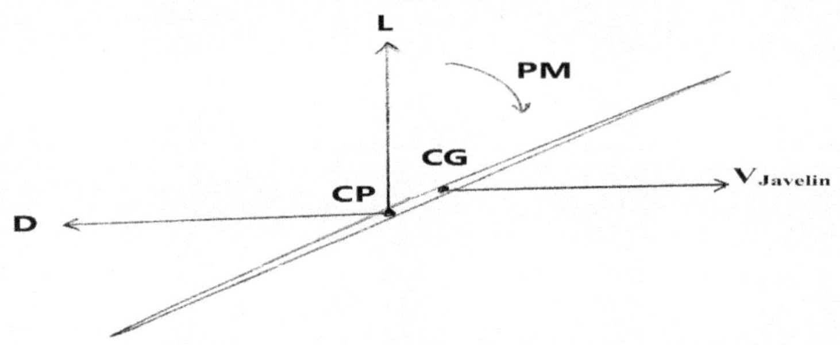

Kinesiological Analysis of Sports Movement

Kinesiology specialists should have broader knowledge about the movements and human physiology to improve the individual's recovery. The range of joint motion and activity of the muscle is analysed. Electromyography and goniometer techniques analyse the muscle activities and range of joint movements, respectively. Knowing these parameters helps in improving the performance of sportsperson. Coaches should also have better knowledge about these physiological movements to guide the athlete in the right way to improve their performance. Here, the role of a kinesiologist is to analyse muscle strength muscle contraction during activities and measure the joint angle. Techniques used in this analysis are electro goniometer and electromyography are discussed in the subsequent section.

TECHNIQUES IN MEASURING THE PERFORMANCE OF ATHLETES

Many computational techniques have been emerged to study sports biomechanics. Some methods that use computational analysis for kinetic and kinematic analysis are video camera, electromyography, force platform, 3D motion capture and inertial motion capture with wearable technologies. Each technology has limitations and advancements over the other technologies. Technologies use sensors, markers, high-speed video cameras, infrared light strobes, force plates to study the kinetics and kinematics involved. Advancement in techniques helps in reducing the error betterment of measurement of kinematic and kinetic data.

Goniometer

A goniometer measures the joint range during various movements (Gronley & Perry, 1984). The conventional goniometer has a movable arm and a stationary arm to measure slow movements. In an electro goniometer, a traditional protractor is replaced with a potentiometer. As the joint movement occurs, the resistance of the potentiometer varies and produces an electrical output. This output is analogous to the joint displacement. The challenges involved are that while knee movement is captured, the knee joint axis and system joint axis have to be correlated. The major drawback is that joint angle measurement can be done for a single plane alone, for which proper calibration for the system joint axis is required. This instrument seems too difficult to fit for joints having more degrees of freedom.

Electromyogram

Electromyogram (EMG) is a technique used to quantify muscle activity. It is measured by placing the sensors on the required muscle. The invasive and non-invasive procedure is available. Based on the type of electrodes used, it is categorised as invasive and non-invasive procedures (Gronley & Perry, 1984). Surface electrodes are placed on the skin's surface in non-invasive procedures, and needle electrodes are inserted in the region of interest for invasive procedures. These electrodes measure muscle contraction and need to be amplified due to low output voltage. The EMG data are used to quantify gait parameters and help in correlating EMG signals with muscle tension during the gait cycle.

Many computational technologies are used to know about physiological movement during sports performance (Baca et al., 2009). Learning about muscle actions and joint activities more deeply help the coaches and kinesiologist increase sports performance and prevent injuries.

Motion Analysis

For motion analysis, 2D and 3D motion capture technology with integrated cameras are available. The prime difference between the 2D and 3D analysis is based on the calibration and reconstruction process

(Colyer et al., 2018). The solitary camera is placed parallelly to the subject plane of motion of interest in 2D camera analysis. Since in 3D analysis, movement in all the planes has to be focused, multiple cameras are placed in complex format (Akhtaruzzaman et al., 2016). Also, reconstruction must be performed at every point of interest on the subject. For this purpose, markers are used to identify the point of interest, and it has to be captured by at least two cameras for processing. Usage of markers causes motion artifacts. So, before the reconstruction process, artifacts have to be eliminated. Usage of markers might lead to the displacement of their initial position leading to errors. To overcome this limitation, markerless technologies were developed using video systems, image features, and algorithms that define the model pose location and shape (Zago et al., 2020). In this markerless motion capture system, two types of cameras are used, namely a depth-sensing camera system where it senses the depth via the projection of light. This is more useful in a real-time system. They are also known as RGB-D cameras; they can capture colour and depth (Zago et al., 2020).

Inertial Sensors

Inertial sensors consist of accelerometers, magnetometers and gyroscopes. The combination of these sensors provides information about the object's orientation, gravitational force, and velocity (Muro-de-la-Herran et al., 2014). The accelerometer working principle is based on Newton's law of motion. According to Newton's law of motion, acceleration of the body is proportional to the net force acting on the body. With a combination of accelerometer and gyroscopes, acceleration and angular velocity of the body is obtained. Velocity is obtained by integrating acceleration, while the position is obtained by combining the body's velocity, and flexion angle is obtained by integrating angular velocity. So, signals received from the accelerometer is filtered, and a classifying algorithm is applied to calculate the number of steps taken in a particular period. The working principle of the gyroscope is based on the Coriolis effect, where all bodies revolve around an axis develop rotational inertia. Moment of inertia is used to determine the rotational inertia of a body. Thus, gyroscopes provide a body's angular velocity, joint angle, and segment velocity. Magnetometers measure the direction and strength of the magnetic field. In biomechanics, the system's reliability can be increased by using a magnetometer in combination with an accelerometer and gyroscopes (Surer & Kose, 2011). Inertial Measurement Unit (IMU) sensors are made of inertial sensors, which is used to measure the kinematics of the subject. It is mainly used in gait analysis to study gait profiling. The sensors are widely used in wearable technologies (Muro-de-la-Herran et al., 2014). Acquiring data using accelerometers and gyroscopes helps determine spatial orientation with respect to an inertial frame. Though the gyroscope produces relative orientation, it has drift and distortion errors. The accelerometer eliminates these drift errors in the horizontal plane, and the magnetic sensors compensate for drift error in the vertical plane (Fong & Chan, 2010). This technique helps to study the locomotion of individuals with injuries and helps provide proper treatment and guidance to recovery fast and also used in gait analysis, which helps improve performance. This technique is more suitable for studying the kinematics of the lower limb.

Force Plate

Force plates are used to measure the ground reaction force and ground reaction momentum in gait profiling (Higginson, 2009). Load cells are placed in the corners of the force plate. These load cells consist of either piezoelectric element, strain gauge or force sensing resistors based on the sensitivity, pressure range and linearity. Force plate with piezoelectric elements on experiencing force gets distorted and provides a measurable voltage directly proportional to the applied force (Anton & Sodano, 2007). Similarly, force plates with force sensing resistors work based on contact resistance. Force sensing resistors are conductive polymer film, which changes its resistance to experiencing pressure. The conductive polymeric film consists of conducting particles and non-conducting particles in matrix form. When an external force is applied on the surface, this article comes in contact with the sensor electrode, which changes its resistance. The amount of change in resistance is directly proportional to the applied force (Sadun et al., 2016). Likewise, the working of the piezoelectric sensor is based on the piezoelectric effect. When the piezoelectric material experiences force, microscopic structures are distorted, resulting in dipole and an electric field. These electric charges are analogous to the applied force, which is measured using a charge amplifier. Generally, force plates are fixed in the ground for having easy calculations of the centre of pressure (Higginson, 2009).

Force plates are used for gait profiling of different actions and exercises. Gait analysis is more helpful in determining the heel strike and toe-off of walking and running patterns (Sinclair et al., 2011). Gait profiling of running helps us understand the various biomechanical aspects of running or other sports activity acutely. It allows the coaches to give the athlete proper training to change their pattern and techniques to improve their performance. Sometimes, sports injuries are fatal, and it is a challenging process in rehabilitation to bring back the original function. Rehabilitation centres require many innovative approaches to solve the problem (Jena et al., 2020). So, gait analysis also helps in rehabilitation engineering to study how the prosthesis helps athletes and improve prosthesis can be done based on the study. Different regression techniques are used to determine gait asymmetry by validating symmetry indices (Jena et al., 2021).

CONCLUSION

The main objective of sports biomechanics is to reduce injury and improve sports performance by studying various physiological movements. Kinetics and kinematics involved in each sports activity and exercise are studied to understand better muscle activity, joint actions, and musculoskeletal functions. Coaches and kinesiologists need to understand it more profoundly to provide suitable training and good treatment for faster recovery of injuries. Computational techniques help study physiological movements during sports activity and help the coaches provide required training for the individual athlete to improve their performance and reduce the injuries during training sessions. The study of sports biomechanics analyses the body loading condition and helps prevent injury by performing exercise more safely.

REFERENCES

Akhtaruzzaman, M., Shafie, A. A., & Khan, M. R. (2016). Gait analysis: Systems, technologies, and importance. *Journal of Mechanics in Medicine and Biology*, *16*(7), 1630003. doi:10.1142/S0219519416300039

Allen, A. A., Caldwell, G. L., & Fu, F. H. (1995). Functional anatomy and biomechanics of the meniscus. *Operative Techniques in Sports Medicine*, *2*(3), 152–163.

Amis, A. A. (2004). The Biomechanics of Ligaments. *Biomechanics and Biomaterials in Orthopedics*, 550–563.

Anderson, K., Strickland, S. M., & Warren, R. (2001). Hip and groin injuries in athletes. *The American Journal of Sports Medicine*, *29*(4), 521–533. doi:10.1177/03635465010290042501 PMID:11476397

Anderson, L. M., Bonanno, D. R., Hart, H. F., & Barton, C. J. (2020). What are the benefits and risks associated with changing foot strike pattern during running? A systematic review and meta-analysis of injury, running economy, and biomechanics. *Sports Medicine (Auckland, N.Z.)*, *50*(5), 885–917. doi:10.100740279-019-01238-y PMID:31823338

Anton, S. R., & Sodano, H. A. (2007). A review of power harvesting using piezoelectric materials (2003-2006). *Smart Materials and Structures*, *16*(3), R1–R21. doi:10.1088/0964-1726/16/3/R01

Ashkavand, Z., Malekinejad, H., & Vishwanath, B. S. (2013). The pathophysiology of osteoarthritis. *Journal of Pharmacy Research*, *7*(1), 132–138. doi:10.1016/j.jopr.2013.01.008

Baca, A., Dabnichki, P., Heller, M., & Kornfeind, P. (2009). Ubiquitous computing in sports: A review and analysis. *Journal of Sports Sciences*, *27*(12), 1335–1346. doi:10.1080/02640410903277427 PMID:19764000

Bahr, R. (2009). No injuries, but plenty of pain? On the methodology for recording overuse symptoms in sports. *British Journal of Sports Medicine*, *43*(13), 966–972. doi:10.1136/bjsm.2009.066936 PMID:19945978

Bartlett, R. (1999). *Sports Biomechanics : Reducing injury and improving performance*. E & FN Spon.

Bartlett, R. (2014). *Introduction to sports biomechanics: Analysing human movement patterns*. Routledge. doi:10.4324/9781315889504

Bhosale, A. M., & Richardson, J. B. (2008). Articular cartilage: Structure, injuries and review of management. *British Medical Bulletin*, *87*(1), 77–95. doi:10.1093/bmb/ldn025 PMID:18676397

Bonnel, F., Toullec, E., Mabit, C., & Tourné, Y. (2010). Chronic ankle instability: Biomechanics and pathomechanics of ligaments injury and associated lesions. *Orthopaedics & Traumatology: Surgery & Research*, *96*(4), 424–432. doi:10.1016/j.otsr.2010.04.003 PMID:20493797

Bridgett, L. A., & Linthorne, N. P. (2007). Changes in long jump take-off technique with increasingrun-up speed. *Journal of Sports Sciences*, 37–41.

Buckwalter, J. A. (1998). Articular cartilage: Injuries and potential for healing. *The Journal of Orthopaedic and Sports Physical Therapy*, *28*(4), 192–202. doi:10.2519/jospt.1998.28.4.192 PMID:9785255

Chen, J., Kim, J., Shao, W., Schlecht, S. H., Baek, S. Y., Jones, A. K., Ahn, T., Ashton-Miller, J. A., Banaszak Holl, M. M., & Wojtys, E. M. (2019). An Anterior Cruciate Ligament Failure Mechanism. *The American Journal of Sports Medicine*, *47*(9), 2067–2076. doi:10.1177/0363546519854450 PMID:31307223

Colyer, S. L., Evans, M., Cosker, D. P., & Salo, A. I. T. (2018). A review of the evolution of vision-based motion analysis and the integration of advanced computer vision methods towards developing a markerless system. *Sports Medicine - Open*, *4*(1), 24. doi:10.118640798-018-0139-y PMID:29869300

Daoud, A. I., Geissler, G. J., Wang, F., Saretsky, J., Daoud, Y. A., & Lieberman, D. E. (2012). Foot strike and injury rates in endurance runners: A retrospective study. *Medicine and Science in Sports and Exercise*, *44*(7), 1325–1334. doi:10.1249/MSS.0b013e3182465115 PMID:22217561

Dixon, S., Fleming, P., James, I., & Carré, M. (2015). *The Science and Engineering of Sport Surfaces*. Routledge. doi:10.4324/9780203133385

Edwards, W. B. (2018). Modelling Overuse Injuries in Sport as a Mechanical Fatigue Phenomenon. *Exercise and Sport Sciences Reviews*, *46*(4), 224–231. doi:10.1249/JES.0000000000000163 PMID:30001271

Farrington, T., Onambele-Pearson, G., Taylor, R. L., Earl, P., & Winwood, K. (2012). A review of facial protective equipment use in sport and the impact on injury incidence. *British Journal of Oral & Maxillofacial Surgery*, *50*(3), 233–238. doi:10.1016/j.bjoms.2010.11.020 PMID:21295384

Finch, C. F., & Cook, J. (2014). Categorising sports injuries in epidemiological studies: The subsequent injury categorisation (SIC) model addresses multiple, recurrent and exacerbation injuries. *British Journal of Sports Medicine*, *48*(17), 1276–1280. doi:10.1136/bjsports-2012-091729 PMID:23501833

Fong, D. T. P., & Chan, Y. Y. (2010). The use of wearable inertial motion sensors in human lower limb biomechanics studies: A systematic review. *Sensors (Basel)*, *10*(12), 11556–11565. doi:10.3390101211556 PMID:22163542

Gearity, B. T., & Murray, M. A. (2011). Athletes' experiences of the psychological effects of poor coaching. *Psychology of Sport and Exercise*, *12*(3), 213–221. doi:10.1016/j.psychsport.2010.11.004

Gee, C. J. (2010). How does sporsportschology actually improve athletic performance? A framework to facilitate athletes' and coaches' understanding. *Behavior Modification*, *34*(5), 386–402. doi:10.1177/0145445510383525 PMID:20935240

Geiringer, S. R. (1985). The biomechanics of running: Implications for the prevention of foot injuries. *Sports Medicine (Auckland, N.Z.)*, *2*(2), 144–153. doi:10.2165/00007256-198502020-00006 PMID:2860714

Getgood, A., Hoshino, Y., & Roessler, P. P. (2019). Biomechanics of musculoskeletal injuries. *The Sports Medicine Physician*, 27-35.

Gronley, J. K., & Perry, J. (1984). Gait analysis techniques. *Physical Therapy*, *64*(12), 1831–1838. doi:10.1093/ptj/64.12.1831 PMID:6505028

Guisasola, I., James, I., Stiles, V., & Dixon, S. (2010). Dynamic behaviour of soils used for natural turf sports surfaces. *Sports Engineering*, *12*(3), 111–122. doi:10.100712283-010-0036-1

Hauser, R. A., & Dolan, E. E. (2011). Ligament Injury and Healing : An Overview of current clinical concepts. *Journal of Prolotherapy*, *3*, 836–846.

Higginson, B. K. (2009). Methods of running gait analysis. *Current Sports Medicine Reports*, *8*(3), 136–141. doi:10.1249/JSR.0b013e3181a6187a PMID:19436169

Jastifer, J. R., McNitt, A. S., Mack, C. D., Kent, R. W., McCullough, K. A., Coughlin, M. J., & Anderson, R. B. (2019). Synthetic Turf: History, Design, Maintenance, and Athlete Safety. *Sports Health*, *11*(1), 84–90. doi:10.1177/1941738118793378 PMID:30096021

Jena, S., Arunachalam, T., & Panda, S. K. (2020). Experimental and numerical investigation of a polypropylene orthotic device for assistance in level ground walking. *Proceedings of the Institution of Mechanical Engineers. Part H, Journal of Engineering in Medicine*, *234*(4), 356–369. doi:10.1177/0954411919894091 PMID:31854229

Jena, S., Sakhare, G. M., Panda, S. K., & Thirugnanam, A. (2021). Implementation of multiple regression technique for detection of gait asymmetry using experimental gait data. *Journal of Medical and Biological Engineering*, *41*(1), 1–10. doi:10.100740846-020-00533-8

Katkat, D., Bulut, Y., Demir, M., & Akar, S. (2009). Effects of different sports surfaces on muscle performance. *Biology of Sport*, *26*(3), 285–296. doi:10.5604/20831862.894793

Knudson, D. (2007). Qualitative biomechanical principles for application in coaching. *Sports Biomechanics*, *6*(1), 109–118. doi:10.1080/14763140601062567 PMID:17542182

Lin, T. W., Cardenas, L., & Soslowsky, L. J. (2004). Biomechanics of tendon injury and repair. *Journal of Biomechanics*, *37*(6), 865–877. doi:10.1016/j.jbiomech.2003.11.005 PMID:15111074

Longo, U. G., Ronga, M., & Maffulli, N. (2009). Acute ruptures of the Achilles tendon. *Sports Medicine and Arthroscopy Review*, *17*(2), 127–138. doi:10.1097/JSA.0b013e3181a3d767 PMID:19440140

Mears, A. C., Osei-Owusu, P., Harland, A. R., Owen, A., & Roberts, J. R. (2018). Perceived links between playing surfaces and injury: A worldwide study of elite association football players. *Sports Medicine - Open*, *4*(1), 40. doi:10.118640798-018-0155-y PMID:30128862

Menzel, H.-J. (1986). Biomechanics of javelin throwing. *Biomechanics of Sports*, *1*(3), 85–98.

Miller, M., & Thompson, S. (2014). Anatomy and Biomechanics of the Knee. *Operative Techniques in Sports Medicine*, *11*(3), 172–186.

Mishra, E., Jena, S., Bhoi, C., Arunachalam, T., & Panda, S. K. (2019). Effect of high heel gait on hip and knee-ankle-foot rollover characteristics while walking over inclined surfaces— A pilot study. *The Foot*, *40*, 8–13. doi:10.1016/j.foot.2019.03.004 PMID:30981083

Mithoefer, K., McAdams, T. R., Scopp, J. M., & Mandelbaum, B. R. (2009). Emerging options for treatment of articular cartilage injury in the athlete. *Clinics in Sports Medicine*, *28*(1), 25–40. doi:10.1016/j.csm.2008.09.001 PMID:19064163

Morio, C., Lake, M. J., Gueguen, N., Rao, G., & Baly, L. (2009). The influence of footwear on foot motion during walking and running. *Journal of Biomechanics*, *42*(13), 2081–2088. doi:10.1016/j.jbiomech.2009.06.015 PMID:19643421

Muro-de-la-Herran, A., García-Zapirain, B., & Méndez-Zorrilla, A. (2014). Gait analysis methods: An overview of wearable and non-wearable systems, highlighting clinical applications. *Sensors (Basel)*, *14*(2), 3362–3394. doi:10.3390140203362 PMID:24556672

Murrell, G. A. C., Maddali, S., Horovitz, L., Oakley, S. P., & Warren, R. F. (2001). The effects of time course after anterior cruciate ligament injury in correlation with meniscal and cartilage loss. *The American Journal of Sports Medicine*, *29*(1), 9–14. doi:10.1177/03635465010290012001 PMID:11206263

Nigg, B. M., Nurse, M. A., & Stefanyshyn, D. J. (1999). Shoe inserts and orthotics for sport and physical activities. *Medicine and Science in Sports and Exercise, 31*(Supplement), 31. doi:10.1097/00005768-199907001-00003 PMID:10416543

Nourissat, G., Berenbaum, F., & Duprez, D. (2015). Tendon injury: From biology to tendon repair. *Nature Reviews. Rheumatology, 11*(4), 223–233. doi:10.1038/nrrheum.2015.26 PMID:25734975

Nurse, M. A., Hulliger, M., Wakeling, J. M., Nigg, B. M., & Stefanyshyn, D. J. (2005). Changing the texture of footwear can alter gait patterns. *Journal of Electromyography and Kinesiology, 15*(5), 496–506. doi:10.1016/j.jelekin.2004.12.003 PMID:15935961

Prochazka, A., & Ellaway, P. (2012). Sensory systems in the control of movement. *Comprehensive Physiology, 2*(4), 2615–2627. doi:10.1002/cphy.c100086 PMID:23720260

Reinschmidt, C., & Nigg, B. M. (2000). Current Issues in the Design of Running and Court Shoes. *Sportverletzung Sportschaden, 14*(3), 71–81. doi:10.1055-2000-7866 PMID:11081243

Sadun, A. S., Jalani, J., & Sukor, J. A. (2016). Force Sensing Resistor (FSR): a brief overview and the low-cost sensor for active compliance control. *First International Workshop on Pattern Recognition.*

Schepsis, A. A., Jones, H., & Haas, A. L. (2002). Achilles tendon disorders in athletes. *The American Journal of Sports Medicine, 30*(2), 287–305. doi:10.1177/03635465020300022501 PMID:11912103

Schoukens, G. (2009). *Developments in textile sports surfaces.* Advances in Carpet Manufacture.

Sinclair, J., Edmundson, C., Brooks, D., & Hobbs, S. (2011). Evaluation of kinematic methods of identifying gait events during running. *International Journal of Sports, 05*(03), 188–192.

Smith, M. (2005). A review of the initial management of soft tissue sports injuries. *Journal of Orthopaedic Nursing, 9*(2), 103–107. doi:10.1016/j.joon.2004.09.011

Smoljanovic, T., Bojanic, I., Hannafin, J. A., Hren, D., Delimar, D., & Pecina, M. (2009). Traumatic and overuse injuries among international elite junior rowers. *The American Journal of Sports Medicine, 37*(6), 1193–1199. doi:10.1177/0363546508331205 PMID:19299531

Stiles, V. H., James, I. T., Dixon, S. J., & Guisasola, I. N. (2009). Natural turf surfaces: The case for continued research. *Sports Medicine (Auckland, N.Z.), 39*(1), 65–84. doi:10.2165/00007256-200939010-00005 PMID:19093696

Surer, E., & Kose, A. (2011). Methods and technologies for Gait analysis. *Computer Analysis of Human Behavior,* 105-123.

Takeda, H., Nakagawa, T., Nakamura, K., & Engebretsen, L. (2011). Prevention and management of knee osteoarthritis and knee cartilage injury in sports. *British Journal of Sports Medicine, 45*(4), 304–309. doi:10.1136/bjsm.2010.082321 PMID:21357577

Thacker, S. B., Stroup, D. F., Branche, C. M., Gilchrist, J., Goodman, R. A., & Kelling, E. P. (2003). Prevention of knee injuries in sports: A systematic review of the literature. *The Journal of Sports Medicine and Physical Fitness, 43*(2), 165–179. PMID:12853898

Thelin, N., Holmberg, S., & Thelin, A. (2006). Knee injuries account for the sports-related increased risk of knee osteoarthritis. *Scandinavian Journal of Medicine & Science in Sports*, *16*(5), 329–333. doi:10.1111/j.1600-0838.2005.00497.x PMID:16978252

Twist, C., & Eston, R. (2005). The effects of exercise-induced muscle damage on maximal intensity intermittent exercise performance. *European Journal of Applied Physiology*, *94*(5–6), 652–658. doi:10.100700421-005-1357-9 PMID:15887020

Tyler, T. F., Silvers, H. J., Gerhardt, M. B., & Nicholas, S. J. (2010). Groin injuries in sports medicine. *Sports Health*, *2*(3), 231–236. doi:10.1177/1941738110366820 PMID:23015943

Vormittag, K., Calonje, R., & Briner, W. W. (2009). Foot and ankle injuries in the barefoot sports. *Current Sports Medicine Reports*, *8*(5), 262–266. doi:10.1249/JSR.0b013e3181b9e3be PMID:19741354

Walker, A., & Subic, A. (2013). Advances in design and materials for indoor sports surfaces. *Advanced Materials Research*, *633*, 47–61. doi:10.4028/www.scientific.net/AMR.633.47

Wei, W., & Yalong, L. (2021). Study on treatment and rehabilitation training of ligament injury of javelin throwers based on sports biomechanics. *Measurement: Journal of the International Measurement Confederation*, *171*, 108757. doi:10.1016/j.measurement.2020.108757

Weiss, K., & Whatman, C. (2015). Biomechanics associated with patellofemoral pain and ACL injuries in sports. *Sports Medicine (Auckland, N.Z.)*, *45*(9), 1325–1337. doi:10.100740279-015-0353-4 PMID:26130304

Zago, M., Luzzago, M., Marangoni, T., Cecco, D. M., Tarabini, M., & Galli, M. (2020). 3D tracking of human motion using visual skeletonization and stereoscopic vision. *Frontiers in Bioengineering and Biotechnology*, *8*, 1–11. doi:10.3389/fbioe.2020.00181 PMID:32195243

Chapter 8
Computational Modeling of the Mechanics of Tissue Engineering Fibrous Scaffolds

Dhruba Jyoti Mech
Indian Institute of Technology, Hyderabad, India

Mohd Suhail Rizvi
Indian Institute of Technology, Hyderabad, India

ABSTRACT

Tissue engineering fibrous scaffolds play a crucial role in regenerative medicine. This chapter discusses the underlying mechanics and various approaches presently available to model these scaffolds and their limitations. The fibrous scaffolds are subjected to forces or deformation at two very different length scales (i.e., macroscopic forces during mechanical characterization and microscopic cellular forces). The scaffold behaves very differently under these two loading conditions, and very few computational frameworks capture the true nature of the scaffold under microscopic loading. The authors have also briefly discussed the two different ways a cell can sense the stiffness of the underlying scaffold and how it differs from the macroscopic stiffness.

INTRODUCTION

The fibrous systems are quite ubiquitous in the biological world. They are found in various forms ranging from purely fiber networks to composites which are characterized by an isotropic protein phase along with structural protein fibers (Dunlop & Fratzl 2010). From the length scale view point the cytoskeletal elements form the smallest fibrous structures in biological systems followed by extracellular matrix (ECM) fibers and finally the supra-cellular fibers such as silk and animal hairs (Lodish et al. 2008; Heim et al. 2010). The principal responsibility of these materials is to deliver structural integrity to the cell or tissue or organism as a whole and it is achieved by the appropriate mechanical properties of individual fibers in the system along with their arrangement (*Figure 1*) (Dunlop & Fratzl 2010; Lodish et al. 2008).

DOI: 10.4018/978-1-7998-9078-2.ch008

The arrangement of the constituent fibers is usually dependent on the biological context of the system. These fibrous structures are also under a regular renewal or recycling, although at different time scales depending on the biological context (*Figure 1*). This turnover at different levels is essential to several physiological functions (Lodish et al., 2008; Tabassi & Garnero, 2007).

For the interest of this chapter the fibrous systems at the extracellular matrix level play a central role. In the ECM fibrous structures are formed by the proteins secreted by the cells of various load bearing tissues (Frantz et al., 2010; Muizniek & Keeley, 2013; Han et al., 2011). Collagen fibers are one of the central players in these systems. They are classified in several categories based on their sequences and post-translational modifications (Frantz, 2010). Some of the important examples where collagen fibers are central to ECM are articular cartilage, bone, skin and other structural tissues. In these example tissues the collagen fibers are arranged in a very specific architecture including aligned and random configurations. As an example, in articular cartilage (AC), collagen fibers are organized in a three-layered structure with alternative aligned and random structures (known as Beninghorff architecture) (Schipper et al., 2008; Clark, 1990; Muir et al., 1970; WEISS et al., 1968; Minns & Steven, 1997; Bullough & Goodfellow, 1968). On the other hand, in ligaments these fibers take an aligned but wavy arrangement (Strocchi et al., 1992; Yahia & Drouin, 1989). Apart from collagen, elastin is another fiber which is an integral element of soft connective tissues such as skin and elastic cartilage (Mithieux & Weiss, 2005; Keith et al., 1977). Elastin fibers also take different structural morphologies depending on tissue functions (Keith et al., 1977; Murakumo et al., 1995; Cheney et al., 2015), for example random arrangement in alveolar structures, and flexor tendon (Mercer & Crapo, 1990; Ritty et al., 2002) and aligned arrangement in ligaments (Grant et al., 2013). Apart from providing structural strength to the tissue, these ECM fibers also facilitate cell adhesion and growth which has led to their use as substrates in tissue engineering applications where they are known as tissue engineering scaffolds.

FIBROUS SCAFFOLD IN TISSUE ENGINEERING

Owing to their ubiquity and role in the physiological functions, fibrous materials are extensively utilized for the application of tissue engineering scaffolds. Depending on the source of the scaffold the fibrous scaffolds used in these applications can be divided into two categories- (i) protein fibrous matrices, which are derived from the biological tissue by processes such as decellularization, (ii) synthetic fiber materials, which are produced in labs with many polymeric materials such as poly (L-lactic acid) (PLLA), poly (lactic acid) (PLA), poly (lactic-co-glycolic acid) (PLGA) etc. One challenge with the synthetic fibrous scaffolds is their lack of cell adhesion molecules. For this these materials are to be modified by some surface treatment which lets cells adhere onto them easily.

In tissue engineering applications, cells are added to these scaffolds of desired architecture and mechanical properties. The effect of the scaffold mechanics on the response of the cells like differentiation, proliferation, migration, adhesion, etc, are widely studied and known by the researchers. Some of the key findings in this domain are summarised and tabulated in the *Table 1* below.

Since the fibrous scaffolds are utilized for the application of tissue engineering extensively, an understanding of their mechanics is essential for an optimal designing of scaffold for any given application.

Figure 1. Fiber diameters and their arrangement in biological systems.

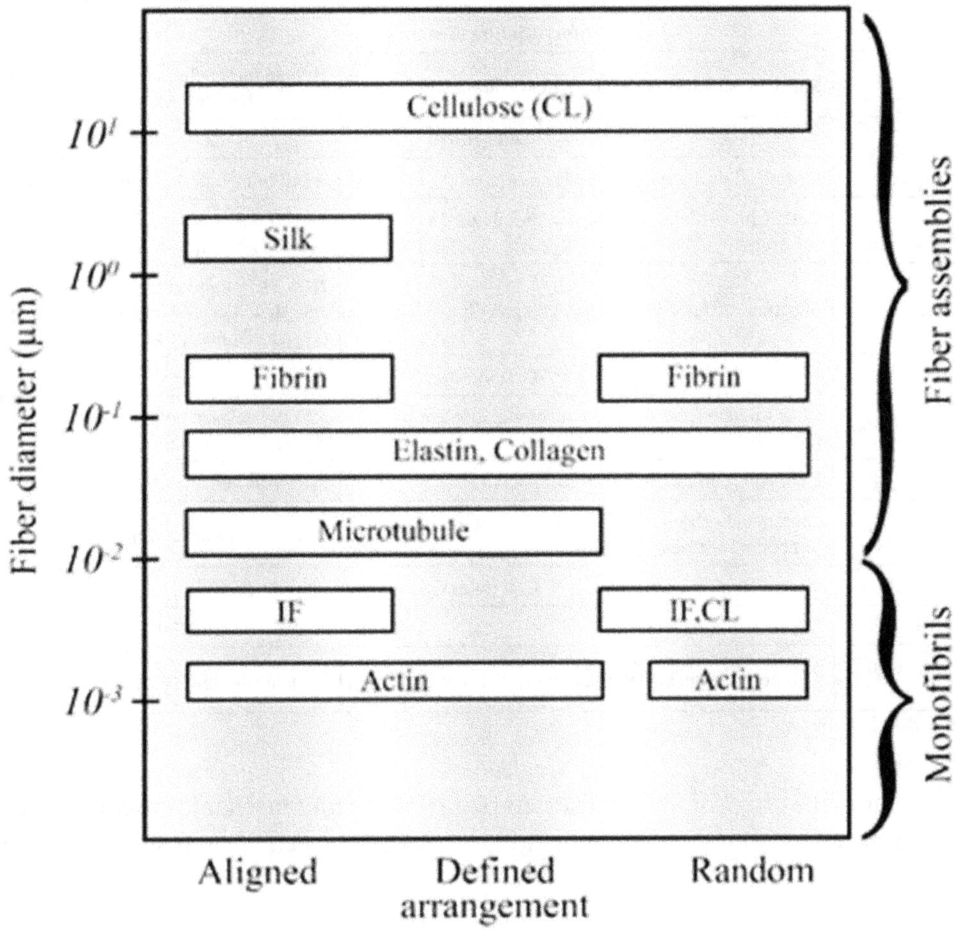

MECHANICS OF FIBROUS SCAFFOLDS

Given the importance of the scaffold mechanics, as described above, there has been a huge amount of experimental investigations and mathematical/computational modeling works to understand their mechanics. In experiments, researchers have looked at the response of single fibers under elongation and bending (Carlisle et al., 2009; Tan & Lim, 2006; Carlisle et al., 2010; Chen et al., 2008) and also that of the fibrous scaffolds under uniaxial (Gentleman et al., 2003; Nerurkar et al., 2007; Hadjizadeh et al., 2011; Hong et al ., 2011; Mauck et al., 2009; Stella et al., 2010; Pu et al., 2014)/ biaxial (Gilbert et al., 2008; Knezevic et al., 2002; Sander et al., 2009) elongations, and shear (Driscoll et al., 2011; Gittes & Mackintosh, 1998). In the experiments, the mechanical tests on the fibrous scaffolds have demonstrated a non-linear nature of stress-strain curve, irrespective of the fiber material. This nonlinearity has been characterized by the presence of a small J-shaped "toe" region in the stress-strain curves (*Figure 2*). Furthermore, at high elongation strains, these scaffolds have been shown to fracture due to the breaking of constituent fibers.

Table 1. Compendium of key results in this domain.

Reference	Conclusion	Substrate and cell type
Differentiation of stem cells		
Engler et al., 2006	Substrate stiffness determines stem cell lineage	Human mesenchymal stem cells (hM-SCs), polyacrylamide gels
Cell growth		
Hadjipanayi et al., 2009a	Faster cell proliferation on stiff substrates	Adult human dermal fibroblasts, collagen scaffold
Ghosh et al., 2007	Cells with stiff and organized cytoskeleton on stiff substrates	Adult human dermal fibroblasts, hyaluronan and fibronectin gels
Yeung et al., 2005	Substrate stiffness dependent cell spreading	NIH-3T3 mouse fibroblasts, bovine aortic endothelial cells, Human blood neutrophils, polyacrylamide gels
Cell motility		
Lo et al., 2000	Cells migrate towards stiffer substrate, also known as durotaxis	3T3 fibroblasts, polyacrylamide gels coated with type-I collagen
Ng et al., 2012	Collective cell migration on stiff substrates	MCF10A cells, PAA gels
Hadjipanayi et al., 2009b	Directional cell migration on scaffolds with directional stiffness	Adult human dermal fibroblasts, collagen scaffold
Cell polarity		
Zemel et al., 2010	Stress fiber alignment	hMSCs, polyacrylamide and hyaluronan gels
Prager-Khoutorsky et al., 2011	Cell polarization and elongation on stiff scaffolds	Human fibroblasts, PDMS

However, as far as the effect of the fiber structure on the scaffold mechanics is concerned an increase in the fiber alignment has been shown to coincide with a stiffer mechanical response (Yan et al., 2012; Ramasamy & Akkus, 2007).

Figure 2. Nonlinear response of polystyrene fiber scaffold (left panel) and plantaris tendon of sheep as reported in (Ker, 1999) (right panel).

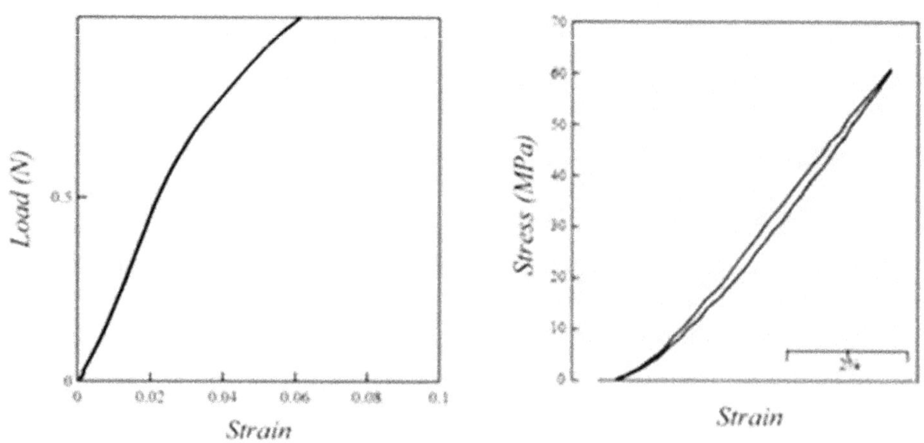

MATHEMATICAL/COMPUTATIONAL MODELING OF SCAFFOLD MECHANICS

In addition to the experiments, several mathematical and computational modeling approaches have also been used to study the fibrous scaffold mechanics. Roughly speaking, one of the main goals of the mathematical/computational models has been to establish connections between structure and mechanics of these scaffold materials (Breuls et al., 2002; Bass & Kuiper, 2008; Lacroix et al., 2006; Stylianopoulos et al., 2008; Argento et al., 2012). One can divide these "*in-silico*" works based on the approaches they take towards this goal, as described below.

Modeling Approaches

Different approaches formulate models of the mechanical response of fibrous scaffolds based on different physical properties of the fibers and their arrangement, and therefore, provide understanding at different levels. These approaches also make several different assumptions (which also put limitations on these approaches) about the scaffolds which we will be discussing next.

Lattice Based Models

In this approach the fibrous scaffolds are considered as a regular arrangement of lattice or unit cells (Stylianopoulos et al., 2008; Stylianopoulos & Barocas, 2008; Kabla & Mahadevan, 2007) where lattice sites are connected with each other via the fibers. Therefore, this approach links the microscopic mechanics of individual fibers with that of the scaffold as a whole. This method has also been used to look at the failure behavior of the fibrous scaffolds under large deformations (Hadi eta l., 2012). As the fiber arrangement in the 'unit cells' can be specified, this approach also has the capability to establish a relationship between the structure and the mechanics of the fibrous scaffolds.

This computational approach has been used to unravel many fundamental properties of the mechanics of the fibrous networks. In one remarkable work (Broedersz et al., 2011), fibrous networks were modeled using lattice base framework with variable connectivity to look at the role of the single-fiber mechanics and network structure on scaffold response. In this approach the energy of the whole system is formulated in terms of the stretching of fibers

$$E_s = \frac{1}{2}\frac{\mu}{l^2}\sum_{ij} g_{ij}\left(u_{ij}\cdot r_{ij}\right)^2 \tag{1}$$

and bending of fiber-fiber junctions

$$E_b = \frac{1}{2}\frac{\kappa}{l^3}\sum_{ijk} g_{ij}g_{jk}\left[\left(u_{jk}-u_{ij}\right)\times r_{ij}\right]^2 \tag{2}$$

where μ and κ are the elastic moduli of fibers, l is the lattice unit length, $\boldsymbol{u}$ is the displacement field, and $\boldsymbol{r}$ are the vector connecting the two lattice sites. The response of the scaffold is obtained by the minimization of the system energy along with the specified boundary conditions. These works show that at low connectivity of the lattice the scaffold response is mainly due to the fiber bending whereas at high connectivity it is due to the stretching of the fibers.

In another work, the same modeling approach was used to investigate the response of the fibrous scaffold against localized forces. In this work it was shown that the network nature of the scaffolds can even amplify the mechanical active stresses which are applied by the cells. These amplification in the mechanical stress can be thought of as a mechanism of the long range mechanical interaction between the cells.

Although this approach has immense utility and application, it also suffers from a limitation which is due to their fundamental (not always justifiable) assumption of structural periodicity. Since experimental techniques for the fabrication of these scaffolds does not allow sufficient control over their structure the 'periodic structure assumption' has a very limited scope.

Discrete Fiber Networks

In this approach, as the name suggests, the scaffold is modeled as a collection of discrete fiber elements. The model considers each and every fiber in the system separately and estimates the collective response of the whole matrix from that of individual fibers (Abhilash et al., 2014; Ma et al., 2013). Here, the structure of the fibrous scaffolds are virtually generated using computational means or obtained from the imaging of physical samples. Since one can generate a computational structure, this method provides a very fine control over the scaffold structure. This approach also requires the knowledge of the constitutive response of the single fibers which is usually modeled as either an elastic beam (high flexural rigidity) or string (low flexural rigidity) (Abhilash et al., 2014; Ma et al., 2013), depending on the application.

In one recent work (Abhilash et al., 2014) on the remodeling of the fibrous scaffold due to cellular forces the system was modeled as discrete fiber network. The response of the fibers was written in terms of the total energy of the system

$$E = \frac{1}{2}\sum_{i=0}^{N}\int\left[EI\left(\frac{\partial\psi_i(s)}{\partial s}\right)^2 + EA\left(\frac{\partial u_i(s)}{\partial s}\right)^2 + \lambda GA\left(\frac{\partial v_i}{\partial s} - \psi_i(s)\right)^2\right]ds \tag{3}$$

where the individual fibers were modeled as flexible Timoshenko beam. Here N is the total number of fibers, E is the Young's modulus, G is the shear modulus, I is the second moment of area, $\partial u/\partial s$ is the axial strain, $\partial v/\partial s$ is the rotation of the fiber cross-section, $\psi(s)$ is the rotation of the plane perpendicular to the normal axis of the fiber, λ is the shear correction factor, u is the axial displacement along the fiber axis s and v is the displacement along the transverse axis. The system response under quasistatic loading was obtained by minimization of this energy functional. This model shows alignment of the scaffold fibers due to the cellular forces and a heterogeneous deformation of the scaffold. Similar to the lattice-based models this modeling approach also shows long distance transmission of the forces in the fibrous scaffolds.

The details of the scaffold architecture this approach can incorporate makes this approach a very powerful method for the study of the mechanics of the scaffold. This strength can, however, sometimes also become the weakness of this approach in the sense of the requirement of large computational resources. Further, the exact architecture of the fibrous scaffolds is not always known in as much details as required by this method. Most of the imaging methods which give us information about the fiber arrangement in the scaffold are of destructive nature and the structural information obtained from one sample is not necessarily applicable for a different sample. This limitation of the approach is sometimes handled by not incorporating very detailed scaffold structure information in the model but only its statistical description which is usually translatable from sample to sample. We are going to look at such statistical modeling approaches next.

Statistical Models

This method is somewhere in between discrete fiber models and continuum formulation (shown below). In this approach the fiber arrangement is described not in its complete detailed information but in terms of the statistical probability density functions (PDFs). As an example, the arrangement of the fibers in a fibrous scaffold made up of straight fibers can be described in terms of the fiber angle probability density function $p_\theta(\theta)$ with

$$\int p_\theta(\theta)d\theta = 1 \tag{4}$$

If the fibers in the scaffold are curved then the information about the degree of the fiber curvature can also be described in terms of another PDF $p_l(l)$. Following this structural description, the response of the fibrous scaffold is written in as the collective response of all the fibers in the system. In one work (Rizvi & Pal, 2014), the shear and normal components of the scaffold response against scaffold shear deformation are written as

$$F_s = N\iint \sigma(\gamma_f)\frac{\pi a^2}{4}\frac{\omega + L\varepsilon}{\sqrt{L^2 + (\omega + L\varepsilon)^2}}p_l(l)\,p_\theta(\theta)\,dld\theta, \tag{5}$$

and

$$F_n = N\iint \sigma(\gamma_f)\frac{\pi a^2}{4}\frac{L}{\sqrt{L^2 + (\omega + L\varepsilon)^2}}p_l(l)\,p_\theta(\theta)\,dld\theta, \tag{6}$$

respectively. Here ε is the macroscopic scaffold shear strain, $\sigma(\gamma_f)$ is the constitutive relation of the fiber material with γ_f as the fiber strain, a is the cross-sectional area of the fiber. This modeling approach has been utilized to look at the relationship between the scaffold structure and its macroscopic mechanical response (Rizvi & Pal, 2014; Rizvi et al., 2012).

One common drawback in all of the three aforementioned approaches is the undefined nature of the mechanical stresses. Since these approaches consider each fiber the conventional measures of mechanical stresses are not well defined in these works. This, sometimes, becomes a big limitation of these approaches while comparing the response of the scaffold with the regular continuum materials. For this we will look at the continuum approach which is also used to study the mechanical response of fibrous scaffolds.

Continuum Mechanics Formulation

This approach is driven by the assumption of the microscopic length scale of the scaffold being considerably smaller than that of the experimental investigation. This assumption leads to the formulation which consider fibrous matrices as elastic continua with fibers embedded in an isotropic phase (Holzapfel et al., 2000). The continuum assumption also puts a constraint on the system that fibers are rigidly attached to the isotropic phase. This puts a limitation on this approach for the study of the fibrous scaffolds if the scaffold does not have the isotropic phase. With these assumptions, this approach obtains the scaffold response as the collective response of its phases- isotropic and fibrous (Holzapfel et al., 2000; Federico & Gasser, 2010; Wang et al. 2014). These methods are also capable to incorporate the structural characteristics of the fibers in the model in the form of degree of fiber alignment to obtain the structure-mechanics relationship (Federico & Gasser, 2010).

In one recent work this approach was used to look at the long-range transmission of the cellular forces in a fibrous collagen scaffold (Wang et al. 2014). In this work researchers used the discrete fiber network model to look at the effect of the fiber alignment, applied force, strain stiffening, cell aspect ratio etc. on scaffold response. For this response of the whole system was written in terms of the energy density function

$$E = \frac{\mu}{2}\left(\bar{I}_3 - 3\right) + \frac{k}{2}\left(J - 1\right)^2 + \sum_{a=1}^{3} f(\lambda_a) \tag{7}$$

where the first two terms correspond to the response of the isotropic phase and the final term stands for the response of the fibers. With this description, one can define the Cauchy stress as

$$\sigma = \frac{2}{J} F \cdot \frac{\partial W}{\partial C} \cdot F^T \tag{8}$$

where $\mathbf{C}=\mathbf{F}^T\mathbf{F}$ is the right Cauchy-Green deformation tensor. Here function $f(\lambda_a)$ describes the response of the individual fibers. In this work the fibers are shown to get aligned due to the cellular forces and also along range force transmission is observed. There have been some extensions (Federico & Gasser, 2010) of this modeling approach where the fiber arrangement is described in terms of the probability density functions, akin to that in the statistical models.

With this description of the modeling approaches utilized for the study of the scaffold mechanics, we now focus on the mechanical response of the fibrous scaffolds and its dependence on the experimental length scale.

Dependence of Fibrous Scaffold on Size

In experiments, mechanical behavior of the scaffold can be characterized either by the uniaxial elongation or shear mechanical testing or by techniques which study the localized mechanical properties, such as atomic probe microscopy. Similarly, the cells seeded on the scaffold also apply very localized forces to the scaffold to sense their stiffness and other mechanical characteristics. This shows that the application of forces/deformation to the scaffolds can be applied at two length scales. We focus on this aspect now.

Macroscopic Mechanics

In the conventional mechanical test in the scaffold of size typically 1cm×1cm is held between two ends of mechanical testing machine and uniaxially elongated. Given this large sample size relative to the fiber diameter the response obtained from this approach can be called "macroscopic mechanical response" of the scaffold. In modeling works also the elastic moduli of the scaffold are estimated by performing such an experiment "virtually". As described earlier, these modeling works have shown that the scaffold response shows a nonlinear nature. Most of the modeling approaches are in qualitative agreement about the macroscopic response of the fibrous scaffold. In few recent modeling works (Rizvi et al., 2012), however, it has been shown that the macroscopic response of the fibrous scaffold can also depend on the size of the sample used in the mechanical testing. This implies that the mechanical characterization of the fibrous scaffolds requires an additional care in measurement and reporting for the reproducibility of the elastic moduli values. Furthermore, using the statistical modeling approach it has also been demonstrated that the two fibrous scaffolds with different structural arrangement of fibers can have same macroscopic stress-stress response (Rizvi & Pal, 2014). Despite this recent observation, the elastic moduli obtained from the macroscopic mechanics of the fibrous scaffold is often taken to be the same as the mechanical stiffness sensed by the cells seeded on scaffold. This can sometimes lead to contradictory experimental outcomes.

This limitation of the use of macroscopic elastic moduli as surrogate of the mechanical properties sensed by cells also makes it important that we should characterize the microscopic mechanical properties of the scaffolds.

Microscopic Mechanics

For the characterization of the microscopic mechanical response of the scaffold we apply a localized force or displacement to the scaffold and estimate the resulting displacement or resistant force, respectively. This is akin to the forces exerted by the cells to the scaffold via focal adhesion. The microscopic mechanical response of the fibrous scaffolds have not been looked at that extensively till now. One reason for this is the lack of any straightforward experimental setup for the application of localized forces in the manner of cells. AFM can apply localized forces but the nature of AFM forces is different from that by cells since the forces applied by the cell are in the plane of contact between cell and scaffold and for AFM forces it is not the case.

In a recent computational work (Mech & Rizvi, 2021), however, the microscopic mechanics of the fibrous scaffold has been characterized using discrete fiber modeling approach. This discrete fiber model estimates the displacement field in the fibrous scaffold due to localized point force and demonstrates an anisotropic behavior. Further, it was shown that the displacement field in the fibrous scaffold decays exponentially with respect to the distance, from the point of force application. This result is in contrast to the known solutions to the Kelvin problem in three-dimensional elasticity where the displacement decays as a power law. Furthermore, the work also looks at the stiffness sensed by a cell under two conditions. First, when cell applies a constant strain and estimates scaffold mechanics in terms of the resistance displayed by the scaffold. Second, when cell exerts a constant magnitude of force and scaffold compliance is sensed in terms of its deformation. It was seen that under constant deformation application scenario, the stiffness sensed by the cell is proportional to the macroscopic elastic moduli, such as Young's modulus.

However, in the second scenario the stiffness sensed by the cell also depends on the stiffness of the cell itself (*Figure 3*). Researchers showed that the stiffness of the same fibrous scaffold sensed by two different cells can differ by many orders of magnitude. This result also implies that one fibrous scaffold can provide very different mechanical microenvironment to different cell types. This implies that an immense care is required in the experimental characterization tests and during modeling the mechanical response of the fibrous scaffold. This also shows that the reporting of only macroscopic elastic moduli, such as Young's or shear moduli, may not always be sufficient for the complete description of the scaffold mechanics.

Figure 3. Scaffold stiffness as sensed by different cell types.

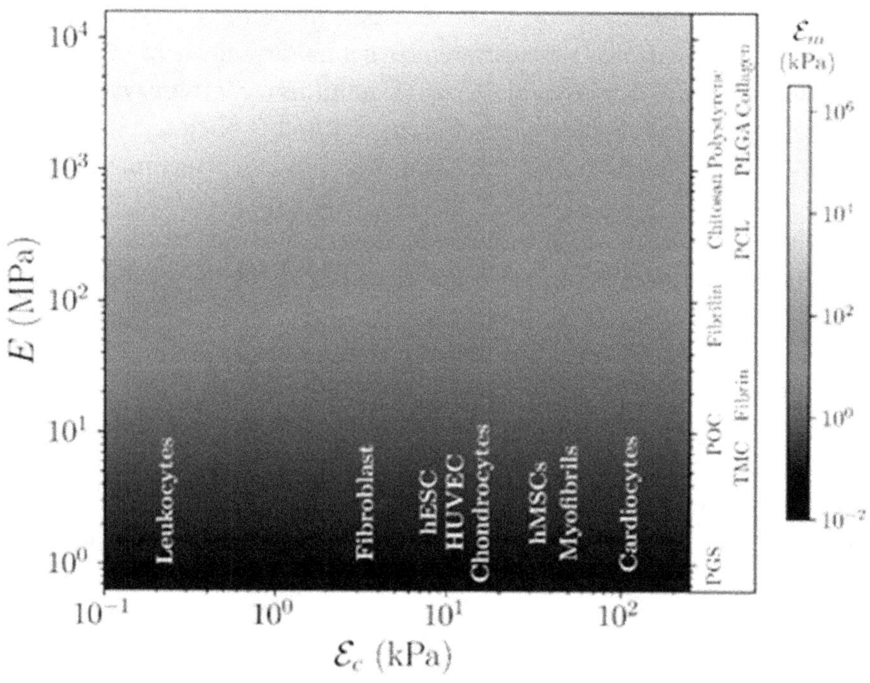

CONCLUSION

In this chapter we have attempted to give a brief overview of different modeling approaches which are utilized for the study of the mechanics of the tissue engineering scaffolds. All of these approaches are based on different assumptions and, therefore, are limited in their scope. Furthermore, authors have also not looked at the transport of the fluid through these scaffolds which is also an extremely important aspect of the scaffold design. Another aspect not covered in here is the viscoelastic nature of the fibrous scaffolds and its modeling. Due to the nature of the materials used for the application of tissue engineering, the viscoelastic response can also be very important in some application where a dynamical mechanical loading is applied to the scaffold. For this, readers are referred to papers such as Rizvi et al., 2016. The information covered in this chapter can be used to study and investigate the mechanical response of the tissue engineering fibrous scaffold for their better design and optimized structure.

ACKNOWLEDGMENT

We would like to thank IIT Hyderabad for the resources and financial support.

REFERENCES

Abhilash, A. S., Baker, B. M., Trappmann, B., Chen, C. S., & Shenoy, V. B. (2014). Remodeling of fibrous extracellular matrices by contractile cells: Predictions from discrete fiber network simulations. *Biophysical Journal*, *107*(8), 1829–1840. doi:10.1016/j.bpj.2014.08.029 PMID:25418164

Argento, G., Simonet, M., Oomens, C. W. J., & Baaijens, F. P. T. (2012). Multi-scale mechanical characterization of scaffolds for heart valve tissue engineering. *Journal of Biomechanics*, *45*(16), 2893–2898. doi:10.1016/j.jbiomech.2012.07.037 PMID:22999107

Baas, E., & Kuiper, J. H. (2008). A numerical model of heterogeneous surface strains in polymer scaffolds. *Journal of Biomechanics*, *41*(6), 1374–1378. doi:10.1016/j.jbiomech.2008.01.018 PMID:18353333

Breuls, R. G., Sengers, B. G., Oomens, C. W., Bouten, C. V., & Baaijens, F. P. (2002). Predicting local cell deformations in engineered tissue constructs: A multilevel finite element approach. *Journal of Biomechanical Engineering*, *124*(2), 198–207. doi:10.1115/1.1449492 PMID:12002129

Broedersz, C. P., Mao, X., Lubensky, T. C., & MacKintosh, F. C. (2011). Criticality and isostaticity in fibre networks. *Nature Physics*, *7*(12), 983–988. doi:10.1038/nphys2127

Bullough, P., & Goodfellow, J. (1968). The significance of the fine structure of articular cartilage. *The Journal of Bone and Joint Surgery. British Volume*, *50*(4), 852–857. doi:10.1302/0301-620X.50B4.852 PMID:5706888

Carlisle, C. R., Coulais, C., & Guthold, M. (2010). The mechanical stressâ€“strain properties of single electrospun collagen type I nanofibers. *Acta Biomaterialia*, *6*(8), 2997–3003. doi:10.1016/j.actbio.2010.02.050 PMID:20197123

Carlisle, C. R., Coulais, C., Namboothiry, M., Carroll, D. L., Hantgan, R. R., & Guthold, M. (2009). The mechanical properties of individual, electrospun fibrinogen fibers. *Biomaterials*, *30*(6), 1205–1213. doi:10.1016/j.biomaterials.2008.11.006 PMID:19058845

Chen, F., Peng, X., Li, T., Chen, S., Wu, X. F., Reneker, D. H., & Hou, H. (2008). Mechanical characterization of single high-strength electrospun polyimide nanofibres. *Journal of Physics. D, Applied Physics*, *41*(2), 025308. doi:10.1088/0022-3727/41/2/025308

Cheney, J. A., Konow, N., Bearnot, A., & Swartz, S. M. (2015). A wrinkle in flight: The role of elastin fibres in the mechanical behaviour of bat wing membranes. *Journal of the Royal Society, Interface*, *12*(106), 20141286. doi:10.1098/rsif.2014.1286 PMID:25833238

Clark, J. M. (1990). The organisation of collagen fibrils in the superficial zones of articular cartilage. *Journal of Anatomy*, *171*, 117. PMID:2081698

Driscoll, T. P., Nerurkar, N. L., Jacobs, N. T., Elliott, D. M., & Mauck, R. L. (2011). Fiber angle and aspect ratio influence the shear mechanics of oriented electrospun nanofibrous scaffolds. *Journal of the Mechanical Behavior of Biomedical Materials*, *4*(8), 1627–1636. doi:10.1016/j.jmbbm.2011.03.022 PMID:22098865

Dunlop, J. W., & Fratzl, P. (2010). Biological composites. *Annual Review of Materials Research*, *40*(1), 1–24. doi:10.1146/annurev-matsci-070909-104421

Engler, A. J., Sen, S., Sweeney, H. L., & Discher, D. E. (2006). Matrix elasticity directs stem cell lineage specification. *Cell*, *126*(4), 677–689. doi:10.1016/j.cell.2006.06.044 PMID:16923388

Federico, S., & Gasser, T. C. (2010). Nonlinear elasticity of biological tissues with statistical fibre orientation. *Journal of the Royal Society, Interface*, *7*(47), 955–966. doi:10.1098/rsif.2009.0502 PMID:20053655

Frantz, C., Stewart, K. M., & Weaver, V. M. (2010). The extracellular matrix at a glance. *Journal of Cell Science*, *123*(24), 4195–4200. doi:10.1242/jcs.023820 PMID:21123617

Fratzl, P. (2008). Collagen: structure and mechanics, an introduction. In *Collagen* (pp. 1–13). Springer. doi:10.1007/978-0-387-73906-9_1

Gentleman, E., Lay, A. N., Dickerson, D. A., Nauman, E. A., Livesay, G. A., & Dee, K. C. (2003). Mechanical characterization of collagen fibers and scaffolds for tissue engineering. *Biomaterials*, *24*(21), 3805–3813. doi:10.1016/S0142-9612(03)00206-0 PMID:12818553

Ghosh, K., Pan, Z., Guan, E., Ge, S., Liu, Y., Nakamura, T., Ren, X.-D., Rafailovich, M., & Clark, R. A. (2007). Cell adaptation to a physiologically relevant ECM mimic with different viscoelastic properties. *Biomaterials*, *28*(4), 671–679. doi:10.1016/j.biomaterials.2006.09.038 PMID:17049594

Gilbert, T. W., Wognum, S., Joyce, E. M., Freytes, D. O., Sacks, M. S., & Badylak, S. F. (2008). Collagen fiber alignment and biaxial mechanical behavior of porcine urinary bladder derived extracellular matrix. *Biomaterials*, *29*(36), 4775–4782. doi:10.1016/j.biomaterials.2008.08.022 PMID:18801572

Gittes, F., & MacKintosh, F. C. (1998). Dynamic shear modulus of a semiflexible polymer network. *Physical Review. E*, *58*(2), R1241–R1244. doi:10.1103/PhysRevE.58.R1241

Grant, T. M., Thompson, M. S., Urban, J., & Yu, J. (2013). Elastic fibres are broadly distributed in tendon and highly localized around tenocytes. *Journal of Anatomy*, *222*(6), 573–579. doi:10.1111/joa.12048 PMID:23587025

Hadi, M. F., Sander, E. A., & Barocas, V. H. (2012). *Multiscale model predicts tissue-level failure from collagen fiber-level damage*. Academic Press.

Hadjipanayi, E., Mudera, V., & Brown, R. A. (2009a). Close dependence of fibroblast proliferation on collagen scaffold matrix stiffness. *Journal of Tissue Engineering and Regenerative Medicine*, *3*(2), 77–84. doi:10.1002/term.136 PMID:19051218

Hadjipanayi, E., Mudera, V., & Brown, R. A. (2009b). Guiding cell migration in 3D: A collagen matrix with graded directional stiffness. *Cell Motility and the Cytoskeleton*, *66*(3), 121–128. doi:10.1002/cm.20331 PMID:19170223

Hadjizadeh, A., Ajji, A., & Bureau, M. N. (2011). Nano/micro electro-spun polyethylene terephthalate fibrous mat preparation and characterization. *Journal of the Mechanical Behavior of Biomedical Materials*, *4*(3), 340–351. doi:10.1016/j.jmbbm.2010.10.014 PMID:21316622

Han, L., Grodzinsky, A. J., & Ortiz, C. (2011). Nanomechanics of the cartilage extracellular matrix. *Annual Review of Materials Research*, *41*(1), 133–168. doi:10.1146/annurev-matsci-062910-100431 PMID:22792042

Heim, M., Römer, L., & Scheibel, T. (2010). Hierarchical structures made of proteins. The complex architecture of spider webs and their constituent silk proteins. *Chemical Society Reviews*, *39*(1), 156–164. doi:10.1039/B813273A PMID:20023846

Holzapfel, G. A., Gasser, T. C., & Ogden, R. W. (2000). A new constitutive framework for arterial wall mechanics and a comparative study of material models. *Journal of Elasticity and the Physical Science of Solids, 61*(1), 1-48.

Hong, Y., Huber, A., Takanari, K., Amoroso, N. J., Hashizume, R., Badylak, S. F., & Wagner, W. R. (2011). Mechanical properties and in vivo behavior of a biodegradable synthetic polymer microfiberâ€"extracellular matrix hydrogel biohybrid scaffold. *Biomaterials*, *32*(13), 3387–3394. doi:10.1016/j.biomaterials.2011.01.025 PMID:21303718

Kabla, A., & Mahadevan, L. (2007). Nonlinear mechanics of soft fibrous networks. *Journal of the Royal Society, Interface*, *4*(12), 99–106. doi:10.1098/rsif.2006.0151 PMID:17015287

Keith, D. A., Paz, A., Gallop, P. M., & Glimcher, M. J. (1977). *Histologic and biochemical identification and characterization of an elastin in cartilage*. Academic Press.

Ker, R. F. (1999). The design of soft collagenous load-bearing tissues. *The Journal of Experimental Biology*, *202*(23), 3315–3324. doi:10.1242/jeb.202.23.3315 PMID:10562514

Knezevic, V., Sim, A. J., Borg, T. K., & Holmes, J. W. (2002). Isotonic biaxial loading of fibroblast-populated collagen gels: A versatile, low-cost system for the study of mechanobiology. *Biomechanics and Modeling in Mechanobiology*, *1*(1), 59–67. doi:10.100710237-002-0005-0 PMID:14586707

Lacroix, D., Chateau, A., Ginebra, M. P., & Planell, J. A. (2006). Micro-finite element models of bone tissue-engineering scaffolds. *Biomaterials*, *27*(30), 5326–5334. doi:10.1016/j.biomaterials.2006.06.009 PMID:16824593

Lo, C. M., Wang, H. B., Dembo, M., & Wang, Y. L. (2000). Cell movement is guided by the rigidity of the substrate. *Biophysical Journal*, *79*(1), 144–152. doi:10.1016/S0006-3495(00)76279-5 PMID:10866943

Lodish, H., Berk, A., Kaiser, C. A., Kaiser, C., Krieger, M., Scott, M. P., & Matsudaira, P. (2008). *Molecular cell biology*. Macmillan.

Ma, X., Schickel, M. E., Stevenson, M. D., Sarang-Sieminski, A. L., Gooch, K. J., Ghadiali, S. N., & Hart, R. T. (2013). Fibers in the extracellular matrix enable long-range stress transmission between cells. *Biophysical Journal, 104*(7), 1410–1418. doi:10.1016/j.bpj.2013.02.017 PMID:23561517

Mauck, R. L., Baker, B. M., Nerurkar, N. L., Burdick, J. A., Li, W. J., Tuan, R. S., & Elliott, D. M. (2009). Engineering on the straight and narrow: The mechanics of nanofibrous assemblies for fiber-reinforced tissue regeneration. *Tissue Engineering. Part B, Reviews, 15*(2), 171–193. doi:10.1089/ten.teb.2008.0652 PMID:19207040

Mech, D. J., & Rizvi, M. S. (2021). *In-silico modeling of the micromechanics of fibrous scaffolds and stiffness sensing by cells.* arXiv preprint arXiv:2112.06064.

Mercer, R. R., & Crapo, J. D. (1990). Spatial distribution of collagen and elastin fibers in the lungs. *Journal of Applied Physiology, 69*(2), 756–765. doi:10.1152/jappl.1990.69.2.756 PMID:2228886

Minns, R. J., & Steven, F. S. (1977). The collagen fibril organization in human articular cartilage. *Journal of Anatomy, 123*(Pt 2), 437. PMID:870478

Mithieux, S. M., & Weiss, A. S. (2005). Elastin. *Advances in Protein Chemistry, 70*, 437–461. doi:10.1016/S0065-3233(05)70013-9 PMID:15837523

Muir, H., Bullough, P., & Maroudas, A. (1970). The distribution of collagen in human articular cartilage with some of its physiological implications. *The Journal of Bone and Joint Surgery. British Volume, 52*(3), 554–563. doi:10.1302/0301-620X.52B3.554 PMID:4247851

Muiznieks, L. D., & Keeley, F. W. (2013). Molecular assembly and mechanical properties of the extracellular matrix: A fibrous protein perspective. Biochimica et Biophysica Acta (BBA)-. *Molecular Basis of Disease, 1832*(7), 866–875. doi:10.1016/j.bbadis.2012.11.022 PMID:23220448

Murakumo, M., Ushiki, T., Abe, K., Matsumura, K., Shinno, Y., & Koyanagi, T. (1995). Three-dimensional arrangement of collagen and elastin fibers in the human urinary bladder: A scanning electron microscopic study. *The Journal of Urology, 154*(1), 251–256. doi:10.1016/S0022-5347(01)67289-6 PMID:7776441

Nerurkar, N. L., Elliott, D. M., & Mauck, R. L. (2007). Mechanics of oriented electrospun nanofibrous scaffolds for annulus fibrosus tissue engineering. *Journal of Orthopaedic Research, 25*(8), 1018–1028. doi:10.1002/jor.20384 PMID:17457824

Ng, M. R., Besser, A., Danuser, G., & Brugge, J. S. (2012). Substrate stiffness regulates cadherin-dependent collective migration through myosin-II contractility. *The Journal of Cell Biology, 199*(3), 545–563. doi:10.1083/jcb.201207148 PMID:23091067

Prager-Khoutorsky, M., Lichtenstein, A., Krishnan, R., Rajendran, K., Mayo, A., Kam, Z., Geiger, B., & Bershadsky, A. D. (2011). Fibroblast polarization is a matrix-rigidity-dependent process controlled by focal adhesion mechanosensing. *Nature Cell Biology, 13*(12), 1457–1465. doi:10.1038/ncb2370 PMID:22081092

Pu, J., & Komvopoulos, K. (2014). Mechanical properties of electrospun bilayer fibrous membranes as potential scaffolds for tissue engineering. *Acta Biomaterialia, 10*(6), 2718–2726. doi:10.1016/j.actbio.2013.12.060 PMID:24434536

Ramasamy, J. G., & Akkus, O. (2007). Local variations in the micromechanical properties of mouse femur: The involvement of collagen fiber orientation and mineralization. *Journal of Biomechanics*, *40*(4), 910–918. doi:10.1016/j.jbiomech.2006.03.002 PMID:16678186

Ritty, T. M., Ditsios, K., & Starcher, B. C. (2002). Distribution of the elastic fiber and associated proteins in flexor tendon reflects function. *The Anatomical Record: An Official Publication of the American Association of Anatomists*, *268*(4), 430–440. doi:10.1002/ar.10175 PMID:12420291

Rizvi, M. S., Kumar, P., Katti, D. S., & Pal, A. (2012). Mathematical model of mechanical behavior of micro/nanofibrous materials designed for extracellular matrix substitutes. *Acta Biomaterialia*, *8*(11), 4111–4122. doi:10.1016/j.actbio.2012.07.025 PMID:22842037

Rizvi, M. S., & Pal, A. (2014). Statistical model for the mechanical behavior of the tissue engineering non-woven fibrous matrices under large deformation. *Journal of the Mechanical Behavior of Biomedical Materials*, *37*, 235–250. doi:10.1016/j.jmbbm.2014.05.026 PMID:24956158

Rizvi, M. S., Pal, A., & Das, S. L. (2016). Structure-induced nonlinear viscoelasticity of non-woven fibrous matrices. *Biomechanics and Modeling in Mechanobiology*, *15*(6), 1641–1654. doi:10.100710237-016-0788-z PMID:27090523

Sander, E. A., Stylianopoulos, T., Tranquillo, R. T., & Barocas, V. H. (2009). Image-based multiscale modeling predicts tissue-level and network-level fiber reorganization in stretched cell-compacted collagen gels. *Proceedings of the National Academy of Sciences of the United States of America*, *106*(42), 17675–17680. doi:10.1073/pnas.0903716106 PMID:19805118

Schipper, H., Kranenbarg, S., van Leeuwen, J., Gijsen, M., Haazelager, M., & van Turnhout, M. (2008). Quantitative description of collagen structure in the articular cartilage of the young and adult equine distal metacarpus. *Animal Biology (Leiden, Netherlands)*, *58*(4), 353–370. doi:10.1163/157075608X383674

Stella, J. A., Wagner, W. R., & Sacks, M. S. (2010). Scaleâ€ dependent fiber kinematics of elastomeric electrospun scaffolds for soft tissue engineering. *Journal of Biomedical Materials Research Part A: An Official Journal of The Society for Biomaterials, The Japanese Society for Biomaterials, and The Australian Society for Biomaterials and the Korean Society for Biomaterials*, *93*(3), 1032–1042. PMID:19753623

Strocchi, R., De Pasquale, V., Gubellini, P., Facchini, A., Marcacci, M., Buda, R., ... Ruggeri, A. (1992). The human anterior cruciate ligament: Histological and ultrastructural observations. *Journal of Anatomy*, *180*(Pt 3), 515. PMID:1487443

Stylianopoulos, T., & Barocas, V. H. (2007). Volume-averaging theory for the study of the mechanics of collagen networks. *Computer Methods in Applied Mechanics and Engineering*, *196*(31-32), 2981–2990. doi:10.1016/j.cma.2006.06.019

Stylianopoulos, T., Bashur, C. A., Goldstein, A. S., Guelcher, S. A., & Barocas, V. H. (2008). Computational predictions of the tensile properties of electrospun fibre meshes: Effect of fibre diameter and fibre orientation. *Journal of the Mechanical Behavior of Biomedical Materials*, *1*(4), 326–335. doi:10.1016/j.jmbbm.2008.01.003 PMID:19627797

Tabassi, N. C. B., & Garnero, P. (2007). Monitoring cartilage turnover. *Current Rheumatology Reports*, *9*(1), 16–24. doi:10.100711926-007-0017-y PMID:17437662

Tan, E. P. S., & Lim, C. T. (2006). Mechanical characterization of nanofibersâ€"a review. *Composites Science and Technology*, *66*(9), 1102–1111. doi:10.1016/j.compscitech.2005.10.003

Wang, H., Abhilash, A. S., Chen, C. S., Wells, R. G., & Shenoy, V. B. (2014). Long-range force transmission in fibrous matrices enabled by tension-driven alignment of fibers. *Biophysical Journal*, *107*(11), 2592–2603. doi:10.1016/j.bpj.2014.09.044 PMID:25468338

Weiss, C., Rosenberg, L., & Helfet, A. J. (1968). An ultrastructural study of normal young adult human articular cartilage. *JBJS*, *50*(4), 663–674. doi:10.2106/00004623-196850040-00002 PMID:5658553

Yahia, L. H., & Drouin, G. (1989). Microscopical investigation of canine anterior cruciate ligament and patellar tendon: Collagen fascicle morphology and architecture. *Journal of Orthopaedic Research*, *7*(2), 243–251. doi:10.1002/jor.1100070212 PMID:2918423

Yan, J., Qiang, L., Gao, Y., Cui, X., Zhou, H., Zhong, S., Wang, Q., & Wang, H. (2012). Effect of fiber alignment in electrospun scaffolds on keratocytes and corneal epithelial cells behavior. *Journal of Biomedical Materials Research. Part A*, *100*(2), 527–535. doi:10.1002/jbm.a.33301 PMID:22140085

Yeung, T., Georges, P. C., Flanagan, L. A., Marg, B., Ortiz, M., Funaki, M., Zahir, N., Ming, W., Weaver, V., & Janmey, P. A. (2005). Effects of substrate stiffness on cell morphology, cytoskeletal structure, and adhesion. *Cell Motility and the Cytoskeleton*, *60*(1), 24–34. doi:10.1002/cm.20041 PMID:15573414

Zemel, A., Rehfeldt, F., Brown, A. E. X., Discher, D. E., & Safran, S. A. (2010). Optimal matrix rigidity for stress-fibre polarization in stem cells. *Nature Physics*, *6*(6), 468–473. doi:10.1038/nphys1613 PMID:20563235

Chapter 9
Biomechanics of the Aortic Valve in Health and Disease

Thirumalai Deepak
National Institute of Technology, Rourkela, India

Patina Yamini
National Institute of Technology, Rourkela, India

Anju R. Babu
National Institute of Technology, Rourkela, India

ABSTRACT

The aortic valve is composed of collagen, elastin, proteoglycan, valvular interstitial cells (VIC), and valvular endothelial cells (VEC). In the open condition, the aorta valve allows blood to leave the heart, and in the closed condition, it prevents the backflow of the blood to the left ventricle. However, when the aortic valve cups become narrow or thickened, cusp motion is impaired and obstructs the blood flow. This chapter investigates the structure and composition of the aortic valve cusp and the role of VIC, VEC, and cross-talk of VEC-VIC. In addition, biomechanical characterization of the aortic cusps such as uniaxial, biaxial, flexure, three-point bending, cantilever bending, and viscoelasticity was discussed. Furthermore, etiology, in vitro cell culture and in vivo animal models, and ex vivo models mimicking aortic stenosis and regurgitation were summarized.

INTRODUCTION

The aortic, mitral, pulmonary, and tricuspid valves are the main component of the heart that regulates the unidirectional flow of blood (Misfeld & Sievers, 2007). The aortic valve contains the cusp, sinotubular junction, sinus, and coronary ostium. During the cardiac cycle, the aorta valve opens and allows the blood to leave the heart, and in closed conditions, it prevents the backflow of the blood (Lindman et al., 2016). Between the years 1990-2017, of the 100 000 death reported due to total valvular heart diseases, 61% is associated with aortic valvular diseases (Bermejo et al., 2021). When the aortic valve cups become narrow

DOI: 10.4018/978-1-7998-9078-2.ch009

or thickened, cusp motion is impaired and obstructs the blood flow, causing diseases like stenosis and regurgitation (Lindman et al., 2016; Misfeld & Sievers, 2007). In aortic stenosis, the aortic valve opening becomes small and inhibits the blood flow from the heart to the aorta. In aortic regurgitation, the aortic root and annulus dilate, leading to cusps malfunction and backflow of blood (Sawaya & Søndergaard, 2018; Zhu et al., 2021). Currently, treatment for aortic stenosis and regurgitation is with the mechanical or bioprosthetic heart valves. Both bioprosthetic and mechanical valves have been associated with health risk complications (Lerman et al., 2015; Otto, 2002; Zakikhani et al., 2019). The mechanical valves are prone to thrombosis, inflammation, and infection. The problem associated with the bioprosthesis valve is calcification, improper closure, and leakage due to the thickening of the valve cusps (Sider et al., 2011; Tsang et al., 2018). Currently, treatment for valvular diseases lacks an understanding of the initiation and progression of calcification of the aortic valvular diseases. The *in vitro* and *in vivo* animal models are robust platforms, and it is a rapid tool in identifying therapeutic interventions and can study early mechanisms of disease progression and pathogenesis (Sider et al., 2011).

This book chapter covers the structure, and mechanical properties, of aortic cusp during health and disease conditions like stenosis and regurgitation and *in vitro, in vivo,* and *ex vivo* models to simulate stenosis and regurgitation. Section 2 reviews the aortic cusp structure and composition, the role of valvular endothelial cells (VEC) and valvular interstitial cells (VIC), and cross-talk between VEC and VIC. Section 3 discusses the biomechanical response of human, porcine, and ovine aortic cusp under various mechanical testing conditions such as uniaxial, biaxial, flexure, three-point bending, and viscoelasticity. Section 4 discusses the diseased conditions of aortic valve stenosis and regurgitation. Further, *in vitro* cell culture model to study calcification process and *in vivo* mice, porcine models, and approaches to mimic aortic stenosis and regurgitation were summarized.

STRUCTURE AND COMPOSITION OF THE AORTIC CUSP

Aortic Cusp Structure

The aortic cusp is composed of four main components: commissural region or hinge, belly region, and lannula with nodule and coapting surface [Figure 1 (A, B, C)] (Misfeld & Sievers, 2007; Tseng, 2011). The central belly region of the cusp is transparent and the lannula appears crescent in shape. The three cusps meet at the coaptation and nodules are present in the midpoint of the coapting surface. Typically, the three aortic valve cusps (left coronary, right coronary, and non-coronary) are thin, flexible in nature, and consist of three layers: ventricularis, spongiosa, and fibrosa [Figure 1 (D, E)] (Hasan et al., 2014; Tseng, 2011). The cusp consists of 13% elastin fibers and 50% collagen by dry weight (Misfeld & Sievers, 2007) (Vesely, 1997). The fibrosa contains more collagen and fewer elastin sheets arranged in the circumferential direction [Figure 1 (F)]. Elastin is made up of elastic protein, and during the loading conditions, it stores energy and, in unloading conditions, it releases to the collagen, as a result, the valve return to its original position (Adham et al., 1996). Among the three layers, the ventricularis is the thinnest and contains more elastin sheets than fibrosa (Hasan et al., 2014). Spongiosa consist of a large amount of glycosaminoglycan (GAG) and a few fibrous proteins (Butcher et al., 2011). The outer layers of the aortic cusp are lined with VEC and VIC are present throughout the aortic cusps layers [Figure 1 (E)] (El-Hamamsy et al., 2010).

*Figure 1. Detailed representation of aortic valve cusps structure and composition(**A**) commissural region or hinge, belly region, and coapting surface of aortic cusp (Tseng, 2011); (**B**) SEM image of aortic cusp in hinge, belly and coaptation region, scale bar at 200 µm (Tseng, 2011); (**C**) Surface morphology of valve leaflet at 1000 X magnification (El-Hamamsy et al., 2010); (**D**) Schematic representation of aortic cusp layers - fibrosa, spongiosa, and ventricularis (Hasan et al., 2014); (**E**) Histological images of fibrosa, spongiosa and ventricularis layers stained with Movat's Pentachrome, scale bar at 200 µm (H Tseng., 2011); (**F**) Arrangements of collagen, elastin, proteoglycan in radial and circumferential direction in aortic cusps layers (Korossis., 2018).*

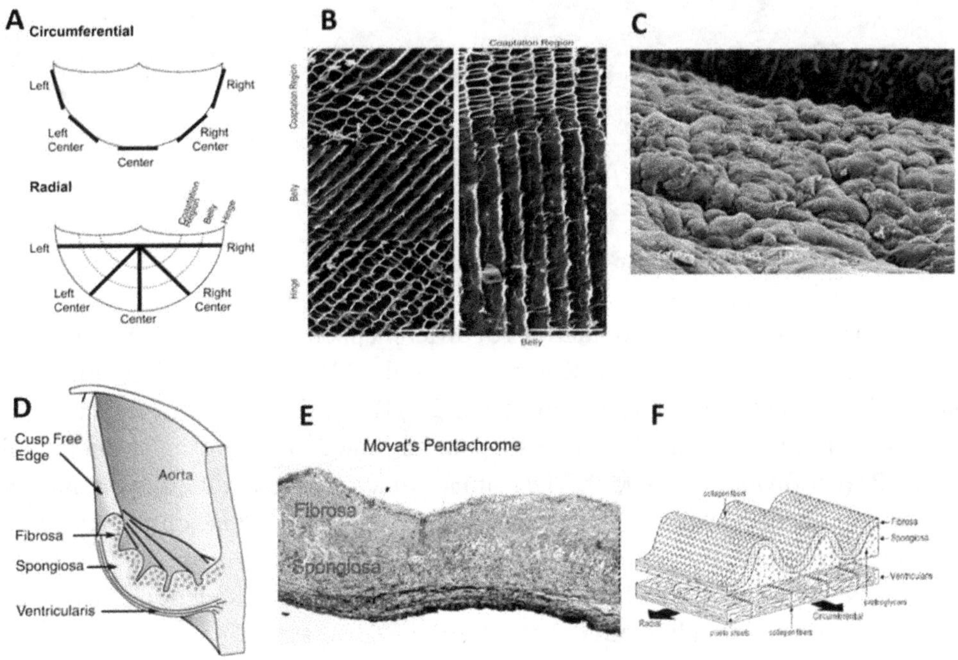

Aortic Valvular Endothelial Cells

The endothelial cells in the aortic valve maintain a non-thrombogenic blood contact surface and transfer the essential nutrients to the VIC (Frater et al., 1992). The VEC appear cobblestone-like morphology, form a monolayer on the cusp surface, and are arranged perpendicular to the direction of blood flow [Figure 2 (A)] (Deck, 1986). Due to the continuous movement of the cusp, the shear stress is challenging to quantify (Butcher et al., 2011). During the progression of stenosis conditions, the shear stress values vary from 30-1500 dyne/cm^2 (Weston et al., 1999). Sucosky et al., (2009) exposed the fibrosa side of VEC of a porcine aortic cusp to an *ex vivo* pulsatile bioreactor to study the shear stress. The result shows that exposing the fibrosa side induces the up-regulation of inflammatory receptors and BMP-4 expression in the VEC.

Figure 2. A full confluent monolayer of VIC (A) & VEC (B) under phase contrast microscope, scale bar 20 μm (Liu et al., 2007)

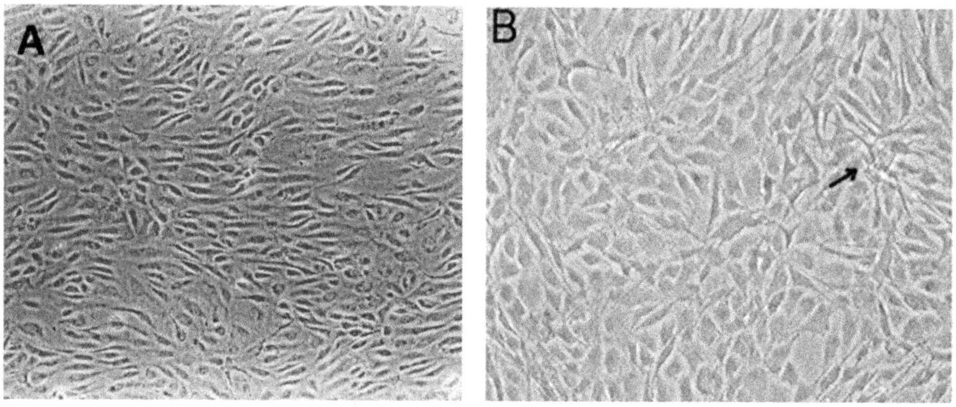

Aortic Valvular Interstitial Cells

Valvular interstitial cells are dynamic populations of different phenotypes such as myofibroblast, smooth muscles, fibroblast, cardiac muscle cells (Rutkovskiy et al., 2017). They exhibit smooth muscles and myofibroblast cells characteristics (Yperman et al., 2004). During culturing of cells, VIC appears cuboidal-like structure and elongated spindle-shaped, and once after confluent, and exhibit swirling patterns [Figure 2 (B)](Taylor et al., 2003). The primary function of the VIC is to organize and remodel extracellular matrix proteins to withstand the tissue strain during the cardiac output (Porras et al., 2017). During the cardiac process, the pressure gradient changes across the cusp cause loading and unloading of the cusps. Weston & Yoganathan (2001) reported that VIC mechanotransduction is affected by pulsatile hemodynamic shear stress and bulk matrix shear stress.

Interaction of VEC-VIC

VEC and VIC synergistically work to maintain the cusp structure and valvular function (Tseng et al., 2014). α-smooth muscle actin (α-SMA) is a marker protein that helps to differentiate between VEC and VIC. Usually, VEC-VIC interaction occurs naturally during cusp tissue homeostasis (Butcher et al., 2011). The damage of VEC causes thrombosis, inflammation, and lipid accumulation in cusps (Rutkovskiy et al., 2017). As a result, activation of VIC calcification and stenosis (Rajamannan, 2011; Tseng et al., 2014). Tseng et al., (2014) designed a three-dimensional co-culture model for VIC-VEC using the magnetic levitation method as an alternate for two-dimensional VEC-VIC studies and suggested that the VEC-VIC co-culturing has anti-thrombotic activity. Butcher & Nerem (2006) also designed three-dimensional co-culture models for VIC-VEC to study the effect of shear stress. The result shows that the co-culture model enhanced the endothelial cell to stabilize the VIC proliferation, increased protein synthesis, decreased glycosaminoglycan.

MECHANICAL PROPERTIES OF AORTIC VALVE

This section discusses the biomechanical properties of the aortic cusp obtained explicitly from the human, porcine and ovine models, reported to load under different mechanical conditions. The different mechanical studies performed on the human, bovine, porcine, and ovine species were summarized in Table 1.

Table 1. Biomechanical testing of the aortic cusp from different species, methods, strain rate, medium, and temperature mentioned below

Species	Method	Strain rate	Temperature	References
Human	Uniaxial	10mm/min	37°C	Vafaee et al., 2018
Porcine	Uniaxial	5mm/min	-	Zhou et al., 2013
Porcine	Uniaxial	10mm/min	37°C	Lee et al., 1984
Human	Uniaxial	10 mm/min	37°C	Desai et al., 2018
Porcine	Biaxial	5mm/min	37°C	Billiar & Sacks, 2000
Human	Biaxial	60N/min	37°C	Martin & Sun, 2012
Porcine	Biaxial	60N/min	37°C	
Ovine	Biaxial	60N/min	37°C	
Porcine	Biaxial	60N/min	-	Stella et al., 2007
Porcine	Biaxial	-	37°C	Borghi et al., 2013
Porcine	Biaxial	60N/min	RT	Liao et al., 2008
Porcine	Flexural	-	22°C	
Porcine	Flexural	25 mm/min	RT	Ragaert et al., 2012
Porcine	Flexural	-	22°C	Lovekamp et al., 2006
Porcine	Three-point bending	-	RT	Brazile et al., 2015
Porcine	Three-point bending	-	RT	Merryman et al., 2006
Porcine	Cantilever bending	Flexural angle of 30°	-	

Uniaxial Testing

The uniaxial stress-strain behavior of the human aortic valve cusps is shown in Figure 3 (Li et al., 2001). The aortic cusps exhibit a non-linear response with an ultimate tensile strength ranging from 2 to 4 MPa. This ultimate strength is significantly higher than the physiological stress in the natural body (Butcher et al., 2011). The aortic cusp Young's modulus is 2-10 kPa and 20-100 kPa in the radial and the circumferential directions, respectively (Billiar & Sacks, 2000). Vafaee et al., (2018) studied the mechanical testing on the human cadaveric cryopreserved aortic cusp at a displacement rate of 10 mm/min at 37 °C. The failure load in the circumferential direction was 6.25±0.89 N, and in the radial direction was 0.47±0.35 N. Stradins et al., (2004) performed comparative studies on the aortic and pulmonary valves from 11 human cadaveric hearts. In the circumferential direction, the pulmonary valve has higher tensile strength than the aortic valve. Koch et al., (2010) reported the anisotropy behavior of porcine aortic cusp during diastolic valve conditions up to 10% strain.

Figure 3.The diagram illustrates the pattern of tissue specimen mechanically stretch under (a) uniaxial and (b) biaxial testing

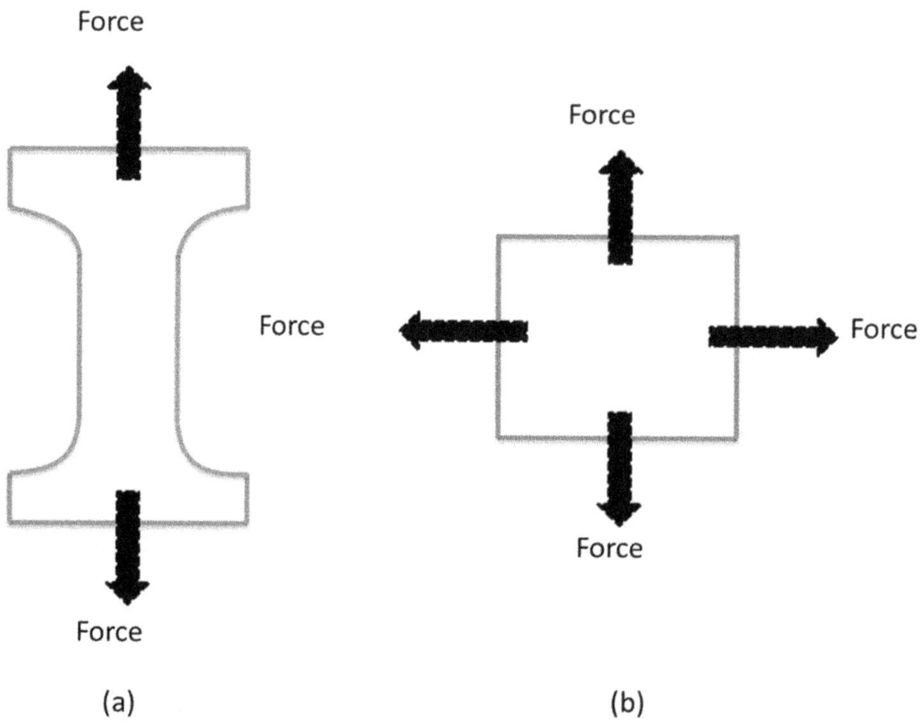

Biaxial Testing

Martin & Sun (2012) compared the biomechanical properties of aortic cusps obtained from human, porcine, and ovine models. The biaxial response shows that aortic cusps exhibit anisotropic behavior. Specifically, the human aortic cusp was stiffer than porcine and ovine cusps, and there was no significant difference between the non-coronary, left coronary, and right coronary cusps within any species. Stella & Sacks (2007) performed the regional separation of fibrosa and ventricularis from the porcine aortic cusp by microdissection technique. The biaxial test result shows that the fibrosa and ventricularis exhibit anisotropic behavior. Similarly, Chaparro et al., (2020) investigated the biaxial testing of mouse aortic cusp, and the study concluded that the cusp is stiffer in the circumferential direction. Billiar & Sacks, (2000) performed a comparative study between the native porcine aortic cusp and glutaraldehyde fixed aortic cusp. The comprehensive study indicates that the native cusps were more extensible than glutaraldehyde-fixed cusps.

Figure 4. Different methods for studying the flexural properties of the aortic cusp. **(a)** *Three-point bending; the sample is fixed between the two brace and the point (P) load applied by the third beam (Merryman et al., 2006);* **(b)** *Cantilever bending; the aortic cusp tissue attached to the fixed end and the point in which (P) bending to the free side with the help of reference post (Mirnajafi et al., 2006);* **(c)** *Stress-strain behavior of aortic cusp in circumferential and radial directions (Li et al., 2001).*

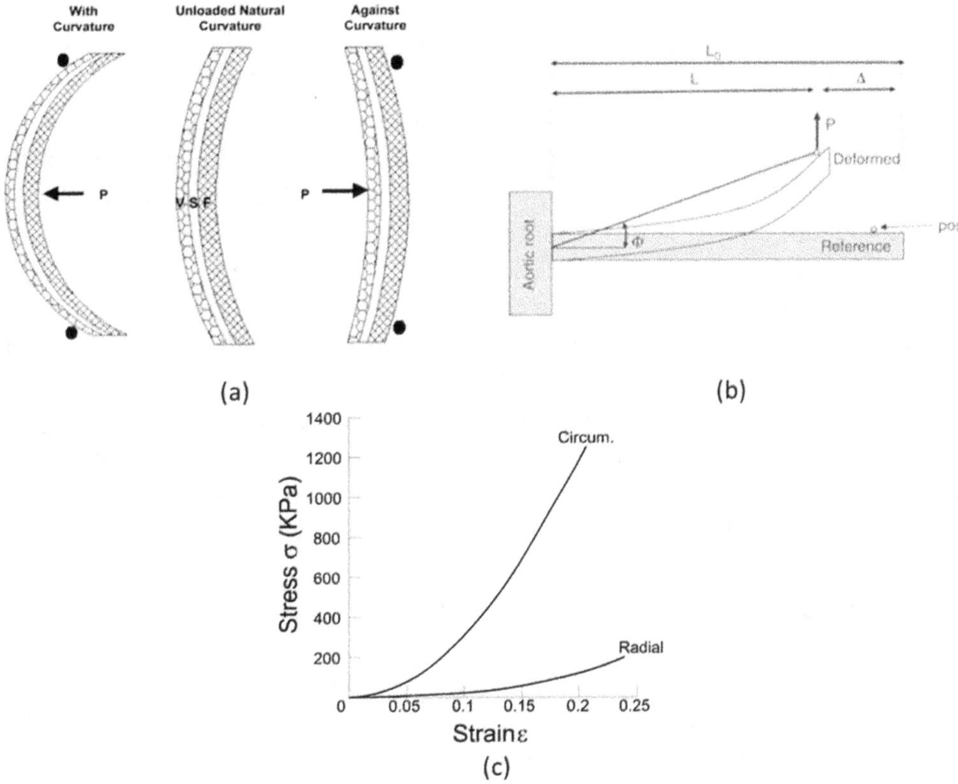

Flexural Testing

Flexure is a distortion method that occurs in the aortic cusp during the cardiac cycle, and it helps to study the biomechanics of the valve (Mirnajafi et al., 2006). Different methods for studying the flexural properties of the aortic cusp are shown in Figure 4. The flexure experiment was performed using three-point bending tests and cantilever bending tests. The flexural properties of the aortic cusp are calculated by using Bernoulli–Euler equation. In most of the flexural study experiments were performed on the central or belly region of the aortic cusps (Brazile et al., 2015; Merryman et al., 2006; Ragaert et al., 2012) These studies reported flexural properties such as extension break, maximum load, and stiffness parameter precisely on the preconditioned porcine aortic cusps. The maximum load is 14.76±4.39 N for the left coronary cusp, 14.45±3.04 N for the right coronary cusp, 16.04±5.68 N for the non-coronary cusp. The stiffness is 7.25±1.48 N/mm on the left coronary cusp, 7.74±1.95 N/mm on the right coronary cusp, 7.64±1.65 N/mm on the non-coronary cusp, respectively. The overall result suggests that there is no significant difference in the flexural study in the cusps. Lovekamp et al., (2006) studied the contribution of GAG to the mechanical properties of the porcine aortic cusp. The flexural rigidity of the cusp determines by bending the specimen in the curvature direction and with and without enzymatic treatment. During the relaxed state, the curvature of the tissue increased, and the stiffness of native and enzyme digested tissue decreased significantly.

Three-Point Bending

The three-point bending mechanism works on the principle of linear beam theory. Using three-point bending testing, the fatigue properties of aortic valve cusps can be obtained. In the three-point bending test, the rectangular specimens were attached with the sleeves and placed between the two stationary rods. The black graphite and cyanoacrylate are used as markers in the specimen to determine the sample's deformation and length (Gloeckner et al., 1999). Brazile et al., (2015) reported the effective bending modulus of the aortic cusp, mainly focused on the momentum-curvature relationship of cusp against the bending and curvature bending in the radial and circumferential directions. Initially, the momentum-curvature relationship is determined by using the Euler-Bernoulli equation. The effective bending modulus of curvature at 0.025 mm, in the circumferential direction, is 231.548±28.467 kPa against the curvature and is 159.169±36.606 kPa with curvature. Whereas in the radial direction, 421.776±72.193 kPa against the curvature, and 97.227±35.124 kPa with curvature. Merryman et al., (2006) studied the effect of cellular contraction on the flexural stiffness of the aortic valve in the circumferential direction. The result shows a 48% increase in cusp stiffness with VIC when bent against curvature. In contrast, curvature direction resulted only in a 5% increase in cusp stiffness with active VIC.

Cantilever Bending

The cantilever bending is based on the principle of simple beam theory. In the test, the aortic root is fixed to the calibrated bending bar and a known load was applied to the cusp using a cantilever beam configuration (Mirnajafi et al., 2006). These experimental conditions almost mimic the deformation occurring in the cusp's commissural region connected to the aortic root. The aortic wall's bending and flexion stresses are caused by the aortic cusps' reversal curvature and flexion (Deck, 1986). Mirnajafi et al., (2006) quantified the stiffness of the porcine aortic cusp in the commissural region. The modulus was greater in the reverse direction than the forward direction at a flexure angle of 30°. Gloeckner et al., (1999) studied the flexural rigidity in the belly region of glutaraldehyde fixed porcine aortic cusp. In the circumferential direction, the stiffness was higher against the cusp curvature.

Viscoelastic Properties

Stella et al., (2007) studied the biomechanical properties of porcine aortic cusp and mitral leaflets. The stress relaxation percentage of the aortic valve in the radial direction is 27.51% and, in the circumferential direction is 33.28%. There are no significant differences between the stress relaxation response of the aortic cusp and the mitral valve leaflets. Tseng et al., (2013) studied the role of GAG in the viscoelastic properties of an aortic cusp by digesting GAG with different concentrations of hyaluronidase. The result indicates that GAG plays a major role in lubricating tissue motion and decreases the cusps stress. Merryman et al., (2009) explored the viscoelastic properties of aortic VIC during diastole conditions using a micropipette aspiration technique. During diastole conditions, the creep was 4.65%, and stress relaxation was 4.39%. Bhatia & Vesely (2005) experimented with the role of GAG on the porcine aortic cusp by using 0.1 M NaOH. The removal of GAG reduces the water content and the viscoelasticity response of the aortic cusp. Similarly, Borghi et al., (2013) used hyaluronidase, chondroitinase, and keratanase enzymes to study the role of GAG on the porcine cusp. The presence and depletion of GAG significantly influence the viscoelastic properties, but there are no changes in the stiffness of cusps before and after enzymatic treatments. Lee et al., (1984) examine the viscoelasticity properties of the porcine native aortic cusp. The cusp is stiffer in the circumferential direction than radial direction as the collagen fibers are alignment is randomly in the radial direction and well-aligned in the circumferential direction.

ETIOLOGY OF AORTIC STENOSIS AND REGURGITATION

In vivo model is essential to study the initiation and progression, molecular mechanism, biomechanics, and cellular interaction during aortic stenosis. The most common animal model to study cardiac aortic valvular diseases are mouse, rabbit, and porcine. This section discusses the causes and progression of aortic stenosis and regurgitation, in vitro model to imitate stenosis conditions and experimental results, and in vivo animal models deliberately to induce calcifications and ex vivo model to create experimental regurgitation results.

Aortic Stenosis

In aortic stenosis, the aortic valve opening becomes thin and inhibits the blood flow from the left ventricle to the aorta. During these conditions, there is an increase in pressure in the left ventricle and aorta to maintain the cardiac output (Weinberg & Kaazempur Mofrad, 2008; Zakikhani et al., 2019). The clinical factor associated with aortic stenosis includes smoking, diabetics, age factor, and hypertension (Otto, 2002). In healthy individuals, VIC is responsible for maintaining and remodeling the extracellular matrix. The patient with a bicuspid aortic wall has more likely to cause calcified aortic stenosis (Weinberg & Kaazempur Mofrad, 2008). One of the reasons for calcified aortic stenosis is that a decrease in the number of VIC in the human aortic valve cusps leads to degeneration of collagen fibers. Another reason is that abnormal mechanical signals in valve cusps lead to VIC dysfunctions (Taylor et al., 2003; Weinberg & Kaazempur Mofrad, 2007).

VICs play a crucial role in the progression of the calcification process (Bogdanova et al., 2019). In order to study aortic stenosis, the VIC cells from the human, porcine, sheep, rat, and bovine models were isolated from the coronary cusps using collagenase enzymes and magnetic assisted cell sorting (Bogdanova et al., 2019; Latif et al., 2015; Rajamannan, 2011; Tseng et al., 2014; Zabirnyk et al., 2020). For calcification studies, the VICs specifically supplemented with osteogenic medium, pro-calcifying medium, to induce the cells calcification process (Rutkovskiy et al., 2017). This medium contains β-glycerophosphate, dexamethasone, and ascorbic acid, essential for calcium crystal formation (Babu et al., 2008). The calcium formation at the molecular level is detected by increasing alkaline phosphatase activity and transcriptional factors like Runx2 and BMP2. At the microscopic level, calcium deposition is detected using alizarin red, Von Kossa staining, and Arsenazo dye (Bogdanova et al., 2019; Chen et al., 2009). Agozzino et al., (1994) studied the gross anatomy of surgically removed aortic cusp from the 210 patients diagnosed with aortic stenosis. The microscopic examination shows calcium deposition, and fibrosis occurs in the cusp. Bogdanova et al., (2019) isolated the VIC from the human calcified aortic cusp and cultured it for 21 days without an osteogenic medium to study the potential of stem cell-like properties. However, the VIC cells cultured for healthy cusp have multipotency compared to the calcified aortic cusp.

The mice model is the most widely used model to study aortic valve stenosis. It has several advantages, including lost cost, easy management, shorter generation time, clone key specific molecule mediators, and matrix Gla protein knockout mice that mimic aortic valve calcification. Another way of studying aortic stenosis is by giving high cholesterol diet plan; this increases valve thickness, activation of osteoblasts, macrophage accumulation, and mineralization (Matsumoto et al., 2010). In human beings, congenital disability occurs due to mutation in the transcriptional regulator of NOTCH 1 (Garg et al., 2005). This NOTCH 1 level is usually expressed in mice during postnatal growth. Deleting a periostin protein-coding gene negatively regulated NOTCH 1 in mice and developed with bicuspid-like morphology (Tkatchenko et al., 2009). The porcine model is an excellent model for studying calcified aortic valvular disease due to its hemodynamics, cusp biomechanical properties, heart anatomy, lipid profile, and lipid metabolism similarity to the human being (Gerrity et al., 2001). A high cholesterol diet plan can induce calcification by feeding animal model supplements with 10-20% fat, 1.5-2% cholesterol, 0.7-1.5% sodium cholate. Another way of studying the calcification process is by bringing the mutation in LDLR and apolipoprotein genes, leading to atherosclerotic lesions without a diet plan.

Aortic Regurgitation

Aortic regurgitation is due to improper closure of the aortic valve leading to backflow of blood. The cause of aortic regurgitation is a combination of factors, including dilation of the aortic root, annulus, and malfunction of valve cusps (Maurer, 2006; Sawaya & Søndergaard, 2018). Clinically, aortic regurgitation is classified as acute and chronic conditions. In acute aortic regurgitation, a sudden increase in volume loading of the left ventricle leads to reduced ejection fraction. This disease condition occurs in infectious endocarditis, disruption of the valve after transcatheter aortic valve replacement surgery. There is a progressive increase in left ventricle stroke volume during chronic aortic regurgitation, causing chamber dilation and hypertrophy. In this condition, elevation occurs in the wall stress of the aortic cusp and the left ventricle response with compensatory eccentric hypertrophy. As a result, the cardiac output is eventually reduced due to eccentric hypertrophy and chamber dilation in the left ventricle. The aortic cusps were repaired by performing Cabrol stitch, re-implantation technique, or ring annuloplasty. However, repair of aortic regurgitation is not widely accepted (Pettersson et al., 2008). Most of the studies performed only on the porcine model for mimicking aortic regurgitation conditions due to its similarity profile with human beings (Dixon et al., 1999; Gerrity et al., 2001). Zhu et al., (2021) studied the aortic regurgitation in a porcine aortic cusp, and the result shows that detaching the two largest cusps can significantly increase the regurgitation volume compared to the controls.

CONCLUSION

Biomechanical characterization studies of the human aortic cusp are limited due to the less availability of fresh specimens. Considerably, there are many biomechanical studies performed on the aortic valve cusps of bovine, porcine, and ovine species. Also, uniaxial and biaxial testing studies dominated the mechanical testing compared to other flexural, three-point bending experiments. Utilizing the existing in vitro, in vivo animal models, and ex vivo models, detailed investigations on aortic stenosis were studied using various approaches. However, in vivo models mimicking the aortic regurgitation conditions is limited. More in vitro, in vivo, and ex vivo calcification studies are required. Studying homeostasis and disease progression in-depth can lead to discovering novel therapies for aortic valve disease.

ACKNOWLEDGMENT

This study has been supported by the National Institute of Technology, Rourkela.

REFERENCES

Adham, M., Gournier, J.-P., Favre, J.-P., De La Roche, E., Ducerf, C., Baulieux, J., Barral, X., & Pouyet, M. (1996). Mechanical Characteristics of Fresh and Frozen Human Descending Thoracic Aorta. *The Journal of Surgical Research*, *64*(1), 32–34. doi:10.1006/jsre.1996.0302 PMID:8806470

Agozzino, L., de Vivo, F., Falco, A., de Luca, L., Schinosa, T., & Cotrufo, M. (1994). Surgical pathology of the aortic valve: Gross and histological findings in 1120 excised valves. *Cardiovascular Pathology*, *3*(3), 155–161. doi:10.1016/1054-8807(94)90024-8 PMID:25990991

Babu, A. N., Meng, X., Zou, N., Yang, X., Wang, M., Song, Y., Cleveland, J. C., Weyant, M., Banerjee, A., & Fullerton, D. A. (2008). Lipopolysaccharide Stimulation of Human Aortic Valve Interstitial Cells Activates Inflammation and Osteogenesis. *The Annals of Thoracic Surgery*, *86*(1), 71–76. doi:10.1016/j.athoracsur.2008.03.008 PMID:18573401

Bermejo, J., Postigo, A., & Baumgartner, H. (2021). The year in cardiovascular medicine 2020: Valvular heart disease. *European Heart Journal*, *42*(6), 647–656. doi:10.1093/eurheartj/ehaa1060 PMID:33388778

Bhatia, A., & Vesely, I. (2005). *The Effect of Glycosaminoglycans and Hydration on the Viscoelastic Properties of Aortic Valve Cusps*. Academic Press.

Billiar, K. L., & Sacks, M. S. (2000). Biaxial mechanical properties of the natural and glutaraldehyde treated aortic valve cusp - Part I: Experimental results. *Journal of Biomechanical Engineering*, *122*(1), 23–30. doi:10.1115/1.429624 PMID:10790826

Bogdanova, M., Zabirnyk, A., Malashicheva, A., Enayati, K. Z., Karlsen, T. A., Kaljusto, M. L., Kvitting, J. P. E., Dissen, E., Sullivan, G. J., Kostareva, A., Stensløkken, K. O., Rutkovskiy, A., & Vaage, J. (2019). Interstitial cells in calcified aortic valves have reduced differentiation potential and stem cell-like properties. *Scientific Reports*, *9*(1), 1–13. doi:10.103841598-019-49016-0 PMID:31506459

Borghi, A., New, S. E. P., Chester, A. H., Taylor, P. M., & Yacoub, M. H. (2013). Acta Biomaterialia Time-dependent mechanical properties of aortic valve cusps : Effect of glycosaminoglycan depletion. *Acta Biomaterialia*, *9*(1), 4645–4652. doi:10.1016/j.actbio.2012.09.001 PMID:22963848

Brazile, B., Wang, B., Wang, G., Bertucci, R., Prabhu, R., Patnaik, S. S., Butler, J. R., Claude, A., Brinkman-Ferguson, E., Williams, L. N., & Liao, J. (2015). On the Bending Properties of Porcine Mitral, Tricuspid, Aortic, and Pulmonary Valve Leaflets. *Journal of Long-Term Effects of Medical Implants*, *25*(1–2), 41–53. doi:10.1615/JLongTermEffMedImplants.2015011741 PMID:25955006

Butcher, J. T., Mahler, G. J., & Hockaday, L. A. (2011). Aortic valve disease and treatment: The need for naturally engineered solutions. *Advanced Drug Delivery Reviews*, *63*(4), 242–268. doi:10.1016/j.addr.2011.01.008 PMID:21281685

Butcher, J. T., & Nerem, R. M. (2006). Valvular endothelial cells regulate the phenotype of interstitial cells in co-culture: Effects of steady shear stress. *Tissue Engineering*, *12*(4), 905–915. doi:10.1089/ten.2006.12.905 PMID:16674302

Chaparro, D., Dargam, V., Alvarez, P., Yeung, J., Saytashev, I., Bustillo, J., Loganathan, A., Ramella-Roman, J., Agarwal, A., & Hutcheson, J. D. (2020). A Method to Quantify Tensile Biaxial Properties of Mouse Aortic Valve Leaflets. *Journal of Biomechanical Engineering*, *142*(10), 1–7. doi:10.1115/1.4046921 PMID:32291440

Chen, J., Ying, C., Yip, Y., Sone, E. D., & Simmons, C. A. (2009). Identification and Characterization of Aortic Valve Mesenchymal Progenitor Cells with Robust Osteogenic Calcification Potential. *American Journal of Pathology*, *174*(3), 1109–1119. doi:10.2353/ajpath.2009.080750 PMID:19218344

Deck, J. D. (1986). Endothelial cell orientation on aortic valve leaflets. *Cardiovascular Research*, *20*(10), 760–767. doi:10.1093/cvr/20.10.760 PMID:3791342

Desai, A., Vafaee, T., Rooney, P., Kearney, J. N., Berry, H. E., Ingham, E., Fisher, J., & Jennings, L. M. (2018). In vitro biomechanical and hydrodynamic characterisation of decellularised human pulmonary and aortic roots. *Journal of the Mechanical Behavior of Biomedical Materials, 79*(September), 53–63. doi:10.1016/j.jmbbm.2017.09.019

Dixon, J. L., Stoops, J. D., Parker, J. L., Laughlin, M. H., Weisman, G. A., & Sturek, M. (1999). Dyslipidemia and vascular dysfunction in diabetic pigs fed an atherogenic diet. *Arteriosclerosis, Thrombosis, and Vascular Biology*, *19*(12), 2981–2992. doi:10.1161/01.ATV.19.12.2981 PMID:10591679

El-Hamamsy, I., Chester, A. H., & Yacoub, M. H. (2010). Cellular regulation of the structure and function of aortic valves. *Journal of Advanced Research*, *1*(1), 5–12. doi:10.1016/j.jare.2010.02.007

Frater, R. W. M., Gong, G., Hoffman, D., & Liao, K. (1992). Endothelial covering of biological artificial heart valves. *The Annals of Thoracic Surgery*, *53*(3), 371–372. doi:10.1016/0003-4975(92)90252-Y PMID:1540049

Garg, V., Muth, A. N., Ransom, J. F., Schluterman, M. K., Barnes, R., King, I. N., Grossfeld, P. D., & Srivastava, D. (2005). Mutations in NOTCH1 cause aortic valve disease. *Nature*, *437*(7056), 270–274. doi:10.1038/nature03940 PMID:16025100

Gerrity, R. G., Natarajan, R., Nadler, J. L., & Kimsey, T. (2001). Diabetes-Induced Accelerated Atherosclerosis in Swine. *Diabetes*, *50*(7), 1654–1665. doi:10.2337/diabetes.50.7.1654 PMID:11423488

Gloeckner, D. G., Bihir, K. L., & Sacks, M. S. (1999). Effects of Mechanical Fatigue on the Bending Properties of the Porcine Bioprosthetic Heart Valve. *ASAIO Journal (American Society for Artificial Internal Organs)*, *45*(1), 59–63. doi:10.1097/00002480-199901000-00014 PMID:9952009

Hasan, A., Ragaert, K., Swieszkowski, W., Selimović, Š., Paul, A., Camci-Unal, G., Mofrad, M. R. K., & Khademhosseini, A. (2014). Biomechanical properties of native and tissue engineered heart valve constructs. *Journal of Biomechanics*, *47*(9), 1949–1963. doi:10.1016/j.jbiomech.2013.09.023 PMID:24290137

Koch, T. M., Reddy, B. D., Zilla, P., & Franz, T. (2010). Aortic valve leaflet mechanical properties facilitate diastolic valve function. *Computer Methods in Biomechanics and Biomedical Engineering*, *13*(2), 225–234. doi:10.1080/10255840903120160 PMID:19657802

Korossis, S. (2018). Structure-Function Relationship of Heart Valves in Health and Disease. In *Structural Insufficiency Anomalies in Cardiac Valves* (pp. 1–38). InTech. doi:10.5772/intechopen.78280

Latif, N., Quillon, A., Sarathchandra, P., McCormack, A., Lozanoski, A., Yacoub, M. H., & Chester, A. H. (2015). Modulation of Human Valve Interstitial Cell Phenotype and Function Using a Fibroblast Growth Factor 2 Formulation. *PLoS One*, *10*(6), e0127844. doi:10.1371/journal.pone.0127844 PMID:26042674

Lee, J. M., Courtman, D. W., & Boughner, D. R. (1984). The glutaraldehyde-stabilized porcine aortic valve xenograft. I. Tensile viscoelastic properties of the fresh leaflet material. *Journal of Biomedical Materials Research*, *18*(1), 61–77. doi:10.1002/jbm.820180108 PMID:6699033

Lerman, D. A., Prasad, S., & Alotti, N. (2015). Calcific Aortic Valve Disease: Molecular Mechanisms And Therapeutic Approaches. *European Cardiology Review*, *10*(2), 108. doi:10.15420/ecr.2015.10.2.108 PMID:27274771

Li, J., Luo, X. Y., & Kuang, Z. B. (2001). A nonlinear anisotropic model for porcine aortic heart valves. *Journal of Biomechanics*, *34*(10), 1279–1289. doi:10.1016/S0021-9290(01)00092-6 PMID:11522307

Liao, J., Joyce, E. M., & Sacks, M. S. (2008). Effects of decellularization on the mechanical and structural properties of the porcine aortic valve leaflet. *Biomaterials*, *29*(8), 1065–1074. doi:10.1016/j.biomaterials.2007.11.007 PMID:18096223

Lindman, B. R., Clavel, M.-A., Mathieu, P., Iung, B., Lancellotti, P., Otto, C. M., & Pibarot, P. (2016). Calcific aortic stenosis. *Nature Reviews. Disease Primers*, *2*(1), 16006. doi:10.1038/nrdp.2016.6 PMID:27188578

Liu, A. C., Joag, V. R., & Gotlieb, A. I. (2007). The Emerging Role of Valve Interstitial Cell Phenotypes in Regulating Heart Valve Pathobiology. *American Journal of Pathology*, *171*(5), 1407–1418. doi:10.2353/ajpath.2007.070251 PMID:17823281

Lovekamp, J. J., Simionescu, D. T., Mercuri, J. J., Zubiate, B., Sacks, M. S., & Vyavahare, N. R. (2006). Stability and function of glycosaminoglycans in porcine bioprosthetic heart valves. *Biomaterials*, *27*(8), 1507–1518. doi:10.1016/j.biomaterials.2005.08.003 PMID:16144707

Martin, C., & Sun, W. (2012). Biomechanical characterization of aortic valve tissue in humans and common animal models. *Journal of Biomedical Materials Research. Part A*, *100A*(6), 1591–1599. doi:10.1002/jbm.a.34099 PMID:22447518

Matsumoto, Y., Adams, V., Jacob, S., Mangner, N., Schuler, G., & Linke, A. (2010). Regular Exercise Training Prevents Aortic Valve Disease in Low-Density Lipoprotein–Receptor–Deficient Mice. *Circulation*, *121*(6), 759–767. doi:10.1161/CIRCULATIONAHA.109.892224 PMID:20124122

Maurer, G. (2006). Aortic regurgitation. *Heart (British Cardiac Society)*, *92*(7), 994–1000. doi:10.1136/hrt.2004.042614 PMID:16775114

Merryman, W. D., Bieniek, P. D., Guilak, F., & Sacks, M. S. (2009). Viscoelastic Properties of the Aortic Valve Interstitial Cell. *Journal of Biomechanical Engineering*, *131*(4), 1–7. doi:10.1115/1.3049821 PMID:19275434

Merryman, W. D., Shadow Huang, H. Y., Schoen, F. J., & Sacks, M. S. (2006). The effects of cellular contraction on aortic valve leaflet flexural stiffness. *Journal of Biomechanics*, *39*(1), 88–96. doi:10.1016/j.jbiomech.2004.11.008 PMID:16271591

Mirnajafi, A., Raymer, J. M., McClure, L. R., & Sacks, M. S. (2006). The flexural rigidity of the aortic valve leaflet in the commissural region. *Journal of Biomechanics*, *39*(16), 2966–2973. doi:10.1016/j.jbiomech.2005.10.026 PMID:16360160

Misfeld, M., & Sievers, H.-H. (2007). Heart valve macro- and microstructure. *Philosophical Transactions of the Royal Society of London. Series B, Biological Sciences*, *362*(1484), 1421–1436. doi:10.1098/rstb.2007.2125 PMID:17581807

Otto, C. M. (2002). Calcification of bicuspid aortic valves. *Heart (British Cardiac Society)*, *88*(4), 321–322. doi:10.1136/heart.88.4.321 PMID:12231576

Pettersson, G. B., Crucean, A. C., Savage, R., Halley, C. M., Grimm, R. A., Svensson, L. G., Naficy, S., Gillinov, A. M., Feng, J., & Blackstone, E. H. (2008). Toward Predictable Repair of Regurgitant Aortic Valves. A Systematic Morphology-Directed Approach to Bicommissural Repair. *Journal of the American College of Cardiology*, *52*(1), 40–49. doi:10.1016/j.jacc.2008.01.073 PMID:18582633

Porras, A. M., van Engeland, N. C. A., Marchbanks, E., McCormack, A., Bouten, C. V. C., Yacoub, M. H., Latif, N., & Masters, K. S. (2017). Robust Generation of Quiescent Porcine Valvular Interstitial Cell Cultures. *Journal of the American Heart Association*, *6*(3). Advance online publication. doi:10.1161/JAHA.116.005041 PMID:28292746

Ragaert, K., De Somer, F., Somers, P., De Baere, I., Cardon, L., & Degrieck, J. (2012). Flexural mechanical properties of porcine aortic heart valve leaflets. *Journal of the Mechanical Behavior of Biomedical Materials*, *13*, 78–84. doi:10.1016/j.jmbbm.2012.04.009 PMID:22842278

Rajamannan, N. M. (2011). Calcific Aortic Valve Disease: Cellular Origins of Valve Calcification. *Arteriosclerosis, Thrombosis, and Vascular Biology*, *31*(12), 2777–2778. doi:10.1161/ATVBAHA.111.237610 PMID:22096095

Rutkovskiy, A., Malashicheva, A., Sullivan, G., Bogdanova, M., Kostareva, A., Stensløkken, K. O., Fiane, A., & Vaage, J. (2017). Valve interstitial cells: The key to understanding the pathophysiology of heart valve calcification. *Journal of the American Heart Association*, *6*(9), 1–23. doi:10.1161/JAHA.117.006339 PMID:28912209

Sawaya, F., & Søndergaard, L. (2018). Aortic Regurgitation. In Diagnosis and Management of Adult Congenital Heart Disease (3rd ed., pp. 387–394). Elsevier. doi:10.1016/B978-0-7020-6929-1.00037-X

Sider, K. L., Blaser, M. C., & Simmons, C. A. (2011). Animal Models of Calcific Aortic Valve Disease. *International Journal of Inflammation, 2011*(Ldl), 1–18. doi:10.4061/2011/364310

Stella, J. A., Liao, J., & Sacks, M. S. (2007). Time-dependent biaxial mechanical behavior of the aortic heart valve leaflet. *Journal of Biomechanics*, *40*(14), 3169–3177. doi:10.1016/j.jbiomech.2007.04.001 PMID:17570376

Stella, J. A., & Sacks, M. S. (2007). On the Biaxial Mechanical Properties of the Layers of the Aortic Valve Leaflet. *Journal of Biomechanical Engineering*, *129*(5), 757–766. doi:10.1115/1.2768111 PMID:17887902

Stradins, P., Lacis, R., Ozolanta, I., Purina, B., Ose, V., Feldmane, L., & Kasyanov, V. (2004). Comparison of biomechanical and structural properties between human aortic and pulmonary valve. *European Journal of Cardio-Thoracic Surgery*, *26*(3), 634–639. doi:10.1016/j.ejcts.2004.05.043 PMID:15302062

Sucosky, P., Balachandran, K., Elhammali, A., Jo, H., & Yoganathan, A. P. (2009). Altered shear stress stimulates upregulation of endothelial VCAM-1 and ICAM-1 in a BMP-4- and TGF-β1-dependent pathway. *Arteriosclerosis, Thrombosis, and Vascular Biology*, *29*(2), 254–260. doi:10.1161/ATVBAHA.108.176347 PMID:19023092

Taylor, P. M., Batten, P., Brand, N. J., Thomas, P. S., & Yacoub, M. H. (2003). The cardiac valve interstitial cell. *The International Journal of Biochemistry & Cell Biology*, *35*(2), 113–118. doi:10.1016/S1357-2725(02)00100-0 PMID:12479860

Tkatchenko, T. V., Moreno-Rodriguez, R. A., Conway, S. J., Molkentin, J. D., Markwald, R. R., & Tkatchenko, A. V. (2009). Lack of periostin leads to suppression of Notch1 signaling and calcific aortic valve disease. *Physiological Genomics*, *39*(3), 160–168. doi:10.1152/physiolgenomics.00078.2009 PMID:19723774

Tsang, H. G., Cui, L., Farquharson, C., Corcoran, B. M., Summers, K. M., & Macrae, V. E. (2018). Exploiting novel valve interstitial cell lines to study calcific aortic valve disease. *Molecular Medicine Reports*, *17*(2), 2100–2106. doi:10.3892/mmr.2017.8163 PMID:29207136

Tseng, B., Balaoing, L. R., Grigoryan, B., Raphael, R. M., Killian, T. C., Souza, G. R., & Grande-Allen, K. J. (2014). A three-dimensional co-culture model of the aortic valve using magnetic levitation. *Acta Biomaterialia*, *10*(1), 173–182. doi:10.1016/j.actbio.2013.09.003 PMID:24036238

Tseng, K., Kim, E. J., Connell, P. S., Ayoub, S., Shah, J. V., & Grande-Allen, K. J. (2013). The Tensile and Viscoelastic Properties of Aortic Valve Leaflets Treated with a Hyaluronidase Gradient. *Cardiovascular Engineering and Technology*, *4*(2), 151–160. doi:10.100713239-013-0122-1

Tseng. (2011). Elastic fibers in the aortic valve spongiosa: A fresh perspective on its structure and role in overall tissue function. *Acta Biomaterialia*, *7*(5), 2101–2108. doi:10.1016/j.actbio.2011.01.022

Vafaee, T., Thomas, D., Desai, A., Jennings, L. M., Berry, H., Rooney, P., Kearney, J., Fisher, J., & Ingham, E. (2018). Decellularization of human donor aortic and pulmonary valved conduits using low concentration sodium dodecyl sulfate. *Journal of Tissue Engineering and Regenerative Medicine*, *12*(2), 841–853. doi:10.1002/term.2391 PMID:27943656

Vesely, I. (1997). The role of elastin in aortic valve mechanics. *Journal of Biomechanics*, *31*(2), 115–123. doi:10.1016/S0021-9290(97)00122-X PMID:9593204

Weinberg, E. J., & Kaazempur Mofrad, M. R. (2007). Transient, Three-dimensional, Multiscale Simulations of the Human Aortic Valve. *Cardiovascular Engineering (Dordrecht, Netherlands)*, *7*(4), 140–155. doi:10.100710558-007-9038-4 PMID:18026835

Weinberg, E. J., & Kaazempur Mofrad, M. R. (2008). A multiscale computational comparison of the bicuspid and tricuspid aortic valves in relation to calcific aortic stenosis. *Journal of Biomechanics*, *41*(16), 3482–3487. doi:10.1016/j.jbiomech.2008.08.006 PMID:18996528

Weston, M. W., LaBorde, D. V., & Yoganathan, A. P. (1999). Estimation of the Shear Stress on the Surface of an Aortic Valve Leaflet. *Annals of Biomedical Engineering*, *27*(4), 572–579. doi:10.1114/1.199 PMID:10468241

Weston, M. W., & Yoganathan, A. P. (2001). Biosynthetic Activity in Heart Valve Leaflets in Response to In Vitro Flow Environments. *Annals of Biomedical Engineering*, *29*(9), 752–763. doi:10.1114/1.1397794 PMID:11599583

Yperman, J., De Visscher, G., Holvoet, P., & Flameng, W. (2004). Molecular and functional characterization of ovine cardiac valve-derived interstitial cells in primary isolates and cultures. *Tissue Engineering*, *10*(9–10), 1368–1375. doi:10.1089/ten.2004.10.1368 PMID:15588397

Zabirnyk, A., Perez, M. del M., Blasco, M., Stensløkken, K., Ferrer, M. D., Salcedo, C., & Vaage, J. (2020). A Novel Ex Vivo Model of Aortic Valve Calcification. A Preliminary Report. *Frontiers in Pharmacology*, *11*(December), 1–7. doi:10.3389/fphar.2020.568764 PMID:33390945

Zakikhani, P., Ho, R., Wang, W., & Li, Z. (2019). Biomechanical assessment of aortic valve stenosis: Advantages and limitations. *Medicine in Novel Technology and Devices*, *2*(September), 100009. doi:10.1016/j.medntd.2019.100009

Zhou, J., Hu, S., Ding, J., Xu, J., Shi, J., & Dong, N. (2013). Tissue engineering of heart valves: PEGylation of decellularized porcine aortic valve as a scaffold for in vitro recellularization. *Biomedical Engineering Online*, *12*(1), 87. doi:10.1186/1475-925X-12-87 PMID:24006837

Zhu, Y., Imbrie-Moore, A. M., Paulsen, M. J., Priromprintr, B., Park, M. H., Wang, H., Lucian, H. J., Farry, J. M., & Woo, Y. J. (2021). A Novel Aortic Regurgitation Model from Cusp Prolapse with Hemodynamic Validation Using an Ex Vivo Left Heart Simulator. *Journal of Cardiovascular Translational Research*, *14*(2), 283–289. doi:10.100712265-020-10038-z PMID:32495264

Chapter 10
Fluid Mechanics in Arterial Diseases: Computational Study

Dipak Kumar Mandal
Department of Mechanical Engineering, College of Engineering and Management, Kolaghat, India

Partha Goswami
https://orcid.org/0000-0001-8267-3652
Kolaghat Thermal Power Station, West Bengal Power Development Corporation, India

Nirmalendu Biswas
Department of Power Engineering, Jadavpur University, Salt Lake, India

Nirmal K. Manna
Department of Mechanical Engineering, Jadavpur University, Kolkata, India

ABSTRACT

With the advancement of computer power, computational solution is considered as a useful predictive tool for study of the blood flow through a diseased artery. The blood flow is governed by the continuity and Navier-Stokes equations. Womersley number and Reynolds number are physiologically important non-dimensional flow parameters and have existence in the non-dimensional governing equations. In the computational solution, the mathematical flow modelled is established considering some boundary conditions, and then it is solved by numerical simulations. The governing equations are converted to discretised form and then solved numerically by readily available commercial CFD software and/or by in-house developed CFD code using appropriate algorithms. Grid independence test and validation of CFD model are the crucial parts of computational solutions. The chapter delivers the knowledge on impact of fluid mechanics on arterial diseases and computational solution techniques.

DOI: 10.4018/978-1-7998-9078-2.ch010

INTRODUCTION

Motivation on blood flow study through the diseased artery and to write a book chapter is felt by the need in obtaining a better understanding of the influence of flow phenomena on initialization, progression, and formation of atherosclerosis. It is hoped that such type of chapter may provide the comparative necessary information to achieve the basic fluid mechanical phenomena in the arterial circulation system. This chapter helps to enhance the knowledge of hemodynamics reference to cardiovascular response, which may help to design of prosthetic devices rationally. With the increasing performances of computer hardware and software, at present days, a computational study is being increasingly used in biomedical research of diseased arteries. Understanding flow dynamics and flow disorders due to flow restriction called stenosis is essential to study yet now numerically and experimentally. The disease is due to the formation of stenosis commonly called atherosclerosis. It is a complex problem as the artery is elliptical, elastic, porous; the shape of the stenosis is asymmetrical; inlet flow is pulsatile, blood is non-Newtonian and heterogeneous, blood property varies concerning human beings. This study of flow dynamics and hemodynamics parameters may be solved by CFD (due to the invention of commercial software) by using high-speed computer.

Last few decades, researchers are working on this area numerically or experimentally. Among them, Zendehbudi and Moayeri have compared the velocity distributions, streamline contours and wall shear stresses for physiological flow and pulsatile flow through an axisymmetrical constricted (61% area) artery. (Zendehbudi & Moayeri, 1999) They have performed the relevant numerical computations considering the blood as Newtonian and laminar flow. The constitutive equations have been solved by using the SIMPLER (Semi-Implicit Method for Pressure Linked Equations Revised) algorithm (Patankar, 1980). Apart from that, they have also discussed the pressure drop across 56% area reduction (PR=33.33) of the cosine-shaped restriction with for steady flow with Reynolds number varying from 50 to 250 to validate their results with the earlier experimental observation (Young & Tsai, 1973), which was found to be having good agreement. Lee and Xu have analysed the flow behaviour of pulsatile flow through trapezoidal-shaped (45% restriction by area) considering the fluid to be incompressible and Newtonian. (Lee & Xu, 2002) They have used two separate commercial codes CFX 4.2 and ABAQUS7 for the analysis. They have investigated velocity profiles, wall shear stress for the rigid and compliant tube. From their wall shear stress representation, it is obvious that the wall shear stress (*WSS*) curves for the rigid and compliant tube are identical. Pressure and flow separation is investigated numerically for different shaped restrictions (Mandal & Chakraborty, 2007a, 2007b, 2007c).

Reynolds-averaged Navier-Stokes approach is used for modelling of pulsatile and turbulent flow by Varghese and Frankel (2003) in stenotic vessels. CFD, Ansys FLUENT has been used and it has been has noted that the minimum wall shear stress occurs at distal to the stenosis. The effect of symmetrical and asymmetrical bell-shaped stenosis has been investigated numerically on haemodynamic parameters for the progression of the disease in steady and pulsatile situations. (Mandal et al., 2010b, 2011) This study revealed that asymmetrical-shaped stenosis predicts higher impact on flow characteristics as observed from their study. Two-dimensional rigid models by assuming unsteady, incompressible, and homogeneous blood flow with double stenosis is also solved numerically (Rabby et al., 2014), where cosine-shaped stenosis, inlet sinusoidal pulsatile laminar flow conditions has been chosen. The effect of stricture length of stenosis and its flow circulation is studied numerically and it has been found that restriction is the main component of disease progression. (Mandal & Chakraborty, 2008, 2009)

The important hemodynamic parameters, *WSS* and oscillatory shear index (*OSI*) have been examined through realistic stenosis both experimentally and numerically for steady flow conditions (Gupta & Agrawal, 2015). They have shown that the developed *WSS* at the stenotic zone for the irregular model is different relative to the regular stenosis. For the irregular shaped stenotic case, the oscillations of *WSS* at the post-stenotic zone are very strong in comparison to the regular stenosis case. Blood flow study for double bell-shaped stenosed coronary arteries with the impact of one stenosis with the other is studied numerically (Mandal et al., 2010a, 2010b, 2011a). Mamun et al. have simulated numerically the effect of physiological pulsatile flow to investigate flow characteristics considering both Newtonian as well as non-Newtonian fluid and compared the results for Newtonian model, Carreau model, and cross model. (Mamun et al., 2016) They have observed that the cross model gives better results than other models. Pulsatile CFD simulations has been performed considering Womersley number as 2.065 to describe the pulsatile character of flow. (Stiehm et al., 2017) They have concluded that the steady-state simulations can be used for analyses while time-averaged values are only considered. They have also performed two different transient CFD simulations using a physiologic waveform at inlet condition. They have investigated time-averaged wall shear stress for hemodynamic analyses.

Mass transport is the phenomenon that dictates plaque formation from the perspective of initiation of the disease. Many people are working in this area. LDL transport and its accumulation rate at different physiological conditions are investigated (Santra et al., 2017, 2018, 2020). They have shown a significant effect on initiation of the diseases that depend on porosity of wall, percentage of LDL, and hyperthermia conditions.

In this chapter, an attempt has been made to show computational study of pulsatile flow through bell-shaped stenosed artery at the stenotic zone and post-stenotic zone for the Reynolds numbers from 50 to 200, Womersley numbers ranging from 5 to 12.5, and the percentage of restriction from 30% to 70% (by diameter). Three different pulsatile profiles, i.e., simple pulsatile profile, physiological pulsatile profile, and realistic pulsatile profile, have been considered at the inlet of considered geometries of the modelled artery. This computational approach to study the biological problem will help the researchers to get an insight into the flow physics of atherosclerotic arteries and further study can be obtained in realistic artery with respect to the initiation, progression, and formation of the disease, atherosclerosis.

FLUID MECHANICS IN ARTERIAL DISEASES AND HEAMODYNAMICS

Stenosis is the narrowing of an artery by forming a restriction in the inner layer of the artery. The process of atherosclerosis called cardiovascular disease happens due to build-up and infiltration of lipid streams and artery walls. This plaque is formed by cholesterol, calcium, and other substances. Human beings cannot feel small plaque deposition. People can feel the disease at higher restriction in the artery. Arterial diseases can lead to various complications like myocardial infarction, stroke, ulceration, thrombosis, and aneurysm, thus affecting the human life span and quality of life of a large segment of the population. Therefore, understanding the initiation and formation of atherosclerosis in the human arteries is of critical importance.

Hemodynamics is the study of blood flow dynamics; here the blood flow is associated with the stenosed artery. Many fluid mechanical factors are connected to the initiation and formation of cardiovascular disease in the artery. The viscosity of blood and diameter of the conduit of blood flow depends on the physiological condition of the different patients and the location of the artery in the arterial system. Thus, Reynolds numbers are different for different patients and different arteries to be concerned. For a particular artery segment of a particular human being, Womersley number solely depends on pulsatile flow frequency. Womersley number increases or decreases with increase or decrease in pulsatile flow frequency respectively. Pulsatile flow frequency varies among different human beings. Pulsatile flow frequency also varies with the different conditions of human beings such as physical exertion, emotional exertion, smoking, drinking alcohol, some types of illness, etc. So, understanding of the effect of Womersley number on blood flow characteristics is important for a better pathological interpretation.

The important hemodynamic parameters are wall pressure, streamline contour for representing recirculation zone, peak and low (*WSS*), and *OSI* to represent oscillation in wall shear stress. The importance of and physiological significance of all parameters are essential to know before initiating any research work in this area. The wall pressure has a great role in developing atherosclerosis plaque. At low pressure, a suction action is exerted in this zone that may tear the artery by damaging the endothelium layer, in turn, initiating the formation of plaque development (Gessner, 1973). The formulation of flow separation and development of flow recirculation zone in the flow field is considered relevant for the formulation of plaque deposition. At this recirculation zone, the flow stream stagnates and recirculates that in turn raises the concentration of blood constituents, mostly lipid and fibrin (Tu et al., 1992). The trapping of lipid particles aggravates the mass transport process through the arterial wall, which may eventually coalesce to form atheromatous plaque. This raises the stenosis and increases the restriction. *WSS* is another parameter to control the fluid dynamic characteristics for interaction between blood and endothelium layer. The magnitude of WSS is determined numerically from velocity distribution since it is not possible to measure in vivo or in vitro. Both low and high wall shear stress are responsible for the initiation and progression of the disease. At high stress, it damages the arterial wall (Fry, 1968, 1969) that initiates the plaque deposition, whereas low *WSS* progresses the deposition. Low stress at the separation zone is dangerous and aggravates plaque endothelium deposition by rising mass transport through the wall (Caro et al., 1971). The oscillation in wall shear stress in an artery creates malfunction in the endothelium layer for the progression of the disease. It enhances the intimal thickening as an oscillation in wall shear stress rises. The effect is measured by an index called OSI (Ku et al., 1985).

Time-averaged and non-dimensional hemodynamic parameters are considered in this study for considering pulsatile inlet.

The time-averaged wall pressure is defined as follows:

$$\overline{p_w^*} = \frac{1}{T}\int_0^T p_w dt \tag{1}$$

In Eq. (1) the wall pressure is expressed as time-averaged normal force of the flow field acts on the arterial wall per unit area.

The *WSS* is defined as follows:

$$\overline{\tau_w^*} = \frac{1}{T}\int_0^T \tau_w dt \tag{2}$$

In Eq. (2) the *WSS* is expressed as the time average of the flow field force acts on the arterial wall per unit area.

The transient nature of *WSS* is denoted by the Oscillatory shear index (*OSI*). This index denotes the pulsatility and its magnitude varies from 0 and 0.5. This helps to locate separation and reattachment points. The OSI is defined as follows in Eq. (3):

$$OSI = \frac{1}{2}\left[1 - \frac{\left|\int_0^T \tau_w dt\right|}{\int_0^T \left|\tau_w\right| dt}\right] \tag{3}$$

Figure 1. Computation domain of artery model

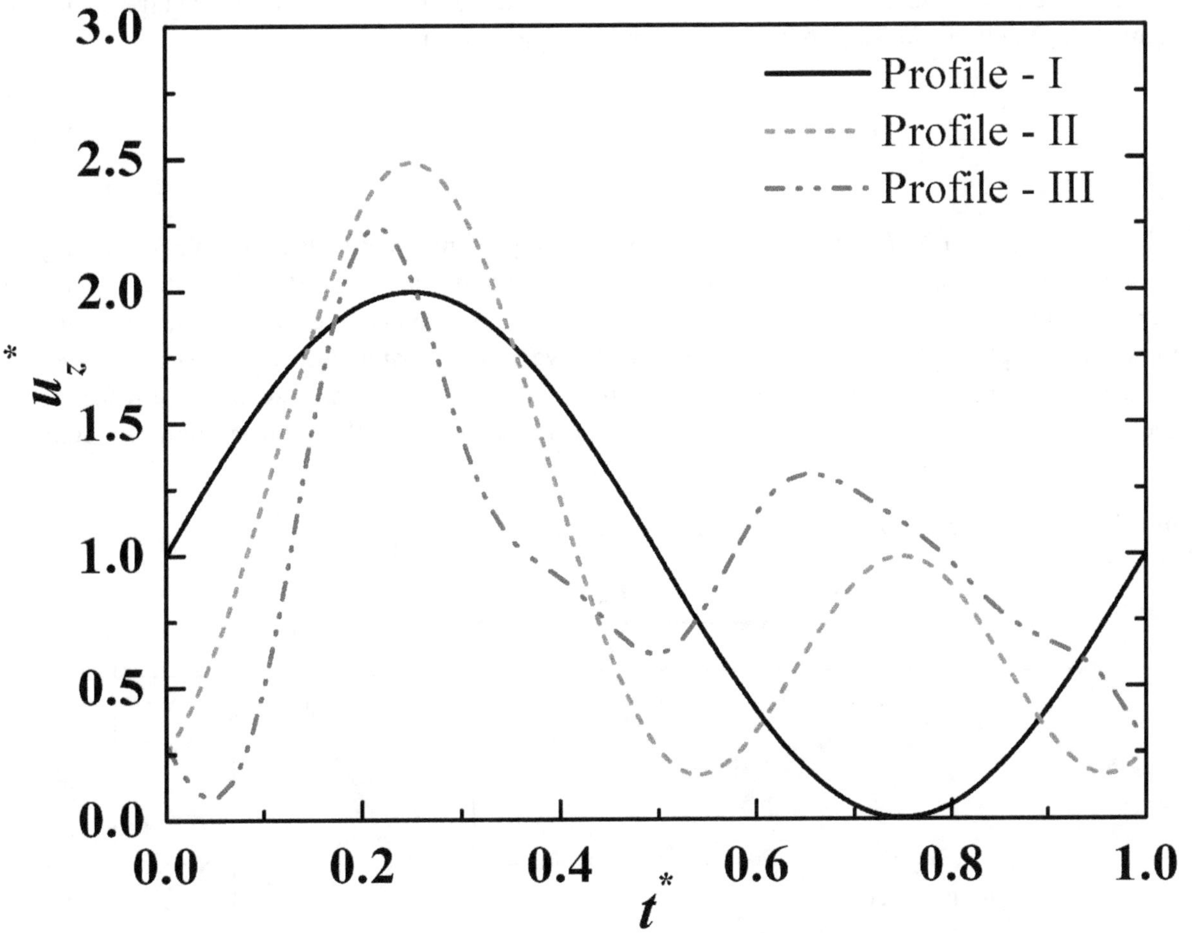

Modelling of Stenosed Artery

The shape of stenosis is difficult to predict that does not follow any rule of geometry. The realistic shape may be like the Gaussian (bell-shaped), or cosine or sinusoidal geometry of constriction. The shape is asymmetrical in the realistic artery. In the present study, axially symmetric but radially asymmetric stenosis has been shown for better understanding. The curvature of the stenosis is chosen as bell-shaped Gaussian distribution profile.

The computational domain is illustrated in Figure 1 (Goswami et al., 2014). This bell-shaped stenosis (Misra and Shit, 2006) geometry is modelled as the following equation.

$$r = 0.5 - h_f \exp\left(-\frac{4m^2 z^2}{L_s^2}\right) \tag{4}$$

Where, m is a parametric constant, which defines the shape of stenosis. In the present study, the value of m is considered as *1*. In general, the degree of stenosis is commonly defined as percentage occlusion by diameter, (Wootton and Ku, 1999), followed by the equation (5).

$$PR = \frac{D - (D/2 - h_f)}{D} \times 100\% \tag{5}$$

The height of stenosis (h_f) rises with an increase in the percentage of restriction. Without changing the distance between the start and throat of stenosis (L_{si}), this is considered that the distance between the throat of stenosis and end of stenosis (L_{se}) increases as the percentage of restriction increases. In this present book chapter, the increment of the distance between the throat and the end of stenosis is equal to the increment of the height of stenosis for all considered variations in the percentage of restrictions. These variations of shape in the percentage of restriction and stenosis length have been shown in Figure 2.

Figure 2. Stenosis models for different percentage of restrictions

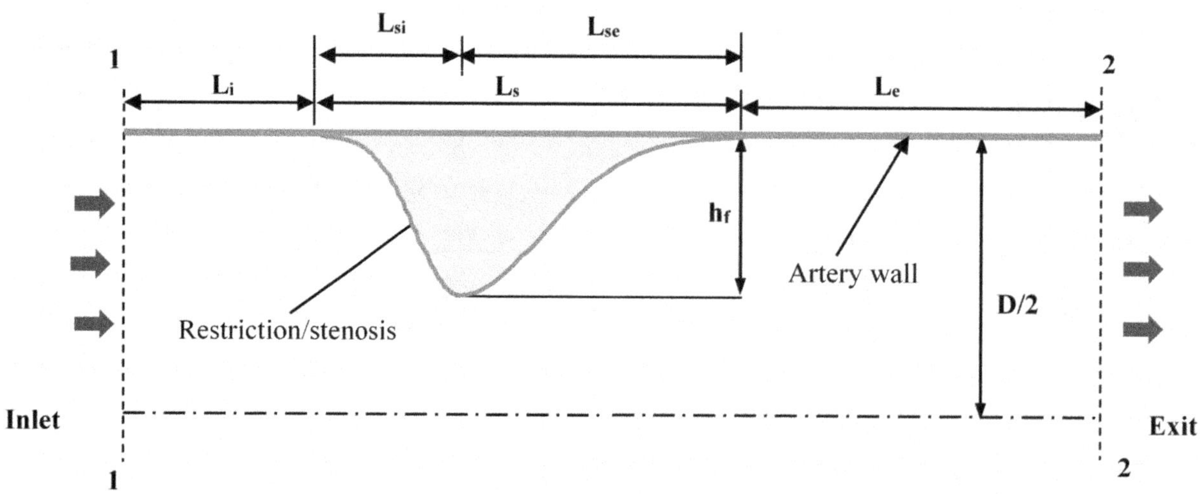

In this work, the percentage of restrictions has been considered to be 30%, 40%, 50%, 60%, and 70% by diameter. The dimensions of stenosis height and stenosis diameter for different stenosis models, in terms of different percentages of restrictions, in the computational domain are shown in Table 1.

Table.1 Different stenosis models with percentage of restrictions

Dimensions	PR 30%	PR 40%	PR 50%	PR 60%	PR 70%
h_f	0.15	0.2	0.25	0.3	0.35
D	1	1	1	1	1
L	200	200	200	200	200
L_i	49.9	49.9	49.9	49.9	49.9
L_S	0.2	0.25	0.3	0.35	0.4
L_{si}	0.1	0.1	0.1	0.1	0.1
L_{se}	0.1	0.15	0.2	0.25	0.3

COMPUTAIONAL SOLUTIONS

With the advancement of computer power, a computational solution is considered a useful predictive tool in studying the blood flow through the diseased artery. Any in-house code or any commercial software like ANSYS, COMSOL can be used to address the problem. The blood flow is governed by the continuity and momentum equations. Womersley number and Reynolds number are physiologically important non-dimensional flow parameters, have existed in the non-dimensional governing equations. In the computational solution, the mathematical flow modelled is established considering some boundary conditions, and then it is solved by numerical simulations. The governing equations are converted to discretized form and then solved numerically by readily available commercial CFD software and or by in-house developed CFD code using different algorithms. The grid independence test and validation of CFD model are the crucial parts of the computational solution. The modelling of the stenosed artery for different percentages of restriction and pulsatile flow simulation of different flow parameters and other features of fluid flow characteristics have been elaborately discussed in this chapter. Following these procedural steps, simulated results have been presented in the subsequent chapter with different focuses.

Assumptions

As mentioned artery is elastic, porous, multiple layers wall, the shape is not circular, presence of different constituents in blood, asymmetrical shaped restriction, and flow is too complicated, therefore, it is difficult to model with the realistic model. So far, almost all the simulations have been carried out under various simplified assumptions. The present numerical study is carried out under the situations of a homogeneous, incompressible, and Newtonian fluid through axially symmetric rigid stenosis. The flow of blood has been considered as two-dimensional, laminar, and pulsatile flow.

Flow of blood usually can be considered as a Newtonian fluid in large arteries (Ku, 1997, Fung, 1997). Cho and Kensey have shown the wall pressure drop at the constricted area and the WSS of blood flow due to consideration of blood as Newtonian fluid are more or less asymptotic with the same due to consideration of blood as non-Newtonian fluid. (Cho & Kensey, 1991) Ishikawa et al. have noted that with the consideration of non-Newtonian property, the strength of the vortex and the vortex size is slightly different from that of consideration of Newtonian property in the analysis. (Ishikawa et al., 1998) Therefore, Newtonian fluid approximation for blood may be regarded as an acceptable condition.

The development of plaque in atherosclerosis in arteries becomes calcified with time that reduces the elastic property of its wall (Liu et al., 2004), therefore rigid boundary may be assumed. However, the effects of elasticity on the results are very small, so an assumption of rigid tube flow is reasonable (Ku, 1997), although the elastic arterial wall affects the flowing blood and a significant impact on the fluid flow dynamics. Therefore, the consideration of rigid wall of an artery in the model is an approximation.

According to some experiments, the maximum Reynolds number can be taken about 500 for stenotic flow that can sustain the flow to be axisymmetric. (Khalifa & Giddens, 1981) It is also to be noted that the arteries are not straight throughout the artery network. Thus the axially symmetric flow can be assumed as a good approximation.

The blood flow in the blood vessel, especially, the flow through a narrow rigid tube can be considered as laminar flow. (Ganong, 2001) The physiological flow range of Reynolds number for the human artery is from 100 to 400, which is considered to be in the laminar zone. (Andersson et al., 2000, Back and Banerjee, 2000)

The diameter of real blood cells in blood flow is about 8×10^{-4} cm. If the vessel diameter is greater than 10^{-3} cm, the homogeneity assumption is valid. The effect red blood cells may be ignored for human arteries as the diameter varies from 0.05-0.3 cm (Lee & Fung, 1970).

Different Pulsatile Flows through Artery

Many investigators have done numerical studies considering blood flow as pulsatile at the inlet of a modelled artery having stenosis and they have compared the flow characteristics for pulsatile flow. Different pulsatile flow profiles are observed in the human arteries depending upon the position in the arterial network, condition of the human body (e.g. physical exertion), and individual to individual (Long et al., 2001, Mills et al., 1970, McDonald, 1974 and Zendehbudi & Moayeri, 1999).

In most of the experimental and numerical studies for blood flow through the artery, researchers have assumed a simple harmonic waveform at the inlet of the stenosis model (Ahmed & Giddens, 1984, Buchanan et al., 2000, Cassanova & Giddens, 1978, Lee & Xu, 2002 and Ojha et al., 1989). In this chapter, the first considered pulsatile flow condition (profile-I) represents simple pulsatile flow, which is defined by the following equation (6).

Profile-I (simple pulsatile flow):

$$u^*_z = 1 + \sin \omega t^* \tag{6}$$

In the arterial system, the waveform is a probable physiological pulsatile flow profile at a downstream location. (Sherwin & Blackburn, 2005) In this present work, the second considered pulsatile flow condition (profile-II) represents physiological pulsatile flow, which is defined by the equation (7) below.

Profile-II (physiological pulsatile flow):

$$u^*_z = 1 + 0.75 \sin \omega t^* - 0.75 \cos 2\omega t^* \tag{7}$$

The third profile represents realistic pulsatile flow condition (profile-III), which is noted in the human right coronary artery (Wiwatanapataphee et al., 2006 and Goswami et al., 2019). The profile-III is defined by the equation (8) below.

Profile-III (realistic pulsatile flow):

$$\begin{aligned} u^*_z =& 1 + 0.29244\cos(\omega t^* + 4.027) - 0.5908\cos(2\omega t^* + 6.509) + 0.2726\cos(3\omega t^* + 1.913) \\ &+0.1980\cos(4\omega t^* + 1.461) + 0.1124\cos(5\omega t^* + 0.074) \end{aligned} \tag{8}$$

The variations of axial velocities for different profiles have been shown in Figure 3.

Figure 3. Distribution of Axial velocities for different pulsatile profiles

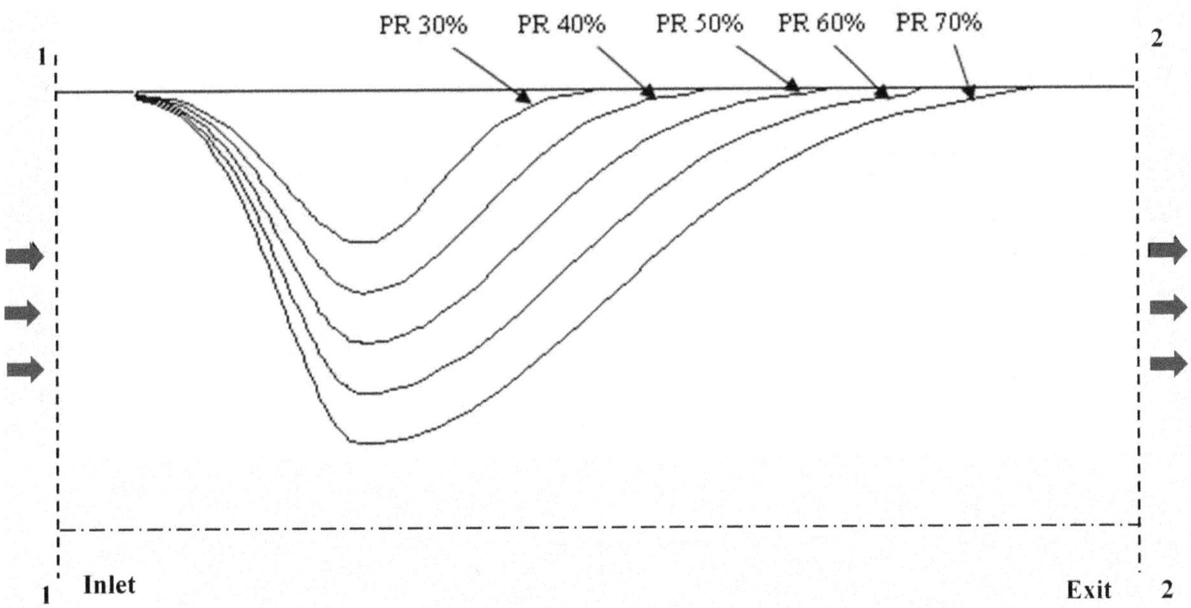

Governing Equations

The different conservation equations in cylindrical co-ordinates (*r*,*z*) can be written as follows:

Continuity equation:

$$\frac{1}{r}\frac{\partial\left(ru_r\right)}{\partial r} + \frac{\partial u_z}{\partial z} = 0 \tag{9}$$

r-direction momentum equation:

$$\frac{\partial u_r}{\partial t} + u_r \frac{\partial u_r}{\partial r} + u_z \frac{\partial u_r}{\partial z} = \frac{1}{\rho}\frac{\partial p}{\partial r} + \frac{\mu}{\rho}\left[\frac{\partial}{\partial r}\left(\frac{1}{r}\frac{\partial\left(r u_r\right)}{\partial r} + \frac{\partial^2 u_r}{\partial z^2}\right)\right] \tag{10}$$

z-direction momentum equation:

$$\frac{\partial u_z}{\partial t} + u_r \frac{\partial u_z}{\partial r} + u_z \frac{\partial u_z}{\partial z} = -\frac{1}{\rho}\frac{\partial p}{\partial z} + \frac{\mu}{\rho}\left[\frac{\partial}{\partial r}\left(\frac{1}{r}\frac{\partial\left(r u_z\right)}{\partial r} + \frac{\partial^2 u_z}{\partial z^2}\right)\right] \tag{11}$$

The above governing equations, (9-11) are nondimensionalized by the following dimensionless variables:

Lengths: $r^*=r/D$, $L^*=L/D$, $z^*=z/D$, $t^*=t/T$.

Velocities: $u_r^* = u_r/U$, $u_z^* = u_z/U$

Pressure: $p^*=p/\rho U2$.

The non-dimensional equations are:

$$\frac{1}{r^*}\frac{\partial\left(r^* u_r^*\right)}{\partial r^*} + \frac{\partial u_z^*}{\partial z^*} = 0 \tag{12}$$

$$\frac{1}{\pi}\frac{W_o^2}{\mathrm{Re}}\frac{\partial u_r^*}{\partial t^*} + u_r^* \frac{\partial u_r^*}{\partial r^*} + u_z^* \frac{\partial u_r^*}{\partial z^*} = -\frac{\partial p^*}{\partial r^*} + \frac{2}{\mathrm{Re}}\left[\frac{\partial^2 u_r^*}{\partial r^{*2}} + \frac{\partial^2 u_r^*}{\partial z^{*2}} + \frac{1}{r}\frac{\partial u_r^*}{\partial r^*}\right] \tag{13}$$

$$\frac{1}{\pi}\frac{W_o^2}{\mathrm{Re}}\frac{\partial u_z^*}{\partial t^*} + u_r^* \frac{\partial u_z^*}{\partial r^*} + u_z^* \frac{\partial u_z^*}{\partial z^*} = -\frac{\partial p^*}{\partial r^*} + \frac{2}{\mathrm{Re}}\left[\frac{\partial^2 u_z^*}{\partial r^{*2}} + \frac{\partial^2 u_z^*}{\partial z^{*2}} + \frac{1}{r}\frac{\partial u_z^*}{\partial r^*} - \frac{u_z^*}{r^{*2}}\right] \tag{14}$$

Where *Wo* is the Womersley number, as described by equation (15) and *Re* is the Reynolds number, as described by equation (16)

$$W_o = R\sqrt{\frac{\omega}{\nu}} \tag{15}$$

$$\mathrm{Re} = \frac{\rho U D}{\mu} \tag{16}$$

Where ω is the radial frequency $2\pi/T$. In this work, four different values of Womersley numbers as 4.0, 7.5, 10.0, and 12.5 from different physiological conditions of the body have been considered in this study (Caro et al., 1978 and Tutty, 1992). In this chapter, the Reynolds number is considered as 50, 100, 150, and 200 respectively.

Boundary Conditions

For numerical modelling of laminar pulsatile flow, it is sufficient to consider Equations (12-14) along with the necessary boundary conditions. For an incompressible fluid, the velocity boundary conditions need to be specified all along the domain boundary. Four different boundary conditions are applied to the considered problem of this work. The boundary conditions are stated as bellows.

1. On the solid wall of the artery, both the components of velocity are zero because no-slip and impregnability conditions on the solid surface have been applied. These may be expressed mathematically as follows:-

$$u_r^* = 0,\ u_z^* = 0$$

2. Axial velocity at the inlet has been applied as pulsatile flow condition (types of considered pulsatile flow have been specified in section of "Different Pulsatile Flows through Artery")

$$u_z^* = specified \text{ and } u_r^* = 0,$$

3. Fully developed flow condition at the exit has been assumed and hence gradients have been set to zero, i.e.

$$\partial u_z^* / \partial z^* = 0,\ \partial u_r^* / \partial z^* = 0$$

4. On the plane or axis of symmetry, the symmetry boundary conditions have been imposed.

$$u_r^* = 0,\ \partial u_z^* / \partial r^* = 0$$

Numerical Simulations

In this study, the constitutive equations have been discretized by our in-house developed CFD code using an integral approach of finite volume approach in the non-uniform staggered grid. The SIMPLER algorithm (Patankar, 1980) is used for the solution.

The distribution of mesh nodes has been considered allowing higher density in the region near to the wall and restriction. For the advective part, the third-order upwind scheme has been used. Tri-diagonal Matrix Algorithm (TDMA) with Alternate Direction Implicit (ADI) scheme has solved the discretized equations iteratively. The convergence of the iterative scheme has been allowed for the mass residue has fallen below 10^{-7}.

The size of numerical mesh for the computational domain has been fixed as *448 x 48* after the grid independence test, which has been given in next section of "Grid Independence Test".

The procedure for numerical calculations is made as per the following steps:

- Step-1: Velocity field is guessed.
- Step-2: Coefficients for the momentum equations are calculated to obtain pseudo velocities.
- Step-3: The pressure equation coefficients are calculated to solve the pressure equation.
- Step-4: Obtained velocity field from momentum equations.
- Step-5: Steps 2 – 4 are continued until the convergence criterion is achieved.
- Step-6: Calculations are repeated for each steps and periods, and comparative study of the previous period until achieving identical results over two successive periods.

Grid Independence Test

Grid independence study is essential for a CFD study to achieve the accuracy of results. One of the methods, Richardson extrapolation (Richardson & Gaunt, 1927) is the discretization of higher order estimate from lower-order discrete values (F_1, F_2,…,F_n). At least three grid solutions are required for the grid independence study (Stern et al., 2001). Roache generalized Richardson extrapolation by introducing the p_e^{th}-order methods, which is defined by equation (15). (Roache, 1994)

$$F_{h=0} = F_1 + \frac{F_1 - F_2}{r_e^{p_e} - 1} \qquad (15)$$

The grid refinement ratio *r* has been adopted as 2 in this study. The accuracy can be estimated (Stern et al., 2001) by using the following equations (16-17).

$$p_e = \frac{\ln\left(\frac{\varepsilon_{32}}{\varepsilon_{21}}\right)}{\ln(r_e)} \qquad (16)$$

Where $\varepsilon_{i+1,i} = F_{i+1} - F_i$ (17)

To evaluate the extrapolated value from the above equations, the convergence conditions of the solutions must be first confirmed. The feasible convergence conditions are as below;

- $0<R_e$; Monotonic convergence;
  - $R_e<0$: Oscillatory convergence;
  - $R_e>1$: Divergence;

Where R_e is the convergence ratio and it is calculated by the equation (18).

$$R_e = \frac{\varepsilon_{21}}{\varepsilon_{32}} \tag{18}$$

The Grid Convergence Index (*GCI*) presents the resolution in achieving the solution to an asymptotic value. It gives a uniform measure of convergence for proper mesh refinement (Roache, 1994). Mathematically it is written as the following equation (19).

$$GCI_{i+i,i} = F_s \frac{\left|\varepsilon_{i+1,i}\right|}{F_i(r_e^{p_e} - 1)} \tag{19}$$

From the recommendation of Wilcox, the safety factor (*Fs*) has been selected as 1.25. (Wilcox, 2006) The asymptotic range of convergence is obtained by the following equation (20)

$$A_c = \frac{GCI_{32}}{r_e^{p_e} GCI_{21}} \tag{20}$$

Table 2. Analysis results for case of M_1, M_2, and M_3 for grid independence study

Flow Characteristic	F_1	F_2	F_3	R_e	P_e	GCI_{32}	GCI_{21}	A_c
Low *WSS*	-690.59	-662.47	-591.05	0.394	1.34	8.69	3.28	1.046
Wall Pressure Drop	29.35	27.66	24.44	0.525	0.93	16.08	79.95	1.061
Recirculation Length (from OSI)	9.15	8.65	7.85	0.62	0.8	0.192	0.113	1.060

Three different mesh sizes, which are represented by *M*3 (112 x 12), *M*2 (224 x 24), and *M*1 (448 x 48), have been arranged as per to their degree of fineness. The grid levels are marked as 3, 2, and 1 respectively. Where, grid-level 1 represents the finest grid while level 3 represents the coarsest grid. The parameters have been compared with respect to maximum low wall shear stress and wall pressure drop, and length of recirculation (calculated from *OSI* value). The results have been presented in Table 2 for physiological pulsatile flow (profile-II) with *PR*= 50%, *Re* =100, and *Wo* = 10. The analysis results show that the calculated results using mesh size as *448 x 48* are in the condition of monotonic convergence as well as in the asymptotic range of accuracy.

The corresponding values of Richardson's Extrapolated data of the exact solution have also been calculated and shown in Table 3.

From the numerical solutions, relative errors are presented in the Table 3 and it is clear that the errors monotonically converge towards zero and are only 1.34% error for maximum low *WSS*, 1.88% error for the wall pressure drop, and 1.96% error for the recirculation length, respectively, when the grid level *1* is adopted. Therefore, this present numerical simulation uses the grid 448 ×48 (*i,j*).

Table 3. Result of grid independence test for PR=50%, Re=100 and W_o=10

Mesh Size (r,z)	M_3 (112x12)	M_2 (224x24)	M_1 (448x48)	Richardson's Extrapolated data
Max. WSS low	-591.05	-662.47	-690.59	-699.96
Error	108.91	37.49	9.3733	
%Error	15.55%	5.36%	1.35%	
WP Drop	24.44	27.66	29.35	29.91
Error	5.47	2.25	0.5633	
%Error	18.29%	7.52%	1.88%	
Length of recirculation (from OSI)	7.95	8.65	9.15	9.33
Error	1.48	0.73	0.1833	
%Error	15.86%	7.82%	1.96%	

Validation of Numerical Code

This is well recognized that the first step of the numerical analysis is the results must be validated with some experimental and numerical work with flow situations. In this present book chapter, pulsatile (i.e., unsteady) laminar flow through the restricted artery (i.e., restricted tube) has been considered by the authors. To validate the developed numerical code for this present work, at first, the comparisons have been made between the exact analytical solution of Womersley and the result of the present numerical simulation taking the same flow situation. The Womersley number and the Reynolds number have taken as 10 and 200, respectively. The sinusoidal pulsatile flow waveforms (i.e., $u_z{*}=1+\sin\omega t^{*)}$ have been considered for this code validation case. The Womersley velocity profile (Womersley, 1955, Taylor et al., 1998) for the axial component of velocity is,

$$u(r,t)=\frac{2B_0}{\pi R^2}\left[1-\left(\frac{r}{R}\right)^2\right]+\sum_{n=1}^{N}\left\{\frac{B_n}{\pi R^2}\left[\frac{1-\dfrac{J_o\left(W_{on}\dfrac{r}{R}i^{3/2}\right)}{J_o\left(W_{on}i^{3/2}\right)}}{1-\dfrac{2J_1\left(W_{on}i^{3/2}\right)}{W_{on}i^{3/2}J_o\left(W_{on}i^{3/2}\right)}}\right]e^{in\omega t}\right\} \tag{21}$$

Where, J_O and J_1 *are* Bessel functions of the first kind of order 0 and 1 respectively, R is the radius of the cylinder, and $W_{on}=R\sqrt{(n\omega)}\,/\,\upsilon$ is Womersley number, where υ is the kinematic viscosity. The pulsatile flow has been simulated over a long time to achieve a periodic solution, namely the solution that has not changed too much from one to the next cycle. The comparison between the axial velocity distribution with the exact solution of Womersley at four different time steps (t*=0.125, 0.375, 0.625 and 0.875) for the sinusoidal pulsatile flow waveforms by the authors. (Goswami et al., 2015b). It shows a good agreement in the results of the present numerical code and the analytical solutions of Womersley.

The present numerical results have been also compared with the experimental works (Ojha et al., 1989) for validation of the results of the present numerical code. The geometry used in this validation is a 45% stenosis with the trapezoidal profile. The considered inlet waveform is a sinusoidal pulse with a mean *Re* of 575 and an amplitude variation of 360, as shown in equation (21).

$$u^*_z = 575 + 360\sin(\omega t^* + 2.23) \tag{22}$$

For validation, the Womersley number has been taken as 7.5 and discretized into 69 time steps. The inlet boundary condition has been considered to be a fully developed. The inlet centerline velocity variation versus time, obtained from the simulation results, has been compared with that of Ojha et al. (1989) and that of Womersley solution in Goswami et al. (2015b). As can be seen from graph of inlet centreline velocity versus time, it is almost similar throughout the pulse.

The axial velocities for different locations along the axial direction in the post-stenotic region at different time steps, obtained from the simulation results, have also been compared with that of Ojha et al. (1989) and the same Womersley solution. The comparison of velocity profiles at the three post stenotic regions ($z^*= 1$, $z^*= 2.5$ and $z^*= 4.3$) for different time steps ($t^*= 0.225$, $t^*= 0.525$ and $t^*= 0.875$) was made in Goswami et al. (2015b). The said comparisons show good agreement between the predictions of present numerical simulation with experimental results that are found at different considered time steps and positions. Further validation of code in different area of work is made in many our research published work (Mandal et al., 2011b, Biswas et al., 2020, 2021a, 2021b).

OUTCOME OF COMPUTER ANALYSIS

The effects of Reynolds number, Womersley number, percentage of restriction, and type of pulsatile flow have been studied from the outcomes of the developed CFD code by the authors (Goswami et al. 2014, 2015a, 2015b, 2016 and 2019) on the wall pressures, streamline contour, peak and low wall shear stress, oscillatory shear index and other various flow characteristics for all the combination cases of mild stenosis, severe stenosis, high pulsatility of flow and low pulsatility of flow respectively for simple pulsatile flow condition. Figures 4 (a-d) and 5 depict how the streamlines vary for different Reynolds number, Womersley number, flow type and percentages of restriction. The streamline contours indicate the area of flow separation as shown in Figure 4(a-d) and Figure 5. The streamline separates from the wall just after the stenosis for every case. The negative pressure gradient termed as favorable pressure gradient enables blood to flow through the artery. The negative pressure gradient counteracts the retarding effect due to viscosity at the boundary. When the blood flow is passed through the stenosis, the pressure drops suddenly from the entry to the throat of the restriction, and the pressure recovers at the exit of the stenosis. Thus positive pressure gradient termed as adverse pressure gradient generates at the distal to the restriction. The separated flow occurs downstream of the stenosis due to adverse pressure gradient. Permanent flow separation is found after the stenosis, regardless of the flow condition as shown in Figure 4(a-d). The streamlines demonstrate the larger flow recirculation zone creates for the simple pulsatile flow while the smaller recirculation zone creates for the realistic pulsatile flow. This has happened because the simple pulsatile flow condition has the lowest flow velocity and the realistic pulsatile flow condition has the highest flow velocity at this time step, just beginning the flow. The size of flow recirculation increases as Reynolds number and percentage of restriction increases in this time step for all the considered pulsatile flow conditions while it decreases as Womersley number increases. Figures 6 to 8 have been adapted here from previous work by the author (Goswami et al., 2019). The variation of wall pressure axially is shown in Figure 6. This shows that the wall pressure drops gradually from entry to exit of the considered length of the artery for different Reynolds numbers and Womersley number. Pressure drops suddenly at the throat of the stenosis and the pressure raises slightly after the throat. The minimum wall pressure corresponds to the maximum mean velocity of the fluid flow. The flow area decreases at the throat of the stenosis from the entry of the stenosis and again the flow area increases after the exit of the stenosis. The flow area has minimum magnitude at the throat. According to the continuity of the flow for an incompressible fluid, the maximum mean velocity of the throat that corresponds to the minimum pressure at the throat. The pressure, after sharp falling at the stenosis region, the wall pressure partially recovers downstream due to the conversion of kinetic energy to pressure energy at the divergence section of stenosis. Wall pressure increases as Reynolds number and Womersley number both rises. This is also studied that wall pressure drop increases slowly at low stenosis, whereas this pressure drop increases markedly from moderate to severe stenosis condition, for any particular combination case of mild stenosis, severe stenosis, low pulsatility of flow, and high pulsatility of flow for all pulsatile flow conditions. From the study on wall pressure and streamline contour, it is felt that the study of *WSS* near the stenosis is important to investigate the causes behind the initiation and progression of arterial stenosis. Blood flow velocity changes in the stenosis zone and post-stenotic zone as the flow pressure gradient changes. It is understood from previous studies that the maximum velocity gradient exists at the throat and the minimum velocity gradient exists at the post stenotic zone. Therefore, it is very relevant to examine the wall shear distribution in the stenotic region as well as post stenotic zone. The variation of the time-averaged *WSS* along the flow direction in the stenotic zone is shown for different *PR* for *Re* 200 and *Wo* 12.5 in Figure 7a to show the effect of *PR* on peak wall shear stress. It is noted that *TAWSS* rises as the flow approaches towards the stenosis or restriction and attains a peak value at the maximum restricted area. Then at the downstream of the stenosis, wall shear stress decreases rapidly to zero, and then it becomes of negative magnitude. As the spatial and temporal magnitude of axial velocity increases as *PR* increases, the flow rate is higher for higher *PR*. The higher flow rate cause the rise in the velocity gradient in the artery wall and therefore *WSS* at the throat attains a higher value for a higher percentage of restriction. From Figure 7a, it is also revealed that peak wall shear stress value rises as Reynolds number increases. Figure 7b shows the variation of wall shear stress along the flow at the post-stenotic region to scrutinize the impact of *PR* on the low wall shear stress. It is revealed that the value of low *WSS* is prone just immediately after stenosis. This is supported to the existence of a recirculation zone downstream of stenosis. Figure 7b depicts that the *WSS* changes its direction from positive to negative when the flow leaves the stenosis and chance vice versa at the downstream flow after stenosis. There are two points of zero wall shear stress. The time-averaged *WSS* zero means the velocity gradient is zero. Thus there are two points of flow stagnation. The first one is the flow separation point and the second one is the flow reattachment point. From Figure 7b, it is noticed that the reattachment length rises with as the value of *PR* increases. It is also observed that the magnitudes of maximum low WSS increase with the increase in *PR* and the area under low WSS also increases as Reynolds number increases. During the study on instantaneous streamline patterns, it has been observed that the flow field always changes with the time steps of a cardiac cycle. The flow reversal is noticed at each time step and the pattern is different. The time-averaged *WSS* is studied in the previous section. The time-averaged WSS is estimated by integrating the instantaneous wall shear stress over a cycle time. Viewing on the instantaneous streamline patterns, it can be revealed that the instantaneous *WSS*s are not the same with varying times of a cardiac cycle at any particular location along the axial direction of arterial length. Thus, there exists oscillatory *WSS* in the blood flow. The oscillatory shear index defines the oscillatory motion of *WSS*. It is observed from Figure 8 that the *OSI* has two maximum values for each of all the considered *Re* flow, one closely after the midpoint of stenosis and other one far from the midpoint. As can be seen in the mathematical definition of *OSI*, it is clear that *OSI* reaches its maximum value when the instantaneous WSS at each time step of a cardiac cycle aligns in the same direction or remains at the same sign. Near the midpoint, the velocity gradient is high and the instantaneous wall shear stress reaches its peak value for each time step and thus *OSI* increases at this location. At the reattachment point, instantaneous *WSS*, which remains in the negative direction, changes the sign. *OSI* again reaches its maximum value at the point of reattachment point. Therefore, the location of the second maximum *OSI* magnitude coincides with the same point of time-averaged reattachment point. The time-averaged reattachment points for all percentage of restrictions are clearly estimated by both *WSS* and *OSI* graphs, and this has a good agreement.

CONCLUSION

This study focuses on a computational study of a biological problem by using CFD. In the present study, flow physics with its haemodynamic factors for initiation and progression of the disease atherosclerosis is investigated in detail. The book chapter tells the details of different arterial diseases its hemodynamic factors and shows a numerical study of the pulsatile flow through a stenosed artery by using our own house CFD code. It is noteworthy to mention here that the process used and outcomes of computational analysis will enrich the knowledge in the area of arterial disease and could also be utilized for better design of a medical tool to detect early atherosclerosis as well as the severity of atherosclerosis. The systematic study of blood flow may help the medical people to make a better decision upon the treatment management of atherosclerosis such as exercise, medication, surgery, etc.

Figure 4. Streamline contours for profile-I, profile-II and profile-III (top to bottom) at time step t/T=0 for a. Re of 50, W_o of 5 and PR of 70%. b. Re of 200, W_o of 5 and PR of 70%, c. Re of 50, W_o of 12.5 and PR of 70% and d. Re of 200, W_o of 12.5 and PR of 70%

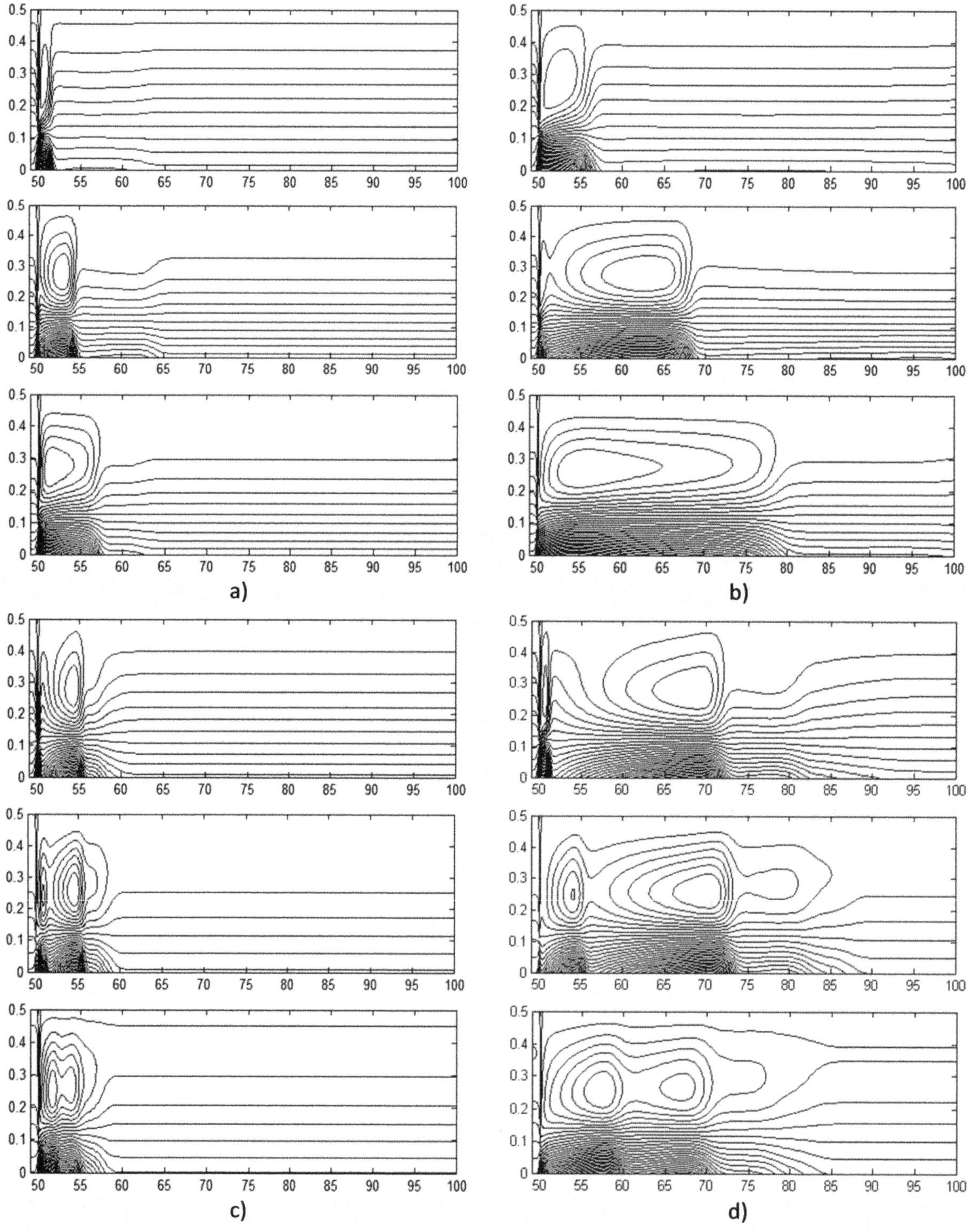

Figure 5. Streamline contour for PR 30%, PR 50% and PR 70% (top to bottom) respectively at t=0 for realistic pulsatile flow*

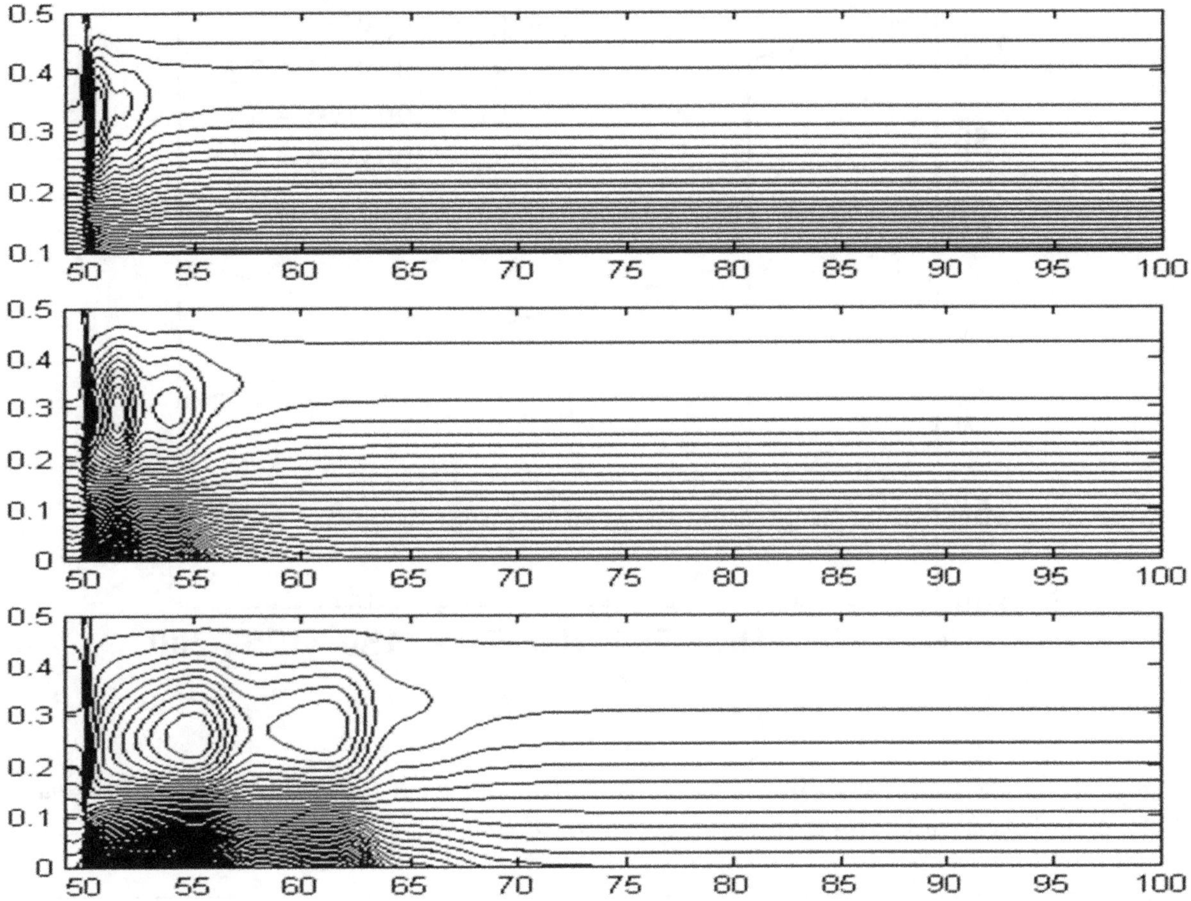

Nomenclature

D Artery diameter, $[m]$
R Artery radius, $[m]$
L Total length of the computational domain, $[m]$
L_i Length of the computational domain before stenosis, $[m]$
L_s Width of stenosis, $[m]$
L_{si} Width of upstream stenosis, $[m]$
L_{se} Width of downstream stenosis, $[m]$
L_e Length of the computational domain after stenosis, $[m]$
h_f Height of stenosis
u_z Velocity in z-direction, $[ms^{-1}]$
u_r Velocity in r-direction, $[ms^{-1}]$
U Average velocity in r-direction at inlet, $[ms^{-1}]$

Figure 6. Distribution of wall pressure along axial direction for PR of 70% with different Re and different Wo for realistic pulsatile flow

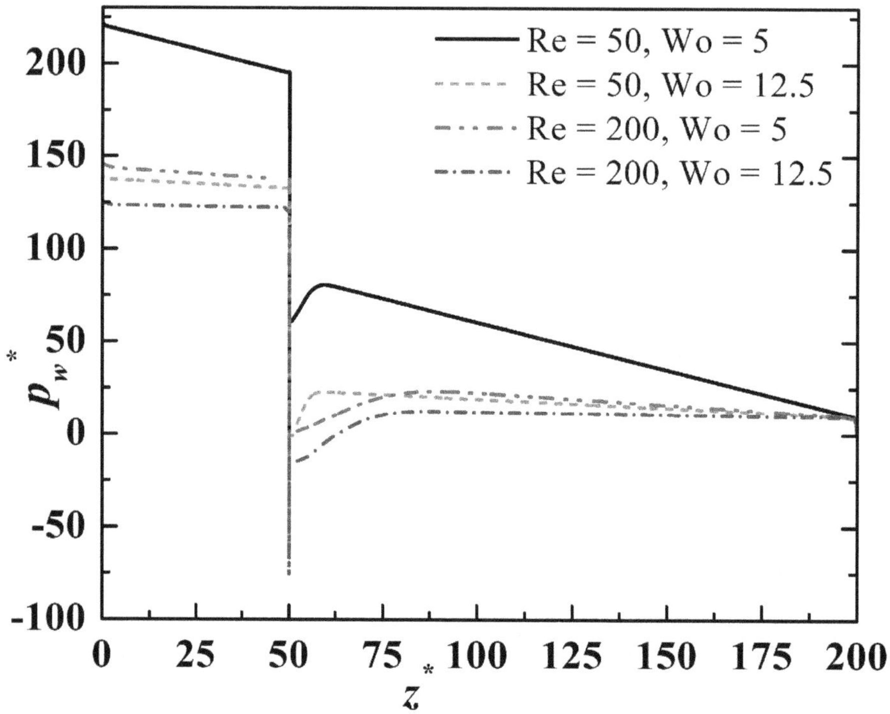

Figure 7. a. Distribution of peak WSS along axial direction for Re of 100 and W_0 of 10. b. Distribution of low WSS along Axial direction for Re of 100 and W_0 of 10

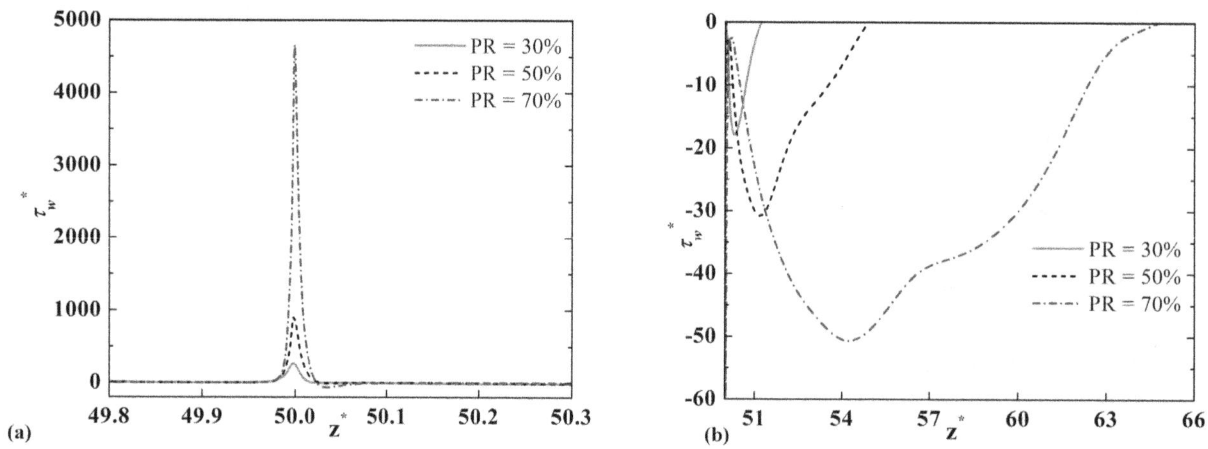

Figure 8. a. Distribution of oscillatory shear index along axial direction for Re of 100 and Wo of 10. b. Distribution of oscillatory shear index at senosis zone for Re of 100 and Wo of 10

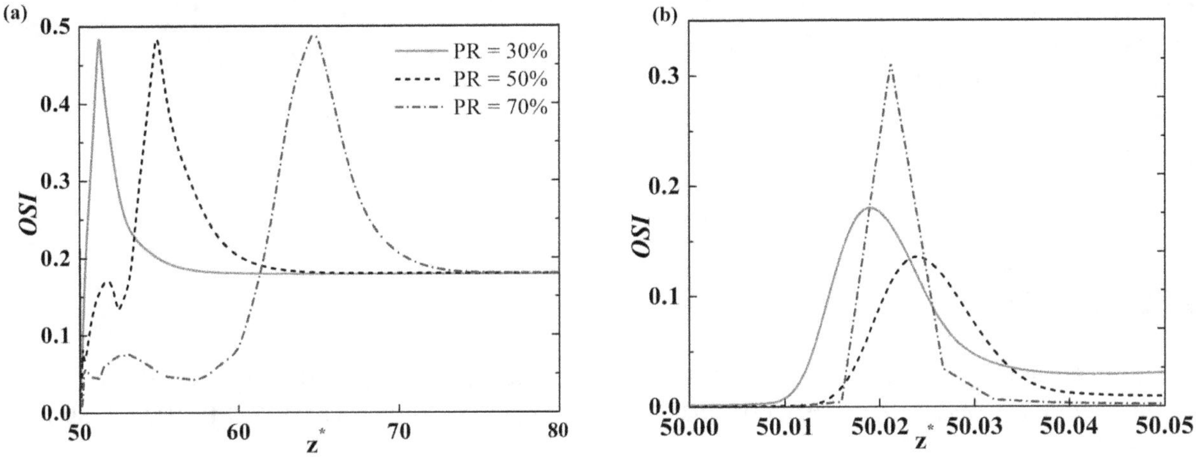

μ Dynamic viscosity, [$kg\ m^{-1}s^{-1}$]
ρ Density, [$kg\ m^{-3}$]
ν Kinematic viscosity, [ms^{-2}]
T Time period of the pulsatile cycle, [s]
T Time, [s]
p Static pressure, [Nm-2]
R_e Reynolds Number $= \dfrac{\rho UD}{\mu}$
W_o Womersley number $= \dfrac{D}{2}\sqrt{\dfrac{2\pi}{\upsilon T}}$
p_w Wall pressure[Nm^{-2}]
τ_w Wall shear stress (WSS), [Nm^{-2}]
PWSS Peak wall shear stress
LWSS Low wall shear stress
TAWSS Time-averaged wall shear stress
OSI Oscillatory shear index
PR Percentage of restriction
r_e Grid refinement ratio
R_e Convergence ratio
GCI Grid convergence index
F_s Factor of safety
A_c Asymptotic range of convergence
p_e Order of accuracy

Superscripts

* Dimensionless terms
- Time-averaged
r, z Cylindrical co-ordinates
1-1 Inlet
2-2 Exit

REFERENCES

Ahmed, S. A., & Giddens, D. P. (1984). Pulsatile poststenotic flow studies with laser Doppler anemometry. *Journal of Biomechanics*, *17*(9), 695–705. doi:10.1016/0021-9290(84)90123-4 PMID:6238968

Andersson, H. I., Halden, R., & Glomsaker, T. (2000). Effects of surface irregularities on flow resistance in differently shaped arterial stenosis. *Journal of Biomechanics*, *33*(10), 1257–1262. doi:10.1016/S0021-9290(00)00088-9 PMID:10899335

Back, L. H., & Banerjee, R. K. (2000). Estimated flow resistance increase in a spiral human coronary artery segment. *Journal of Biomechanical Engineering*, *122*(6), 675–677. doi:10.1115/1.1319661 PMID:11192391

Biswas, N., Manna, N. K., Datta, A., Mandal, D. K., & Benim, A. C. (2020). Role of aspiration to enhance MHD convection in protruded heater cavity. *Progress in Computational Fluid Dynamics*, *20*(6), 363–378. doi:10.1504/PCFD.2020.111408

Biswas, N., Manna, N. K., Mandal, D. K., & Chamkha Ali, J. (2021b). Effect of surface waviness on MHD thermo-gravitational convection of Cu−Al2O3−water hybrid nanofluid in a porous oblique enclosure. *Physica Scripta*, *96*(10), 105002. doi:10.1088/1402-4896/ac0f94

Biswas, N., Manna, N. K., Mandal, D. K., & Gorla, R. S. R. (2021a). Magnetohydrodynamic mixed bioconvection of oxytactic microorganisms in a nanofluid-saturated porous cavity heated with a bell-shaped curved bottom. *International Journal of Numerical Methods for Heat & Fluid Flow*, *31*(12), 3722–3751. doi:10.1108/HFF-10-2020-0668

Buchanan, J. R. Jr, Kleinstreuer, C., & Comer, J. K. (2000, April). (20000). Rheological effects on pulsatile hemodynamics in a stenosed tube. *Computers & Fluids*, *29*(6), 695–724. doi:10.1016/S0045-7930(99)00019-5

Caro, C. G., Fitz-Gerald, J. M., & Schroter, R. C. (1971). Atheroma and arterial wall shear: Observation, correlation, and proposal for a shear dependent mass transfer mechanism for atherogenesis. *Proceedings of the Royal Society of London. Series B, Biological Sciences*, *117*, 109–159. PMID:4396262

Caro, C. G., Pedley, T. J., Schroter, R. C., & Seed, W. A. (1978). The mechanics of the circulation. Oxford Medical.

Cassanova, R. A., & Giddens, D. P. (1978). Disorder distal to modeledstenoses in steady and pulsatile flow. *Journal ofBiomechanics*, *11*(10-12), 441–453. doi:10.1016/0021-9290(78)90056-8 PMID:730759

Cho, Y. I., & Kensey, K. R. (1991). Effects of the non-Newtonian viscosity of blood flows in a diseased arterial vessel, part-1: Steady flows. *Biorheology*, *28*(3-4), 241–262. doi:10.3233/BIR-1991-283-415 PMID:1932716

Fry, D. L. (1968). Acute vascular endothelial changes associated with increased blood velocity gradients. *Circulation Research*, *12*(2), 165–197. doi:10.1161/01.RES.22.2.165 PMID:5639037

Fry, D. L. (1969). Certain histological and chemical responses of the vascular interface to acutely induced mechanical stress in the aorta of the dog. *Circulation Research*, *24*(1), 93–108. doi:10.1161/01.RES.24.1.93 PMID:5763742

Fung, Y. C. (1997). *Biomechanics: Circlulation* (2nd ed.). Springer. doi:10.1007/978-1-4757-2696-1

Ganong, W. F. (2001). *Review of Medical Physiology*. Appleton.

Gessner, F. B. (1973). Haemodynamic theories of atherogenesis. *Circulation Research*, *3*(3), 259–266. doi:10.1161/01.RES.33.3.259

Goswami, P., Mandal, D. K., Manna, N. K., & Chakrabarti, S. (2014). Study on the Effect of Steady, Simple Pulsatile and Physiological Pulsatile Flows through a Stenosed Artery. *Heat and Mass Transfer*, *50*(10), 1343–1352. doi:10.100700231-014-1334-0

Goswami, P., Mandal, D. K., Manna, N. K., & Chakrabarti, S. (2015a). Wall Shear Stress Characteristics for the Progression of the Disease, Atherosclerosis. *Journal of the Institution of Engineers (India): Series C*, *96*(3), 311-323.

Goswami, P., Mandal, D. K., Manna, N. K., & Chakrabarti, S. (2015b). Analysis of Steady and Physiological Pulsatile Flow Characteristics in an Artery with Various Percentages of Restrictions. *International Journal of Fluid Mechanics Research*, *42*(3), 260–280. doi:10.1615/InterJFluidMechRes.v42.i3.60

Goswami, P., Mandal, D. K., Manna, N. K., & Chakrabarti, S. (2016). Analysis of Wall Shear Parameters of Physiological Pulsating Flow through Mild and Severe Arterial Stenosis and Correlation to Atherosclerosis. *International Journal of Science and Technology*, *2*(3), 40–54.

Goswami, P., Mandal, D. K., Manna, N. K., & Chakrabarti, S. (2019). Numerical Investigation of Fluid Mechanical Factors of Realistic Pulsatile Flow through Constricted artery and Various Aspects of Plaque Deposition. *Journal of Mechanical Engineering Science*, *13*(3), 5306–5322. doi:10.15282/jmes.13.3.2019.07.0433

Gupta, A. K., & Agrawal, S. P. (2015). Computational modeling and analysis of the hydrodynamic parameters of blood through stenotic artery. *Procedia Computer Science*, *57*, 403–410. doi:10.1016/j.procs.2015.07.355

Ishikawa, T., Guimaraes, L. F. R., Oshima, S., & Yamane, R. (1998). Effect of non-Newtonian property of blood on flow through a stenosed tube. *Fluid Dynamics Research*, *22*(5), 251–264. doi:10.1016/S0169-5983(97)00041-5

Khalifa, A. M. A., & Giddens, D. P. (1981). Characterization and evolution of poststenotic flow disturbances. *Journal of Biomechanics, 14*(5), 279–296. doi:10.1016/0021-9290(81)90038-5 PMID:7263720

Ku, D. N. (1997). Blood flow in arteries. *Annual Review of Fluid Mechanics, 29*(1), 399–434. doi:10.1146/annurev.fluid.29.1.399

Ku, D. N., Giddens, D. P., Zarins, C. K., & Glagoy, S. (1985). Pulsatile flow and atherosclerosis in the human carotid bifurcation—Positive correlation between plaque location and low and oscillating shear-stress. *Arteriosclerosis (Dallas, Tex.), 5*(3), 293–302. doi:10.1161/01.ATV.5.3.293 PMID:3994585

Lee, J. S., & Fung, Y. C. (1970). Flow in locally constricted tubes at low Reynolds number. *Journal of Applied Mechanics, 37*(1), 9–16. doi:10.1115/1.3408496

Lee, K. W., & Xu, X. Y. (2002). Modelling of flow and wall behaviour in a mildly stenosed tube. *Medical Engineering & Physics, 24*(9), 575–586. doi:10.1016/S1350-4533(02)00048-6 PMID:12376044

Liu, G-T., Wang, X-J., Ai, B-Q., & Liu, L-G. (2004). Numerical study of pulsating flow through a tapered artery with stenosis. *Chinese Journal of Physics, 42*(4-1), 401-409.

Long, Q., Xu, X. Y., Ramnarine, K. V., & Hoskins, P. (2001). Numerical investigation of physiologically realistic pulsatile flow through arterial stenosis. *Journal of Biomechanics, 34*(10), 1229–1242. doi:10.1016/S0021-9290(01)00100-2 PMID:11522303

MacDonald, D. A. (1979). On steady flow through modelled vascular stenosis. *Journal of Biomechanics, 12*(1), 13–20. doi:10.1016/0021-9290(79)90004-6 PMID:762177

Mamun, K., Ali, M., & Akhter, M. N. (2016). Physiological non-Newtonian blood flow through single stenosed artery. *Theoretical and Applied Mechanics, 43*(1), 99–115. doi:10.2298/TAM160322006M

Mandal, D. K., & Chakrabarti, S. (2007a). Effect of Restriction and Reynolds Number on the Pressure of Blood of a Stenotic Artery. *International Journal of Fluid Mechanics Research, 34*(2), 159–178. doi:10.1615/InterJFluidMechRes.v34.i2.50

Mandal, D. K., & Chakrabarti, S. (2007b). Two Dimensional Simulation of Steady Blood Flow through a Stenosed Coronary Artery. *International Journal of Dynamics of Fluids, 3*(2), 187–209.

Mandal, D. K., & Chakrabarti, S. (2007c). Study of Pressure Drop and Flow Characteristics across Rectangular Stenotic Models. *International Journal of Fluid Mechanics Research, 34*(5), 434–461. doi:10.1615/InterJFluidMechRes.v34.i5.40

Mandal, D. K., & Chakrabarti, S. (2008). Effect of Stricture Length on Reattachment Point and Wall Shear Stress through a Stenosed Coronary Artery. *International Journal of Fluid Mechanics Research, 35*(2), 188–202. doi:10.1615/InterJFluidMechRes.v35.i2.70

Mandal, D. K., & Chakrabarti, S. (2009). Effect of Stenosis Length on Flow Characteristics across Rectangular Stenotic Models. *International Journal of Fluid Mechanics, 1*(1), 29–39.

Mandal, D. K., & Chakrabarti, S. (2010). Study on the effect of Different Shaped Stenoses on Blood flow through Coronary Artery. *International Journal of Biomedical Engineering and Technology, 4*(1), 1–17. doi:10.1504/IJBET.2010.034274

Mandal, D. K., Manna, N. K., Bandyopadhyay, S., Biswas, B. P., & Chakrabarti, S. (2011b). A Numerical Study on the Performance of a Sudden Expansion with Multisteps as a Diffuser. *International Journal of Applied Mechanics*, *3*(4), 779–802. doi:10.1142/S1758825111001238

Mandal, D. K., Manna, N. K., & Chakrabarti, S. (2010a). Numerical Study of Blood Flow Through Different Double Bell Shaped Stenosed Coronary Artery During the Progression of the Disease, Atherosclerosis. *International Journal of Numerical Methods for Heat & Fluid Flow*, *20*(6), 670–698. doi:10.1108/09615531011056827

Mandal, D. K., Manna, N. K., & Chakrabarti, S. (2010b). A numerical model study of steady flow through bell-shaped stenoses with and without asymmetry. *International Journal of Experimental and Computational Biomechanics*, *1*(3), 306–331. doi:10.1504/IJECB.2010.035263

Mandal, D. K., Manna, N. K., & Chakrabarti, S. (2011a). Influence of Primary Stenosis on Secondary One and Vice Versa in case of Double Stenoses. *Journal of Applied Fluid Mechanics*, *4*(4), 31–42.

McDonald, D. A. (1974). *Blood Flow in Arteries*. Camelot.

Mills, C. J., Gabe, I. T., Gault, J. H., Mason, D. T., Ross, J., Braunwald, E., & Shillingford, J. P. (1970). Pressure-flow relationships and vascular impedance in man. *Cardiovascular Research*, *4*(4), 405–417. doi:10.1093/cvr/4.4.405 PMID:5533085

Misra, J. C., & Shit, G. C. (2006). Blood flow through arteries in a pathological state: A theoretical study. *International Journal of Engineering Science*, *44*(10), 662–671. doi:10.1016/j.ijengsci.2005.12.011

Ojha, M., Cobbold, C., Johnston, K. W., & Hummel, R. L. (1989). Pulsatile flow through constricted tubes: An experimental investigation using photochromic tracer methods. *Journal of Fluid Mechanics*, *203*, 173–197. doi:10.1017/S0022112089001424

Patankar, S. V. (1980). *Numerical heat transfer and fluid flow*. Hemisphere Publication.

Rabby, M. S., Shupti, S. P., & Molla, M. M. (2014). Pulsatile non-Newtonian laminar blood flows through arterial double stenosis. *Journal of Fluids*, *2014*, 1–13. doi:10.1155/2014/757902

Richardson, L. F., & Gaunt, J. A. (1927). The deferred approach to the limit.Part I. Single lattice. Part II. Interpenetrating lattices. *Philosophical Transactions of the Royal Society of London. Series A, Containing Papers of a Mathematical or Physical Character*, *226*, 299–361.

Roache, P. J. (1994). Perspective: A method for uniform reporting of grid refinement studies. *Journal of Fluids Engineering*, *116*(3), 405–441. doi:10.1115/1.2910291

Santra, S., Mandal, D. K., & Chakrabarti, S. (2017). Assessment of Accumulation Rate of LDL Species in Arterial Wall layers under Hypertension and Hyperlipidemia Conditions. *International Journal of Fluid Mechanics Research*, *44*(1), 79–92. doi:10.1615/InterJFluidMechRes.2017016597

Santra, S., Mandal, D. K., & Chakrabarti, S. (2018). Effect of pulsatile blood flow on LDL transport in arterial layers. *Progress in Computational Fluid Dynamics*, *18*(3), 177–187. doi:10.1504/PCFD.2018.091746

Santra, S., Mandal, D. K., & Chakrabarti, S (2020).A New Approach for Assessing the LDL Species in Arterial Wall Layers. *Journal of The Institution of Engineers (India): Series C, 101*(1), 13–24.

Sherwin, S. J., & Blackburn, H. M. (2005). Three-dimensional instabilities and transition of steady and pulsatile axisymmetric stenotic flows. *Journal of Fluid Mechanics, 533*, 297–327. doi:10.1017/S0022112005004271

Stern, F., Wilson, R. V., Coleman, H. W., & Paterson, E. G. (2001). Comprehensive approach to verication and validation of cfd simulations part 1: Methodology and procedures. *Journal of Fluids Engineering, 123*(4), 793–802. doi:10.1115/1.1412235

Stiehm, M., Wüstenhagen, C., Siewert, S., Grabow, N., & Schmitz, K. (2017). Numerical simulation of pulsatile flow through a coronary nozzle model based on FDA's benchmark geometry. *Current Directions in Biomedical Engineering, 3*(2), 775–778. doi:10.1515/cdbme-2017-0163

Taylor, C. A., Hughes, T. J. R., & Zarins, C. K. (1998). Finite element modeling of blood flow in arteries. *Computer Methods in Applied Mechanics and Engineering, 158*(1-2), 155–196. doi:10.1016/S0045-7825(98)80008-X

Tu, C., Deville, M., Dheur, L., & Vanderschuren, L. (1992). Finite element simulation of pulsatile flow through arterial stenosis. *Journal of Biomechanics, 25*(10), 1141–1152. doi:10.1016/0021-9290(92)90070-H PMID:1400514

Tutty, O. R. (1992). Pulsatile flow in a constricted channel. *Journal of Biomechanical Engineering, 114*(1), 50–54. doi:10.1115/1.2895449 PMID:1491586

Varghese, S. S., & Frankel, S. H. (2003). Numerical modeling of pulsatile turbulent flow in stenotic vessels. *Journal of Biomechanical Engineering, 125*(4), 445–460. doi:10.1115/1.1589774 PMID:12968569

Wilcox, D. C. (2006). *Turbulence modeling for CFD* (3rd ed.). DCW Industries, Inc.

Wiwatanapataphee, B., Poltem, D., Wu, Y. H., & Lenbury, Y. (2006). Simulation of pulsatile flow of blood in stenosed coronary artery bypass with graft. *Mathematical Biosciences and Engineering, 3*(2), 71–383. PMID:20361829

Womersley, J. R. (1955). Method for the calculation of velocity, rate of flow, and viscous drag in arteries when the pressure gradient is known. *The Journal of Physiology, 127*(3), 553–563. doi:10.1113/jphysiol.1955.sp005276 PMID:14368548

Wootton, D. M., & Ku, D. N. (1999). Fluid mechanics of vascular systems, diseases, and thrombosis. *Annual Review of Biomedical Engineering, 1*(1), 299–329. doi:10.1146/annurev.bioeng.1.1.299 PMID:11701491

Young, D. F., & Tsai, F. Y. (1973). Flow characteristics in models of arterial stenoses-I. Steady flow. *Journal of Biomechanics, 6*(4), 395–410. doi:10.1016/0021-9290(73)90099-7 PMID:4732939

Zendehbudi, G. R., & Moayeri, M. S. (1999). Comparison of physiological and simple pulsatile flows through stenosed arteries. *Journal of Biomechanics, 32*(9), 959–969. doi:10.1016/S0021-9290(99)00053-6 PMID:10460133

Chapter 11
Small Intestinal Peristalsis: Biomechanics and Clinical Prominence of Digestion

Jetal Bhanarkar
National Institute of Technology, Rourkela, India

Rashi Singh
National Institute of Technology, Rourkela, India

Anupama Usha Rani Siddhanathi
CWS Hospital, India

Ravi Kant Avvari
https://orcid.org/0000-0001-5586-746X
National Institute of Technology, Rourkela, India

ABSTRACT

Digestion of food forms an essential process in a living organism. Digestion of food can be classified into two categories: (1) mechanical digestion and (2) chemical digestion. Whereas mechanical digestion contributes to the physical breakdown of food using mechanical forces, chemical digestion contributes via breaking down the chemical bonds present in the food. The process is manifested by muscular contraction. The intestine develops undulations via contraction of the muscle fiber that involves contraction and relaxation of muscle fibers (referred to as the peristalsis). As a consequence, the momentum generated by the muscular contraction serves to develop fluid motion in the lumen, which eventually leads to mixing and transport. Whereas the bodily secretions of the digestive enzymes participate in innate digestion, the gut flora, on the other hand, also influence digestion to an extent by means of fermentation. Small intestinal peristalsis serves as the key mediator, whether the digestion proceeds normally during physiology or abnormally leading to pathology.

DOI: 10.4018/978-1-7998-9078-2.ch011

INTRODUCTION

Digestion is the process of breaking down ingested food into a simpler form that can be easily absorbed by the body. It is carried out by two sets of organs, primary and assisting organs. Primary organs include the mouth, esophagus, stomach, small intestine, and large intestine, whereas, assisting organs include salivary glands, pancreas, liver and gall bladder. Primary organs function as storage, mixing and transport and the assisting organs helps in the facilitating the digestive process through secretion of the digestive juices directly in to the lumen of the primary organ

The process of digestion begins in mouth after ingestion of food (Figure 1). It is broken down into smaller particles by teeth (also known as mastication) and mixed with saliva to bind it together in the form of a bolus. It then enters the esophagus and moves towards stomach by the action of peristalsis. Stomach breaks down the food into simpler molecules with the help of enzyme in a highly acidic condition (pH 1-3). Such acidic environment is crucial for digestion as it not only facilitates the mechano-chemical breakdown but also helps in destroying pathogens that come along with the food during ingestion.

After a lapse of 1-2 hours in the stomach, the partially digested food, also known as the chyme, enters the duodenum (duodenum is first segment of the small intestine, followed by jejunum and ileum) through a small opening known as the pylorus. The pH of chyme entering from stomach to small intestine increases from pH of 1.7 to 5 at initial portion of the duodenum and then gradually increases to pH 6 in the distal duodenum to a pH of 7.4 in the ileum (Avvari, 2019a). Duodenum, which lies next to the pyloric valve, is the shortest in length of around 30 cm. It facilitates the digestion of the chyme by allowing the digestive enzymes from pancreas and bile juice from liver to directly enter into the lumen. These enzymes help in digesting carbohydrates and lipids, whereas, secretions from pancreas increase the pH level of duodenum to make it slightly alkaline. Jejunum, connected next to duodenum, is of length 2.5m. It is involved in absorption of sugar, fatty acids and amino acids from digested food with the help of villi. Villi are small finger-like structures present throughout the interior surface of the gut. It helps to the effective surface area of contact for the luminal contents so the digested product can be absorbed. Ileum is the last section of small intestine of length 3m. It also contains small finger like projections for nutrient absorption similar to jejunum. It is responsible for absorption of nutrients that were not absorbed in jejunum. Ileum is then connected to large intestine by ileocecal valve. Food at the terminal ileum is then emptied into the large intestine eventually, where the final transformation of the food takes place via absorption of water to eliminate the waste in the form of feces before excreting it from the body.

The digestion can be classified into two types, mechanical digestion and chemical digestion.

1. Chemical digestion

Chemical digestion involves the chemical breakdown of bonds of ingested food to produce simpler compounds and provide an appropriate pH balance to ease the process of digestion. This is carried out with the help of several enzymes and secretions such as saliva (in mouth), hydrochloric acid and gastric juice (in stomach), digestive enzymes (in pancreas) and bile juice (in liver). Chemical digestion starts in the mouth by digestion of carbohydrates content of the food by salivary amylases. In stomach, secretion of hydrochloric acid and pepsin aids the digestion of protein content and reduces the pH levels to 1.5-3. The walls of stomach secrete mucus and bicarbonate to shield itself from such low pH environment. In small intestine, digestive enzymes from pancreas and bile juice from liver mix together with the chyme coming from stomach to aid digestion during peristalsis and to increase pH levels.

Figure 1. Schematic diagram of the human digestive system showing small intestine as a convoluted tube and its association with pancreas, liver and gall bladder.

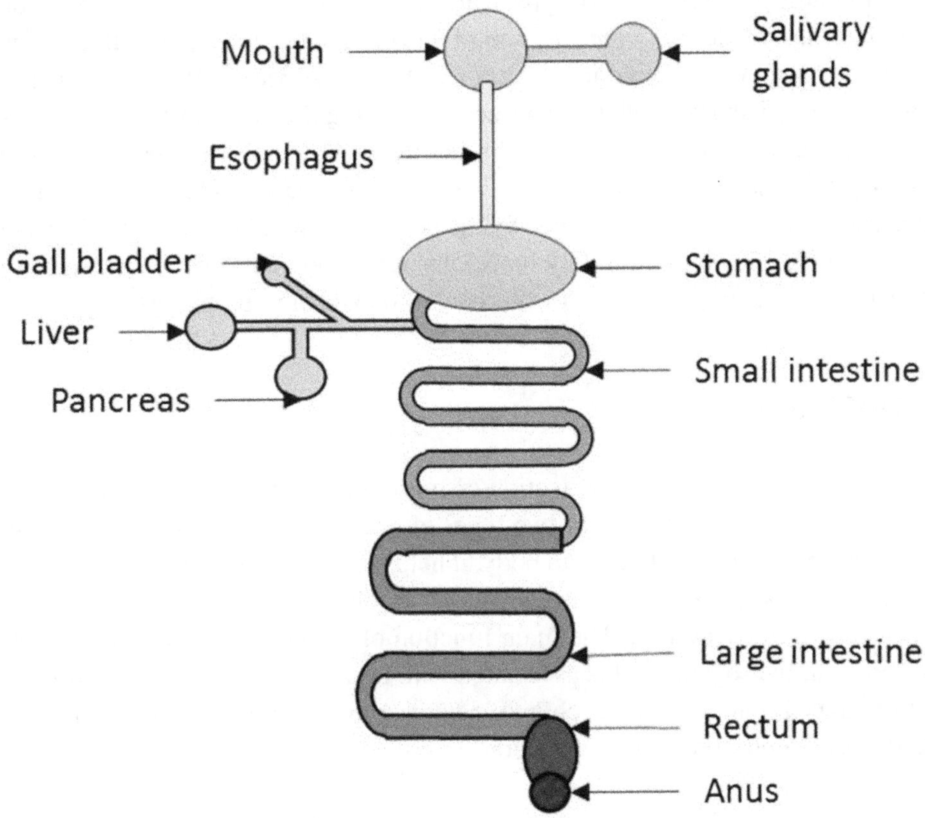

2. Mechanical digestion

It is a purely physical process of breaking down larger particles of food into smaller particles and mixing them to facilitate chemical digestion. The initial breakdown of the food takes place in the mouth via mastication (chewing of food) by teeth and mixing with saliva by tongue. In the stomach, peristalsis promotes mixing of food and physical breakdown of chunks of food into smaller particles of size 2-3 mm so it can pass through pyloric channel. And in the small intestine the peristalsis helps in further mixing and agitation of the contents for increased digestibility by the enzymes; besides shearing.

Pylorus (valve-like structure connecting the stomach to the duodenum) plays an important role in transferring partially digested food from stomach to small intestine. Transport of chyme through pylorus depends on the pressure gradient across pylorus channel. Antegrade (forward moving) and retrograde (backward moving) flow can be observed depending on the direction of pressure gradient across this channel. When the volume of retrograde flow increases, this can lead to physiological condition called duodenogastric reflux (DGR), which is the backflow of fluid from small intestine to the stomach (Avvari, 2015).

In this chapter, we discuss on the digestibility of the food in the small intestine due to peristalsis. Peristalsis confers the mechanical forces require for the digestions, especially by facilitating various process such as – i) digestion by actively mixing the chyme with duodeno-biliary pancreatic (DBP) secretion coming from pancreas, ii) homogenization of partially digested food in small intestine, iii) increasing the pH in duodenal segment, iv) physical breakdown of food into smaller particles through generation of higher shearing forces, v) absorption of nutrients with the help of villi present at the walls of intestine, and vi) transport of chyme for further absorption (Avvari, 2015). Since peristalsis drives and interfere with the digestion of food, the parameters defining the contractility of the muscles plays a key role in the digestibility of the food. We also discuss the mathematical modeling of the peristalsis of relevance to the small intestinal digestion so as to bridge the gap between the mechanical and chemical digestion. Since chemical digestion is governed by the peristalsis, we also review the breakdown of food using digestive enzymes. This is followed by biomechanical prominence of the peristalsis to digestion by discussing as to how the suppression of the peristalsis may lead to pathology.

WHAT IS PERISTALSIS?

Peristalsis is a wave of contraction and relaxation of muscle fibers present in the intestinal walls. The wall of gastrointestinal tract is composed of two types of muscle fibers – longitudinal (outer layer) and circular (inner layer) muscle fibers. In human body, it helps in an involuntary transport of ingested food through the gastrointestinal (GI) tract. Peristalsis has been observed in esophagus, small intestine and large intestine of the digestive system. The main function of peristalsis involves transport and mixing of intraluminal content. In esophagus, it is associated with transfer of bolus from mouth to the stomach, in small intestine, transfer and mixing of chyme, whereas, in the large intestine, it is associated with transport of undigested food for further evacuation.

Circular Contraction

Small intestine can be viewed as a circular tube of certain radius. When the circular muscle fiber of intestinal wall contract radially, the radius of the lumen reduces. The ratio of smallest radius during contraction to the radius of cylindrical tube is called as *'degree of occlusion'*. During circular contraction, the adjacent muscle fibers of occlusion region undergo tension. Let us consider that the circular contraction travels in forward direction to propel the chyme from left to right direction. As the contraction traverses the segment, the muscle fibers present on the right hand side of contraction wave undergoes relaxation and the muscle fibers present on the left hand side undergoes contraction. This can be visualized as a sinusoidal wave of tension in the fiber traveling from left to right, which tries to accept the bolus on head-region and contract at the tail-end to push forward the bolus.

Local Longitudinal Shortening

Longitudinal muscle fibers are present at the outer side of circular muscle fiber. Imagine keeping a thumb and an index finger over your skin. When you try to bring these two fingers closer to each other, the skin beneath accumulates between the two fingers. Local longitudinal shortening (LLS) can be visualized in a similar way. Further, the LLS should be considered as a wave of contraction of the longitudinal muscle fibers over one wavelength of contraction. When a small section of intestine of length 'l_0' undergoes contraction, i.e. the muscle fibers becomes shorter, the length l_0 becomes l. The ratio of l/l_0 is defines the extent of local longitudinal shortening of small intestine. Similar to circular contraction, local longitudinal shortening also travels along the axis of small intestine with the same speed as that of circular contraction. The combination of both circular and longitudinal contraction of intestinal wall is called as peristalsis (Nicosia 2001; Pal & Brasseur 2002; Avvari 2019b; Avvari 2019c).

Combined Effect of Circular Contraction and Local Longitudinal Shortening

A pioneering work from Pal and Brasseur (2002) study suggest that the local longitudinal shortening provides for a mechanical advantage in coordination with the circular contraction to help reduce the amount of power required to perform peristaltic activity (Pal & Brasseur, 2002). As per the author, when observed from the mechanical perspective, local longitudinal shortening can help in gaining mechanical advantage when associated with circular contraction. This can be explained with Laplace equation for a thin walled cylindrical tube by considering force balance equation between net pressure (ΔP) acting wall and the average tension (T_{ave}) developed on the intestinal wall; where the average tension is,

$$T_{ave} = R_{int} \frac{\Delta P}{t}. \tag{1}$$

Where, R_{int} is the internal radius of the circular tube and t is the thickness of circular muscle fiber. Using this equation, difference between intraluminal pressure and external (thoracic) pressure across the wall, ΔP can be determined. The average tension times the effective surface area of the muscles corresponds to the amount of force required to propel the chyme in forward direction.

As per the equation, T_{ave} is inversely proportional to the thickness of circular muscle fiber, which states that with increase in thickness of muscle layer, T_{ave} can be reduced. Shortening of longitudinal muscle at the region of circular contraction, brings circular muscle fibre closer to each other. This accumulation of circular muscle fiber leads to increase in thickness, which eventually reduces the tension required for forward propelling of the chyme.

Apart from mechanical advantage, LLS also helps in transport of bolus by weakening or strengthening the segment in the vicinity of the contraction region. During peristalsis, the longitudinal muscle fiber of gastrointestinal wall present at the ahead of digesta contracts whereas circular muscle fiber in that segment relaxes to its original diameter. This simultaneous action of both the muscle fiber causes expansion of luminal tube. The segment of GI tract present behind of bolus undergoes opposite behavior to that of the segment present ahead, i.e. the circular muscle contracts and the longitudinal muscle relaxes. As a consequence of this, the segment ahead of bolus acts as a receiving segment and the one present behind the bolus acts as propulsive segment. Propulsive segment pushes the bolus to the receiving segment due to simultaneous action of these two muscle fibers. This causes forward movement of luminal content (Wood, 2004). Similar mechanism has been observed in the esophageal segment of opossum with the help of strain gauge. This experiment showed that both the types of muscle fiber coordinate with each other during peristalsis. Coordination of these fibers help in peristalsis activity to be performed at relatively lower power as compared to the amount of power required to perform peristalsis with the help of single type of muscle fiber (Sugarbaker, 1984a; Sugarbaker, 1984b).

The peristalsis in small intestine has not been clinical evaluated for the significance of LLS, however has been studied from a computational perspective. As per the computational modelling, it has been demonstrated that peristalsis is involved in transport as well as mixing of the luminal content in small intestine depending of the degree of occlusion of peristaltic wave (Avvari, 2019b; Avvari, 2019c).

MATHEMATICAL MODELLING

Multiple studies on mathematical modelling of peristalsis have been reported; where the small intestine has been considered as a cylindrical tube of uniform cross section of radius (R, $R = 2D$, where D is the diameter for cylindrical tube) and axial length (L). Peristaltic wave in 2D can be visualized as a sinusoidal wave travelling along the axial length with occlusion (Figure 2). We can apply lubrication theory in the current context where the amplitude of wave is significantly small as compared to the wavelength. In this modelling, a fluid of power index 'n', with viscosity (μ) as 1000 *Pa.s* and density of 1 kg/cm^3 is considered for a wave moving at a velocity of 1 cm/s. For an intestine of diameter 2 cm and ratio of radius to axial length of 0.02, the Reynold's number (*Re*) becomes very small (*Re*<<1) so that the flow can be described as laminar in the lumen (Avvari, 2019b; Avvari, 2021d).

In general, we describe the fluid model for the peristalsis for a wave travelling along small intestine of radius (a), wave speed (c), wavelength (λ) and wavenumber (k), where axial and radial velocity component of fluid in the peristaltic flow being denoted as u_r and u_z in this section as follows, where the subscript 'r' and 'z' represents the radial and axial direction respectively. The parameters (denoted by *) are converted in non-dimensional form with the help of these characteristics parameters.

$$k = \frac{a}{\lambda} \quad t = \frac{t^*}{\lambda / c} \quad z = \frac{z^*}{\lambda} \quad r = \frac{r^*}{a} \quad u_z = \frac{u_z^*}{c} \quad u_r = \frac{u_r^*}{kc} \tag{2}$$

$$P = \frac{P^*}{\mu \lambda c^n / a^{n+1}} \quad \tau = \frac{\tau^*}{\mu \left(c / a \right)^n}.$$

General equation for modelling fluid flow in small intestine in cylindrical coordinate system using lubrication theory is given by (Avvari, 2019b),

$$\frac{\partial P}{\partial r} = k^2 \frac{1}{r} \frac{\partial}{\partial r} \left(r \frac{\partial u_r}{\partial r} \right) + k^4 \frac{\partial^2 u_r}{\partial z^2} - k^2 \frac{u_r}{r^2}. \tag{3}$$

$$\frac{\partial P}{\partial z} = sgn \left(\frac{\partial u_z}{\partial r} \right) \frac{1}{r} \frac{\partial}{\partial r} \left(r \left| \frac{\partial u_z}{\partial r} \right|^n \right) + k^2 \frac{\partial^2 u_r}{\partial z^2}. \tag{4}$$

$$\frac{1}{r} \frac{\partial (r u_r)}{\partial r} + \frac{\partial u_z}{\partial r} = 0. \tag{5}$$

The pressure inside the lumen is denoted as *P.* Applying lubrication theory approximation to this problem, where the radial dimension is assumed to be significantly small compared to axial dimension, the wave number becomes very small ($k<1$), thus neglecting the higher powers of *k* in the above governing equation simplifies to the form,

$$\frac{\partial P}{\partial r}=0. \quad (6)$$

$$\frac{\partial P}{\partial z}=sgn\left(\frac{\partial u_z}{\partial r}\right)\frac{1}{r}\frac{\partial}{\partial r}\left(r\left|\frac{\partial u_z}{\partial r}\right|^n\right). \quad (7)$$

$$\frac{1}{r}\frac{\partial(ru_r)}{\partial r}+\frac{\partial u_z}{\partial r}=0. \quad (8)$$

We have the following boundary conditions,

$$\left.\frac{\partial u_z}{\partial r}\right|_{r=0}=0,\ u_z\big|_{r=R}=U_{z,B}\left(z,t\right).$$

$$u_r\big|_{r=0}=0,\ u_r\big|_{r=R}=U_{r,B}\left(z,t\right)=\frac{dR\left(z,t\right)}{dt}. \quad (9)$$

$$P\big|_{z=0}=P_0,\ P\big|_{z=L}=P_L,$$

Where $U_{z,B}(z,t)$ and $U_{r,B}(z,t)$ denotes the axial and radial velocity of the intestinal wall respectively at position *z* and time *t*.

Local longitudinal shortening of longitudinal muscles, travelling with the same speed long axial direction (z-direction) as that of contraction, can be visualized as a narrow segment of length (l_0) of small intestine contracting to a length of (l). Local longitudinal shortening can be defined as (Pal, 2002),

$$s_{LLS}\left(z,t\right)=\frac{l\left(z,t\right)}{l_0}. \quad (10)$$

Axial velocity of the wall is driven by the local longitudinal shortening by the equation (axial boundary velocity),

$$U_{z,B}\left(z,t\right)=\frac{\partial}{\partial t}\left(\int_0^z s\,dz\right). \quad (11)$$

The wall movement of small intestine due to peristalsis will be governed by axial and radial velocity due to circular and longitudinal contraction by,

$$U_{r,B}(z,t)=U_{z,B}\frac{dR}{dz}+\frac{dR}{dt}. \tag{12}$$

Solving the above governing equations using the boundary conditions gives the solution for axial and radial velocity of fluid flow due to peristalsis. A typical flow pattern is shown by a streamline in figure 2, with a forward transport of the fluid at the far end relative to the wave and a flow in the reverse direction in the occlusion region formed by the wave. As a result of this, there is stagnation of flow at the head-end and the other at the tail-end of the wave. We notice that the propagating contraction gives rise to a recirculation in the lumen, with generation of higher velocity at the center. With higher occlusion this velocity increases to shear any fluid particles that comes in to this occlusion region. This can be looked upon as the strategy employed by the peristalsis to shear the food particles and agitate the luminal contents.

Figure 2. Schematic diagram of a small intestinal segment showing a constriction arising from muscular contraction or a peristalsis wave moving along the left to right direction. The blue dash line gives the streamline indicating local fluid velocity in the lumen. A zoom-in view of the mucosal folds is shown indicating plicae circulares, villi and the enterocyte with brush border enzymes (BBE). BBEs are the enzymes present at the apical surface of the enterocytes and participate in the membrane digestion of the food; shown here is the catalysis of polysaccharides into glucose. M refers to the microbes which collectively contribute to the gut microbiota; drives the fermentation of the undigested products.

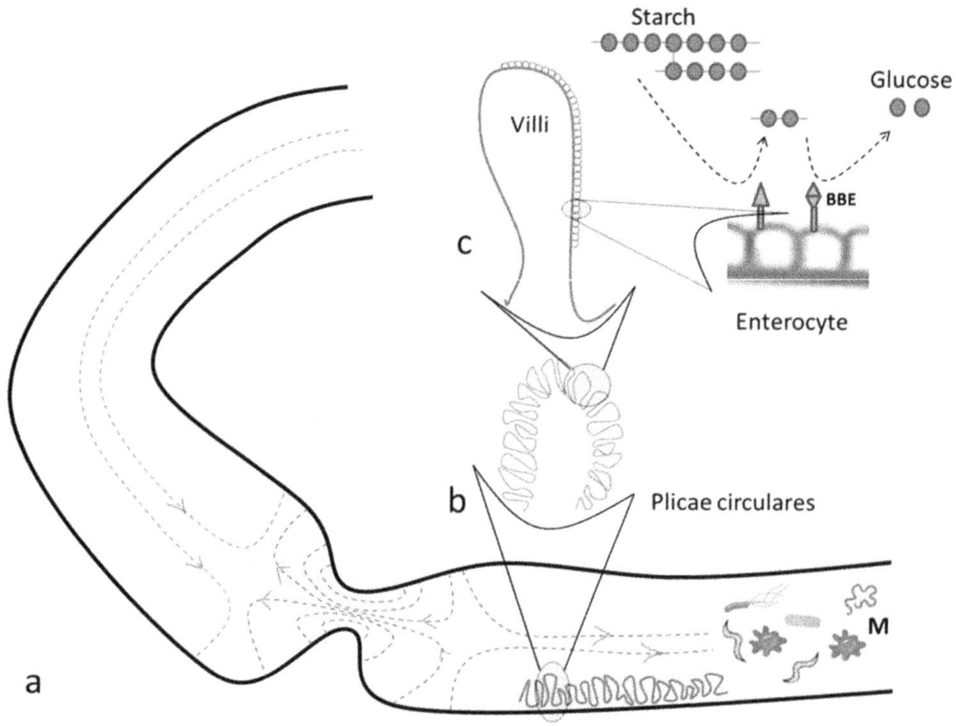

At the surface, the velocity of the fluid is equal to the local velocity of the interior surface of the intestinal wall and satisfies the no-slip boundary condition. With local longitudinal shortening, there is translation of the intestine wall along its surface, which as a consequence leads to the development of a tangential velocity component (Avvari, 2019b).

Shear stress at the wall of small intestine due to fluid flow can be calculated as,

$$(z,t) = \left.\frac{\partial u_z}{\partial t}\right|_{r=R} . \tag{13}$$

FACTORS AFFECTING THE SMALL INTESTINAL DIGESTION

Digestion is majorly driven by mechanical forces, through action of peristalsis, however depends on various factors as discussed in the following.

Nature of Wave

The small intestine shows two types of motor patterns- peristalsis and segmentation. Peristalsis is associated with the movement of digesta, whereas segmentation is associated with the mixing of luminal content to help in absorption of nutrients and water in the further sections of the gastrointestinal tract. There are two types of peristaltic waves observed in the small intestine- antegrade propagating wave (APW) and retrograde propagating wave (RPW). APW is responsible for forward movement of digesta in the anal direction and RPW is responsible for backward movement in the oral direction, which if prominent, can cause Duodenogastric Reflux (DGR). On the other hand, segmentation is a type of contraction which divides the small intestine into several sections and these sections contracts simultaneously. During contraction, the chyme present adjacent to the occlusion tends to move away from the peak of occlusion in both the direction. This causes local mixing of chyme with enzymes and secretions present in the small intestine and also promote nutrient and water absorbtion by the villi present at the walls of intestine (Avvari, 2015; Huizinga et al., 2014). Stationary contractions which do not propagate also appear in the motility (Avvari, 2021a). These contractions facilitate mixing via shearing of the contents while allowing for flow in the either directions of the occlusion zone (Avvari, 2021a).

Occlusion of the Wave

When the wave is traveling with certain degree of occlusion, the momentum is transferred to the immediate particles lying close to the walls of intestine. Due to this momentum transfer, the author demonstrated based on his computational approach, two types of particle trajectory of the luminal content can be observed, radial and axial. When the wave is traveling with degree of occlusion less than 50%, radial trajectory of the particles was dominant, causing mixing of the content, whereas, when it is travelling with more than 80% of occlusion, axial displacement of particles was observed to be dominant, making particle to move forward along the intestine (Avvari, 2019b).

Frequency of the Wave

A standing wave (SW) is mainly responsible for breaking down and mixing the intraluminal content of the small intestine. The change in intensity and volume of mixing for change in wave frequency is not significant compared to APW and RPW wave velocity. Only some mixing was observed in SW at a frequency of 6Hz.

Wave Velocity

Mixing of intraluminal content is very sensitive to the wave velocity of APW and RPW. It has been seen that, the transpyloric flow (i.e. flow from stomach to small intestine via pylorus) increases with wave velocity. Furthermore, it also affects the intensity and volume of mixing in a similar manner. With an increase in velocity from 1cm/s to 4cm/s, degree of mixing increases by ten folds. Mixing index (which determines the degree of dispersion of solute or suspended material to cause homogeneity of solution) in duodenum is observed to be greater than 1.005. For standing wave (SW), wave frequency is considered rather than velocity. (Avvari, 2015)

Multiplicity of the Wave

We now state the effect of an increasing number of waves on pumping, the intensity of mixing, and the volume of mixing for the three contraction types. An increase in the number of waves showed an increase in pumping effect from stomach to duodenum for all three contraction types. In terms of mixing, for APD contraction, the intensity of mixing can wither increase or decrease, on the other hand, for SW and RPW, the intensity of mixing was observed to increase with wave multiplicity. In the case of volume of mixing, SW showed a negligible effect, whereas, for APW and RPW, a direct relationship was observed.

Rheology of the Content

Fluid can be categorized as Newtonian and non-Newtonian fluid (Takahashi, 2011). Rheology of the content also modulates the flow in the lumen (Avvari, 2021b). Considering the peristaltic wave to move from left to right hand side, the left side region in the vicinity of occlusion creates lower pressure due to relaxation of circular muscle fiber, whereas the right side region shows higher pressure due to contraction as compared to the pressure at the occlusion region. Due to lower pressure generated at the left side, this causes backward movement of chyme. Thus, it can be predicted that for shear-thickening fluid, backward flow can be higher as compared to the shear-thinning fluid. For non-Newtonian fluid of flow behavioral index, n, it has been studied that the pressure gradient from left to right side of occlusion increases with increase in the value of n. Similar cases can be correspondingly observed for wall shear stress. There is development of higher shear stress for shear-thickening fluid ($n>1$) as compared to shear-thinning fluid ($n<1$). With increase in wall shearing values, the axial velocity at the occlusion region is also increased. Hence, it can be reported that the rate of global transport of luminal fluid can be seen faster for shear thickening fluid and slower for shear thinning fluid (Avvari, 2019b).

DIGESTION OF FOOD

The human digestive system offers a diverse environment for the digestion of food. In stomach, it encounters a highly acidic environment (pH 1.5–3.5) which eliminates any microorganism that enters into the body via ingestion of food. This ensures that the food is free from any microbes and prevents any kind of infection. There are few exceptional cases where the microorganisms can sustain such harsh environment such as *H. Pylori*, which employs urease dependent and urease independent mechanisms for the survival (Scott et al., 2002; Ansari & Yamaoka, 2017). The stomach however does not support the growth of microorganisms; as evident from the reduced concentration of microbes ~10^3-10^4 number/cc (O'Hara & Shanahan, 2006; Sender et al., 2016). It also implies that the microbial competition for food is hereby suppressed and eliminates any possibility of microbial foraging in the stomach and upper bowels. Stomach facilitates the digestion of food under harsh acidic condition where enzymes such as pepsin and gastric lipase function to breakdown the proteins and fats. Carbohydrates however are not broken down chemically in the stomach due to lack of enzymes to digest it.

As the chyme (food from the stomach) enters into the small intestine it undergoes transition in terms of pH due to neutralizing effect of the alkaline secretions from pancreas and duodenal mucosa. The small intestine provides a platform for digestion of diverse food components through provision of various enzymes and provision of suitable environment for its optimal function. If the enzymes are secreted by the pancreas directly into the lumen, it facilitates the luminal digestion, whereas digestion occurring at the mucosa (by virtue of having catalytic enzymes at the apical surface of the enterocytes, also known as brush border enzymes) facilitates membrane digestion. Both these types of digestion contribute to the metabolism of various food components. To facilitate the process and timely digestion of food, the small intestine provides a large surface area of contact for the intestinal contents with the mucosa through formation of villi and microvilli (Figure 2).

The food ingested comprises of carbohydrates, proteins and fats, with carbohydrates typically accounting for more than 50% of the energy in terms of calories on an average. The chemical breakdown of the food is driven by the digestive enzymes which cleaves specific bond such as the glycosidic bond in carbohydrates and peptide bond in protein. The chemistry of the digestion involves in food breakdown is discussed in the following with reference to its composition.

Carbohydrates

Carbohydrates form the main source of energy for human body. They occur in either simple or complex form – simple carbohydrates comprises of monosaccharides, disaccharides, trisaccharides and complex carbohydrates. Simple sugars exits in form of monomer units such as the glucose, galactose and fructose (monosaccharides), sugars with two monomer units such as sucrose, maltose and lactose (disaccharides) are also commonly found. Complex sugars such as starch, fibers and glycogen have more number of monomer units bounded by 1-4 glycosidic bond and 1-6 glycosidic bonds. Starch contributes to major source of energy as they are very commonly found in stable foods. It occurs in two forms – amylose (glucose units are linked in a straight chain with alpha 1-4 glycosidic bond) and amylopectin (polymer of glucose unit with branching using alpha 1-6 glycosidic bond). Fibers (such as cellulose, hemicellulose, lignin and pectin), which are consumed at lower rate, are made up of many sugar units bonded together by beta glycosidic bond; render them mostly indigestible by the digestive enzymes. Digestion is driven by the breakdown of the bonds. For example, the pancreatic α-amylase enzyme cleaves the alpha 1-4 glycosidic bonds in starch into smaller units such as the maltose, maltotriose, and α-limit dextrins; contributes to luminal digestion. The smaller units such as the disaccharides (maltose) and trisaccharides (maltotriose) are further hydrolyzed by the brush border enzyme (such as maltase-glucoamylase and sucrase-isomaltase) glucose, galactose, and fructose molecules, which can then be absorbed by enterocytes and delivered into the bloodstream (Levin, 1994; Southgate, 1995). There are two active sites in maltase glucoamylase. The maltase site hydrolyzes the maltose and maltotriose's terminal 1,4-linked D-glucose residues to produce D-glucose. In addition to hydrolyzing 1,4 connected glycosidic links when next to a 1,4 linkage, the glucoamylase site hydrolyzes 1,6 glycosidic linkages, resulting in D-glucose. Sucrase isomaltase likewise contains two active sites, each with its own molecular specificity. The sucrase site hydrolyzes sucrose into glucose and fructose, and it can also partially hydrolyze maltose. The isomaltase component catalyses the digestion of 1,6-dextrin bonds.

Before the polysaccharides are further hydrolyzed or transported, any disaccharide or free monosaccharide generated in the bulk phase must diffuse across the enterocyte's surface. At the surface, water becomes more densely structured, forming an unstirred layer that stretches outwards into the bulk phase. The sort of stirring present in the gut lumen determines the thickness of this unstirred layer. If just linear flow is present, the unstirred layer thickness can be rather thick; however, if turbulent flow is present or the villi are motile, the unstirred layer thickness can be quite small. Early research revealed that the unstirred layer in several small intestines was large and thus served as a diffusion barrier for molecules, particularly those that were highly permeant across the mucosa. This could be a crucial component in obtaining a precise quantitative characterization of the kinetics involved in hexose transfer (Levin, 1994)

Active transport, facilitated diffusion, and passive diffusion are three ways for movement through the enterocyte brush boundary membrane. The active transport is driven by the sodium-dependent transporter present on the enterocytes which allows the entry of the glucose into the cell. This glucose eventually enters into the capillaries through diffusion from the epithelial cells. Galactose is delivered similarly to glucose, using the same transporters. Galactose is absorbed predominantly from the breakdown of lactose because it is not found as a monosaccharide in nature. Fructose is transported fully through facilitated diffusion. When fructose enters the enterocytes, it uses a different transporter than glucose; yet, when fructose and glucose escape the enterocyte and enter the capillaries, they use the same transporter. Fructose absorption is significantly slower than glucose absorption and is quantitatively constrained (Levin, 1994)

Proteins

Protein also contributes to significant portion of the diet. The luminal digestion is facilitated by the pancreatic enzymes and digestion at the mucosa by the brush border enzyme. The enzymes for protein digestion are triggered in the small intestine's duodenum with the alkaline secretions from pancreas, resulting in an increase in pH with a consequence of inactivating the gastric enzymes while activating enzymes secreted by the pancreas. Enteropeptidase, a brush boundary enzyme found in duodenal enterocytes, initiates pancreatic enzyme activation. Endopeptidases such as trypsin, chymotrypsin, and elastase break the bond present interior to the protein. The endopeptidases carboxypeptidases A and B convert proteins into oligopeptides. In humans, intestinal protein digestion is more important than gastric digestion for protein assimilation. (Erickson & Kim, 1990; Gary, 1971).

The proteolytic enzymes specializing in the breakdown of the peptide bond can be classified into two – exopeptidase (cleavage at the terminus) and exopeptidase (cleavage within the chain). Some of these enzymes target C-terminal residues, whereas others target N-terminal residues. They produce free dipeptides or amino acids as a result of their activity. Aminopeptidase, for example, has a high specificity. Internally, endopeptidases cut peptide chains, resulting in smaller peptide chains. Enteropeptidase (or enterokinase) is a brush border enzyme that catalyses conversion of trypsinogen to trypsin, a key pancreatic protease. Mostly found in the enterocytes of the duodenum. It helps activate trypsin, chymotrypsin, elastases, and carboxypeptidases A and B by initiating a cascade of proteolytic processes. The most abundant aminopeptidase N removes the N-terminal amino acids from smaller peptides in a sequential manner. There are further four peptidases which provide relatively faster catalysis, especially by cleaving peptides having a proline residue. Prolyl peptides are catalyzed by the enzymes at its amino terminus by dipeptidyl aminopeptidase IV and aminopeptidase P. While at the carboxy terminus the prolyl peptides are catalyzed by the angiotensin-converting enzyme and carboxypeptidase P. The brush border membrane also contains many neutral metallo-endopeptidases and glutamyl aminopeptidase which also facilitate the digestion at the membrane (Erickson & Kim, 1990).

The small intestine absorbs the amino acids (including short peptides) using active transport systems (using transporter such as PepT1) and eventually enters the circulation for the first time. Those entering the enterocytes are catalyzed by the cytoplastic peptidases. Although the small intestine absorbs about 95% of dietary protein, the remaining amino acids, undigested proteins, and unabsorbed peptides, whether dietary or endogenous, enter the large intestine. These components are digested further by the intestinal microflora in the large intestine. The small intestine acts as a link between the gut lumen and the rest of the body, controlling the amount and pace of amino acid transport from dietary protein to the liver and systemic circulation via the portal vein. Because some amino acids are used for local metabolism, not all amino acids that pass through the gut enter the circulation (e.g., oxidation, protein synthesis) (Have, 2007).

Fats

The digestion of fats in the duodenum is facilitated firstly by emulsification, where bile salts participate in the digestion by allowing the conversion of the fat globules into a miniscule, that is, micelles. Micelles provide a large surface area for lipid digesting enzymes to catalyze the reaction; a mechanism which amplifies the digestion of fat globules in comparison to case without emulsification. The three major enzymes involved in lipid digestion are pancreatic lipase, phospholipase A2 and cholesterol ester hydrolase, which participate by hydrolyzing the micelles, breaking them down into fatty acids, monoglycerides, cholesterol and lysolecithin. Lipolysis (lipid triglycerides are hydrolyzed into a glycerol and three fatty acids) takes place in the cytoplasm, where the fatty acids are oxidized by beta-oxidation into acetyl CoA. After lipolysis, the glycerol enters the glycolysis pathway as dihydroxyacetone phosphate (DHAPJ) (Ahmadian et al., 2011). Phospholipase B1, neutral ceramidase, and alkaline sphingomyelinase are the most important brush border enzymes for fat digestion. Pancreatic enzymes do not breakdown sphingolipids. The hydrolysis of these lipids is carried out by the brush border alkaline sphingomyelinase and neutral ceramidase. Phospholipids and sphingolipids can be found in minute levels in almost every meal, but they're especially abundant in eggs, soybeans, cream, and cheese (Hooton et al., 2015).

At the apical membrane, the products are released and diffuse into the enterocyte. Fat digestion is further taken forward in the cell, with further re-esterified to produce lipids, triglycerides, cholesterol, and phospholipids. Inside the cell the fatty acids and monoglycerides are transported into the endoplasmic reticulum which is used for packaging into chylomicrons. Chylomicron makes its way into the lymphatic circulation (Carey et al., 1983; Phan, 2001).

FERMENTATION IN THE SMALL INTESTINE

Microbial foraging gives a new dimension to the digestion in the gut. In physiology the oral route of microbial entry is highly regulated by the harsh acidic conditions of the stomach, whereas the entry from the anal route has not such restriction, except for the case that the microbial growth is regulated by the immune system of the gut. This suggests that the anal route has no much restriction and allows for flourishing of the microbial colonies that establishes a symbiotic relation with the gut. It gets clear that there a gradient of bacterial concentration across the gut. The concentration of microbes is low at the duodenum (but significantly higher in comparison to stomach), however increases few fold as we go down the small intestine. The bacterial concentration of the digestion in colon reaches its maximum of ~10^{12}-10^{14} number/cc (O'Hara & Shanahan, 2006; Sender et al., 2016).

Not all the food ingested undergoes complete digestion. The gut flora participates in the digestion through fermentation of the undigested and indigestible food components (Figure 2). It helps in further metabolism of the food components by provision of enzymes (these are not produced by the body such as the cellulase) to simpler compounds. However, due to relatively lower concentration of the microbes, fermentation occurring in the small intestine amounts to a small fraction in comparison to the colon. The microbial community is very large and majorly dominated by Bacteroidetes, Firmicutes, Actinobacteria, Proteobacteria, and Verrucomicrobia. It also includes fungus Candida. These essentially contribute to further digestion and digestion of indigestible food components such as cellulose. It also contributes to the formation of faeces in the process. While the microbes settle in symbiotic relation with the gut, it also benefits the gut in some way. Colonic bacteria synthesize vitamin K, which is a valuable addition to food supplies and renders clinical vitamin K insufficiency uncommon. On the other sides, it can also affect the digestion, if existing in greater numbers as in case of small intestinal bacterial overgrowth or SIBO. The digestion is preferably anaerobic and helps in conversion of the undigested food into short-chain fatty acids (SCFAs). Along with this, there is an increase in the formation of gaseous matter such as methane, hydrogen, and carbon dioxide as the principal end products of fermentation. Considering the case of hydrogen produced during fermentation, it diffuses into the systemic circulation, makes its way through the pulmonary circulation to the lungs and expelled out via diffusion at the alveoli. It is through this mechanism that the gas produced during fermentation enters the respiration system, enabling the detection by breath test. The gases H_2 and CO_2 are the two important gases for detection of SIBO (Zoetendal et al., 2012; King et al., 1984).

The mechanical relevance of fermentation comes into action via elicitation of peristalsis which drives the fluid motion to cause mixing (buffering of the chyme with alkaline secretions and mixing with digestive enzymes secreted into the lumen), absorption and transport. Together with this, peristalsis also drives the movement of the microbes residing in the bowels (small and large intestine). Transient infections are also moved out of the body by the continual movement of leftover matter through the gastrointestinal tract. The transport also serves the purpose of eliminating any toxins released into the lumen. While the normal microbiota provides a second line of defence against infection through a variety of ways, it also competes with potential infectious microbes for space and nutrients in the intestine.

PATHOPHYSIOLOGY OF SMALL INTESTINAL DYSMOTILITY

Any disorders in the motility patterns (such as myopathy and neuropathy) of the small intestine can lead to several pathophysiological conditions. If the intestinal muscles do not elicit proper contractions of an appropriate strength or motility patterns of reasonable amplitude, speed and occlusion (dysmotility), it may reduce the digestive capacity of the intestine and the efficiency of digestion (Avvari, 2021e). Further motility dysfunctions also give rise to numerous digestion problems leading to indigestion or dyspepsia, discomfit, abdominal pain, and bloating (non-organic symptoms). When motility dysfunctions lead to non-organic symptoms with no damage to the tissue involving the digestive system it is referred to as the functional gastrointestinal disorder or FGID. In functional constipation, there is an infrequent bowel movement leading to hard stool formation. When partially digested food remains in intestine for prolonged period of time, it leads to large water absorption forming hard stools. This can be due to slower velocity of peristaltic wave, longer periods of segmental contraction in small intestine, or lack of high amplitude propagating contraction wave in large intestine. Major symptoms related to constipation include painful defecation and feeling of incomplete emptying after bowel movements (Klaschik et al., 2003).

In another physiological condition of passing of loose and urgent stools or functional diarrhea, there is a rapid movement of peristaltic wave in small or large intestine causing rapid transit of the digesta. This can lead to malabsorption of nutrients and water, causing loose or watery stools. Diarrhea can be caused due to various factors like, change in diet, pancreatic, small or large intestinal disorders, increased flow through ileo-caecal valve (sphincter muscle present at the junction of small and large intestine), and lower water absorption at large intestine. Diarrhea can be classified into two categories depending on the causing mechanism – distension and chemical irritation. The result of these mechanisms leads to motility disorders as a secondary response. Distension diarrhea is caused due to rapid movement of chyme (usually of the rate of 20ml/min) in small intestine because of faster peristaltic activity. During this rapid movement, there is less time available to absorb nutrients in the small intestine. Malabsorption of lactose or higher intake of fructose or sorbital can also lead to this type of diarrhea. On the other hand, chemical irritation diarrhea is a response of body to protect it from the toxins from chemical, bacterial or plants origin. When such toxins are detected in the small intestine, rapid peristaltic activity is elicited in response to it to remove through the formation of loose stools (Spiller, 2006).

If by some reason, the passage of digesta is blocked, either partially or fully, also known as Intestinal pseudo-obstruction, it leads to symptoms like nausea, vomiting and abdominal pain. The movement of digesta in small intestine is governed by the action of nervous system and intestinal muscles to generate the contraction wave. Any abnormality in these systems directly affects the functioning of small intestine and digestion. In case of chronic intestinal pseudo-obstruction (CIP), abnormal and inefficient peristaltic movement is observed and symptoms are very similar to mechanical obstruction (due to tumour or scar tissue formation) in the small intestine, however, no such obstruction is reported; known as pseudo-obstruction. When weak or absence of peristaltic abnormality is observed, it is classified as myopathic (due to abnormalities in the functioning of muscles), whereas, if unsynchronised peristaltic activity is observed it is classified as neuropathic CIP (involving abnormalities in the functioning of nerves) (Colemont & Camilleri, 1989). CIP affects a small section of the small intestine which shows abnormality in the peristaltic activity. In case of myopathic CIP, due to the above-mentioned abnormaltilies, it can be suggested that the digestion and the nutrient absorption process can be affected which gives rise to symptoms like nausea, vomiting and abdominal pain. Because of long time retention of digesta in the intestine with less motility activity, nutrient absorption is drastically affected. If CIP is prominent for a longer duration this can lead to malnutrition in children as well as adults. Such a situation can affect the quality of life of an individual (Gabbard & Lacy, 2013).

In situations where there is reflux of duodenal contents into the stomach or duodeno-gastric reflux (DGR), there is an increased risk of damage to the gastric mucosa due to the action of bile in the highly acidic environment and possibly cancerous formation. Patients with DGR show symptoms of nausea, vomiting, indigestion and abdominal pain. The reflux is driven by the retrograde movement of digesta from the small intestine to stomach. Pylorus, a muscular valve, plays a key role in regulating the flow of chyme into the duodenum. During opening of pylorus, suction is created which pulls the content from either side of the pylorus (stomach and duodenum) and during closure the pressure increase in the vicinity pushes the content on either sides (Avvari, 2021c). During opening, when the pressure gradient in the stomach side of pylorus is higher than the duodenal side, the process of gastric emptying (GE) takes place, i.e. passage of partially digested food from stomach into the duodenum, whereas when the pressure at the duodenal side is greater, the flow is in reversed direction. Similar reversed direction flow can be observed in the closing period, when the high volume of content is pushed back into the stomach. If the reversed flow is prominent through the pylorus, such a condition is called DGR (Avvari, 2020).

Whereas the fermentation is inevitable, the gut microbes are also responsible for causing abdominal bloating, diarrhea and discomfort. When the numbers of bacteria present in the small intestine increases beyond a certain limit we refer to the condition as the small intestinal bacterial overgrowth or SIBO. In healthy individuals, the normal value of small intestinal bacteria count is less than10^4 cfu/ml (cfu refers to colony forming unit). When the count becomes greater than 10^5 cfu/ml, it becomes a pathophysiological problem. One of the major causes of SIBO is reverse peristalsis (retrograde propagating wave) of digesta from large intestine to small intestine. Other causes can be due bacteria coming from ingested food, insufficient pH balance, bacterial adhesion to the walls of small intestine, intestinal pseudo-obstruction, etc. Our body has several defense mechanisms to prevent overgrowth of bacteria such as – eliciting antegrade propagating wave (to prevent bacterial from adhering on to the gut), regulating microbial entry via oral route (by use of gastric secretion of acid), and eliminating the microbes (using gastric acid and bile secretion in small intestine), intact ileo-cecal valve (sphincter muscle present at the junction of small intestine and large intestine to prevent backward movement of digesta form the large intestine rich in microbes), and action of immune system (to prevent overgrowth) (Bures et al., 2010). In irritable bowel syndrome (IBS), patients show symptoms of abdominal pain, abdomen swelling due to gas and infrequent bowel movements. IBS is directly associated with SIBO and motility disorders. Strong intestinal contractions when travelling rapidly causes diarrhea and bloating whereas, and travelling with a slower velocity increase the chances of constipation (Whitehead et al., 1980; Camilleri & Choi, 1997).

Dysmotility can also affect nourishment of an individual through indigestion or unable to digest the food completely. It is advisable to regulate intake of food rich in lactose, gluten, and FODMAP; especially in those individual who cannot digest such food components. In lactose intolerance, the individual cannot digest the food rich in daily products; due to lack of sufficient amount of lactase, an enzyme that catalyzes the hydrolysis of the lactose into glucose and galactose. Lactose is the most abundant carbohydrate present in the milk and catalyzed by lactases which are most abundantly present in the small intestinal brush border lining of the small infants. Lactase expression starts plummeting shortly after weaning. As a consequence, many adults face with the issue of lactose intolerance due to the non-persistence of lactase activity (Hooton et al., 2015). Similarly for food rich in gluten, certain individual show allergic responses to the gluten, most prominent being in celiac disease (gluten intolerance). Food rich in FODMAP constituents such as the short-chain carbohydrates also have trouble in digesting it (FODMAP intolerance).

CONCLUSION

Peristalsis plays an essential role in the digestion of food in the small intestine. The process is mechanical and driven by the smooth muscle contractions of the small intestine, comprising of two muscle fibers – circular and longitudinal smooth muscular fibers. It is the contractility of these muscles which determine as to how the food is digested. In this chapter, we have studied the biomechanics of peristalsis using a mathematical model to define the fluid flow (or flow pattern) resulting from the intestinal motility. Studies indicate that the small intestine employs both the muscle fibers to elicit various types of contraction to facilitate the digestion. For example, at higher occlusion, APW transfers the content in the forward direction, RPW transfer it in backward direction and SW helps in the segmentation. The function is altered with changing geometry of the contraction wave, where the contraction also facilitates the mixing, besides transport as in APW and RPW. We have also reviewed digestion from the biology point of view, the digestion process of carbohydrates, proteins and fats taking place in the intestine – luminal and membrane digestion. The microbes in the small intestine are inevitable, as it contributes to the digestion symbiotically in exchange for a benefit. When the peristalsis cease to function in a normal way or the microbial population increases beyond a limit, it give rise of various pathological condition or motility disorders. In conclusion, we can say that the peristalsis plays an important role in overall functioning of small intestine and any disorder in its motility can cause several medical problems.

REFERENCES

Ahmadian, M., Wang, Y., & Sul, H. S. (2010). Lipolysis in adipocytes. *The International Journal of Biochemistry & Cell Biology*, *42*(5), 555–559. doi:10.1016/j.biocel.2009.12.009 PMID:20025992

Ansari, S., & Yamaoka, Y. (2017). Survival of Helicobacter pylori in gastric acidic territory. *Helicobacter*, *22*(4), e12386. doi:10.1111/hel.12386 PMID:28402047

Avvari, R. K. (2015). *Bio-mechanics of the distal stomach and duodenum: An insight into mechanisms of duodenogastric reflux and duodenal mixing*. Department of Biological Sciences and Bioengineering.

Avvari, R. K. (2019a). Managing motility disorders of the antro-pyloro-duodenal segment: A biomedical engineering perspective. *United J Biochem Biotechnol*, *1*, 1–20.

Avvari, R. K. (2019b). Effect of local longitudinal shortening on the transport of luminal contents through small intestine. *Lixue Xuebao*, *35*(1), 45–60. doi:10.100710409-018-0809-5

Avvari, R. K. (2019c). Biomechanics of the small intestinal contractions. In *Digestive system- Recent advances*. IntechOpen.

Avvari, R. K. (2020). Enteric and Central Nervous System Mediated Control of Digestive Processes in the Small Intestine: A Coprocessor-Processor Paradigm. *Food Science and Engineering*, 39-44.

Avvari, R. K. (2021a). Effects of stationary contraction of the small intestine on digestion. *Journal of Mechanics in Medicine and Biology*, *21*(02), 2150021. doi:10.1142/S0219519421500214

Avvari, R. K. (2021b). Is Rheology a Concern in GI Physiology? *Evolutions in Mechanical Engineering*, *3*(3), 1–4.

Avvari, R. K. (2021c). Theoretical modeling of the resistance to gastric emptying and duodenogastric reflux due to pyloric motility alone, presuming antral and duodenal quiescence. *Journal of Theoretical Biology*, *508*, 110460. doi:10.1016/j.jtbi.2020.110460 PMID:32891592

Avvari, R. K. (2021d). A novel two-indenter based micro-pump for lab-on-a-chip application: Modeling and characterizing flows for a non-Newtonian fluid. *Meccanica*, *56*(3), 569–583. doi:10.100711012-020-01303-1

Avvari, R. K. (2021e). Managing motility disorders of the gastrointestinal segment and obesity through electrical stimulation. *Health and Technology*, *11*(6), 1175–1189. doi:10.100712553-021-00590-2

Bures, J., Cyrany, J., Kohoutova, D., Förstl, M., Rejchrt, S., Kvetina, J., & Kopacova, M. (2010). Small intestinal bacterial overgrowth syndrome. *World Journal of Gastroenterology*, *16*(24), 2978. doi:10.3748/wjg.v16.i24.2978 PMID:20572300

Camilleri & Choi. (1997). Irritable bowel syndrome. *Alimentary Pharmacology & Therapeutics*, *11*(1), 3-15.

Carey, M. C., Small, D. M., & Bliss, C. M. (1983). Lipid digestion and absorption. *Annual Review of Physiology*, *45*(1), 651–677.

Colemont, L. J., & Camilleri, M. (1989, January). Chronic intestinal pseudo-obstruction: Diagnosis and treatment. *Mayo Clinic Proceedings*, *64*(1), 60–70. doi:10.1016/S0025-6196(12)65304-X PMID:2642997

Erickson, R. H., & Kim, Y. S. (1990). Digestion and absorption of dietary protein. *Annual Review of Medicine*, *41*(1), 133–139. doi:10.1146/annurev.me.41.020190.001025 PMID:2184718

Gabbard, S. L., & Lacy, B. E. (2013). Chronic intestinal pseudo-obstruction. *Nutrition in Clinical Practice*, *28*(3), 307–316. doi:10.1177/0884533613485904 PMID:23612903

Hooton, D., Lentle, R., Monro, J., Wickham, M., & Simpson, R. (2015). The secretion and action of brush border enzymes in the mammalian small intestine. *Reviews of Physiology, Biochemistry and Pharmacology*, *168*, 59–118. doi:10.1007/112_2015_24 PMID:26345415

Huizinga, J. D., Chen, J. H., Zhu, Y. F., Pawelka, A., McGinn, R. J., Bardakjian, B. L., & Chen, D. (2014). The origin of segmentation motor activity in the intestine. *Nature Communications*, *5*(1), 1–11. doi:10.1038/ncomms4326 PMID:24561718

King, C. E., Toskes, P. P., & Watkins, J. B. (1984). Breath tests in the diagnosis of small intestine bacterial overgrowth. *CRC Critical Reviews in Clinical Laboratory Sciences*, *21*(3), 269–281. doi:10.3109/10408368409165785 PMID:6439469

Klaschik, E., Nauck, F., & Ostgathe, C. (2003). Constipation—Modern laxative therapy. *Supportive Care in Cancer*, *11*(11), 679–685. doi:10.100700520-003-0525-x PMID:14505158

Levin, R. J. (1994). Digestion and absorption of carbohydrates—From molecules and membranes to humans. *The American Journal of Clinical Nutrition*, *59*(3), 690S–698S. doi:10.1093/ajcn/59.3.690S PMID:8116552

O'Hara, A. M., & Shanahan, F. (2006). The gut flora as a forgotten organ. *EMBO Reports*, *7*(7), 688–693. doi:10.1038j.embor.7400731 PMID:16819463

Pal, A., & Brasseur, J. G. (2002). The mechanical advantage of local longitudinal shortening on peristaltic transport. *Journal of Biomechanical Engineering*, *124*(1), 94–100. doi:10.1115/1.1427700 PMID:11871611

Phan, C. T., & Tso, P. (2001). Intestinal lipid absorption and transport. *Frontiers in Bioscience*, *6*(5), D299–D319.

Scott, D. R., Marcus, E. A., Weeks, D. L., & Sachs, G. (2002). Mechanisms of acid resistance due to the urease system of Helicobacter pylori. *Gastroenterology*, *123*(1), 187–195. doi:10.1053/gast.2002.34218 PMID:12105847

Sender, R., Fuchs, S., & Milo, R. (2016). Revised estimates for the number of human and bacteria cells in the body. *PLoS Biology*, *14*(8), e1002533. doi:10.1371/journal.pbio.1002533 PMID:27541692

Southgate, D. A. (1995). Digestion and metabolism of sugars. *The American Journal of Clinical Nutrition*, *62*(1), 203S–210S. doi:10.1093/ajcn/62.1.203S PMID:7598078

Spiller, R. (2006). Role of motility in chronic diarrhoea. *Neurogastroenterology and Motility*, *18*(12), 1045–1055. doi:10.1111/j.1365-2982.2006.00836.x PMID:17109687

Sugarbaker, D. J., Rattan, S., & Goyal, R. K. (1984a). Mechanical and electrical activity of esophageal smooth muscle during peristalsis. *American Journal of Physiology. Gastrointestinal and Liver Physiology*, *246*(2), G145–G150. doi:10.1152/ajpgi.1984.246.2.G145 PMID:6696111

Sugarbaker, D. J., Rattan, S., & Goyal, R. K. (1984b). Swallowing induces sequential activation of esophageal longitudinal smooth muscle. *American Journal of Physiology. Gastrointestinal and Liver Physiology*, *247*(5), G515–G519. doi:10.1152/ajpgi.1984.247.5.G515 PMID:6496741

Takahashi, T. (2011). Flow behavior of digesta and the absorption of nutrients in the gastrointestine. *Journal of Nutritional Science and Vitaminology*, *57*(4), 265–273. doi:10.3177/jnsv.57.265 PMID:22041908

Whitehead, W. E., Engel, B. T., & Schuster, M. M. (1980). Irritable bowel syndrome. *Digestive Diseases and Sciences*, *25*(6), 404–413. doi:10.1007/BF01395503 PMID:7379673

Wood, J. D. (2004). Peristalsis. Encyclopedia of Gastroenterology, 164-165.

Zoetendal, E. G., Raes, J., Van Den Bogert, B., Arumugam, M., Booijink, C. C., Troost, F. J., & Kleerebezem, M. (2012). The human small intestinal microbiota is driven by rapid uptake and conversion of simple carbohydrates. *The ISME Journal*, *6*(7), 1415–1426. doi:10.1038/ismej.2011.212 PMID:22258098

Chapter 12
Computational Study of In-Vitro Ureter Urine Flow in DJ Stent

Ranjit Barua
https://orcid.org/0000-0003-2236-3876
CHST, Indian Institute of Engineering Science and Technology, Shibpur, India

Pallab Datta
National Institute of Pharmaceutical Education and Research, Kolkata, India

Amit Roy Chowdhury
Indian Institute of Engineering Science and Technology, Shibpur, India

Surajit Das
R. G. Kar Medical College and Hospital, India

ABSTRACT

Many researchers and urologists are presently studying different designs of ureteral stents to advance the feature of their surgeries and the succeeding recovery of the patient. With the aim of help during this design procedure, several computational models have been established to simulate the performance of various biological tissues and deliver an accurate computational environment to estimate the stents. As a result of the high difficulty of the complicated issues, they generally introduce interpretations to create these simulations a smaller amount computationally trying. A DJ stent (double J) is used to improve the blocking of urine in the upper urinary tract while there is ureteral stenosis, which causes the disturbance of normal urine flow and affects renal or kidney failure. The intention of employing a DJ stent is to confirm enough urine flow in the ureter, but the DJ stent performs as a foreign body in the urinary tract and sometimes acts as a difficulty in achieving satisfactory urine flow.

DOI: 10.4018/978-1-7998-9078-2.ch012

INTRODUCTION

At the time of normal urination, urine flows or streams from the kidney to the urinary bladder via ureter in normal fit persons for two causes: (a) the peristaltic movement or peristalsis, and (b) a physical event (Matloubiech et al., 2020). Several usual pathologies which can block or obstruct the ureters are urinary stones or renal stones, infections, tumorous tissues, or inflammations (Park et al., 2020). These difficulties can decrease or even make problems with the normal urination and cause more severe malfunctions or pathologies and threats for patients (Modi et al., 2019). To resolve these difficulties, two dissimilar treatments are generally performed, for example, (a) nephrostomy which provisionally helps to open the urinary tract portion by an external flexible thin tube or catheter, and (b) ureteral stenting technique which attachments a flexible thin tube named stent into the ureter that helps the drain urine from the kidney. Both methods accomplish repairing the urine flow, reducing intra-pelvic stress, and avoiding kidney failure. In the processes while a stent is applied in obstructed ureters, a continuing loss of muscle tone happens due to this ureter stent (Kinn et al., 2002) (Demzik et al., 2021) and the ureter does not create the peristaltic function anymore. Hereafter, stented ureters are merely affected by the pressure gradient.

Figure 1. Refine the 3D model from MRI/CT image

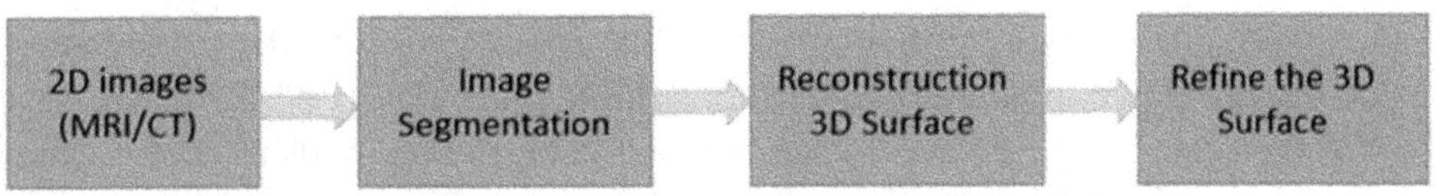

At present, all these conducts are experienced and improvedbyapplying the investigational models that permit mimicking exact precisely the physiological performance under dissimilarboundary circumstances. For urologic investigation, a pig is assumed as the perfect model, because its features (kidney and urethral anatomy and physiology) related to humans (Leonhäuser et al., 2019). Though, a number of studies are complicated or impracticable to be conducted with experimentaltests for dissimilarcauses: dimensionlimitations, ethical concerns, instrumental constraints, etc. Therefore, FEM (finite element models) and another computational software representation have been initiated as a model analysis of the urine flow and for supporting in the design and improvement of innovative ureteral stents (Beysens et al., 2018) (Al et al., 2020) (Oliver et al., 2018). Figure 1 shows flow chart of the procedure of refine the 3D model making. Various experiments have conducted to achieve the biomechanical characteristics of the urethral wall (Taguchi et al., 2018). Different ex-vivo examinations have performedby Yin et al., 1971to find the stress-strainresults of a ureter and describe its biomechanical performs. Afterward, in-vivo characteristics were executed by Sokolis et al., 2012, and presenting the anisotropic performance of ureters that demonstrated with a four-factor Fung-type strain energy function. Figure 2 shows the (a) CT image of urinary system, and the 3D model of (b) front, (c) side view of urinary system. Furthermore, Rassoli et al., 2014 presented important data of human ureter stress-strain results fixed with a four-constraint Fung-type representation and five- constraint polynomial models.

Figure 2. (a) CT scan image of urinary system, and 3D model (b) Front view (d) Side view

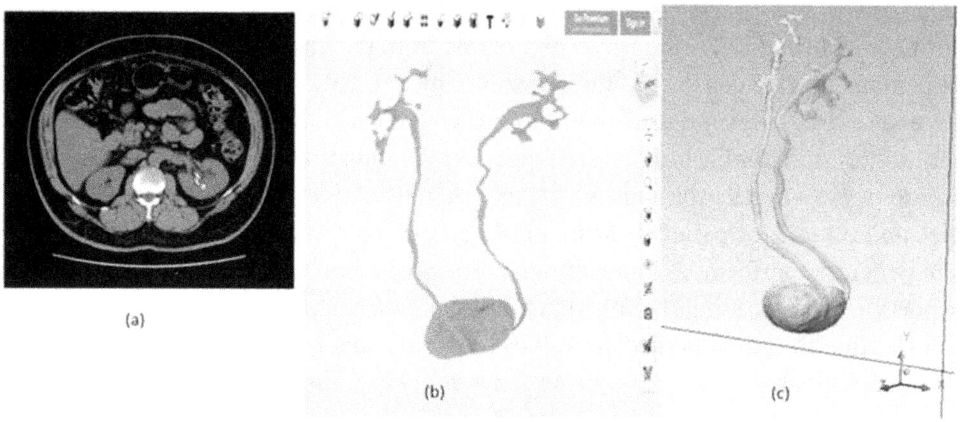

However, Computational Fluid Dynamics (CFD) simulations have been applied by many researchers (Zheng et al., 2015) to simulation analysis of peristaltic activities of a ureter in a two dimensional analysis. Later, Zhang et al., 2021analyzed that the investigation to the three dimensional case using finite element method for the Computational Fluid Dynamics (CFD) result. Other researchers (Kim et al., 2017) (Fitt et al., 2014) focused their works on the investigation of the urine flow through the ureter stent. The background explained investigations the biomechanical performance of a ureter and the urine flow for the period of peristalsis or inureter stent. All the methods to analyze the urine flow, but in Gómez-Blanco et al., 2016 work, were axisymmetrical Computational Fluid Dynamics (CFD) investigations and many of them were about peristalsis. Heexamined the three dimensional movement, but mimicking the peristaltic motion not the urine flow during a ureter stent.

Figure 3. Lithotripsy process

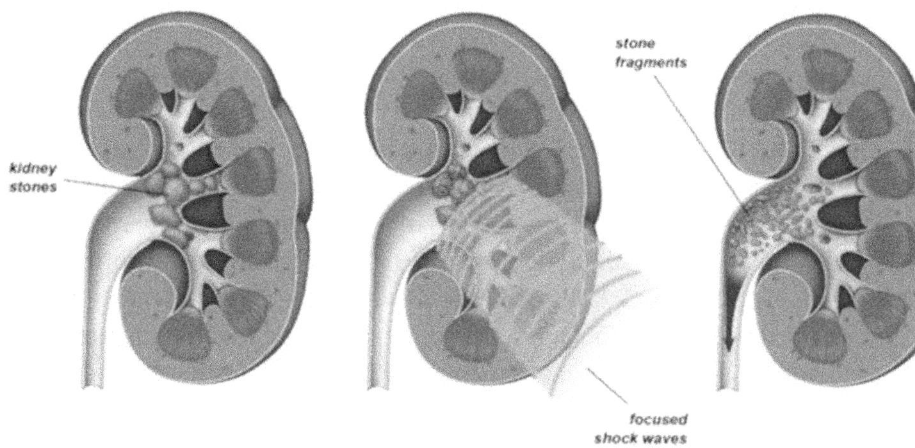

In a ureteral stent process for ureter stones before and after lithotripsy (Figure 3), different types of DJ stents are used, such as 5Fr, 6Fr, 7Fr, and 8-Fr. The selection of stent's size depends on the urology surgeon's estimation based on the probable urine flow rate via the stent. The luminal urine flow (flow inside stent) is directly proportionate to the internal diameter of the stent, whereas, the extra-luminal flow (flow outside of the stent in the ureter) does not connect with the outer diameter of the stent (Islamov et al. 2017). The TFR (total flow rate) is the summation of the both two flow rates.

Here, we examined which size of DJ stent is improved at attaining a satisfactory urine flow via the stented ureter with the help of CFD (computational fluid dynamics) analysis to define the TFR (total flow rate), the sum of luminal and extra luminal flow. We have considered two different types of tubular 4.6 mm diameter of ureter DJ stents (5Fr, 7Fr) for the study the of urine flow in three different processes (Table. 1).

DOUBLE-J (DJ) STENT

A Double-J (DJ) stent (Figure 4a) is applied together with metallic stents in ureteral obstruction or ureteral stenosis (Clavica et al., 2014). The objective of using a DJ stent is to get a passage or track for urine flow and to reach an adequate and satisfactory urine flow via the stented ureter (Figure 4b). The indwelling ureteral stent (pigtail) has confirmed its effectiveness in continuing ureteral patency in numerous cases (Wheeler et al., 2012). It is broadly considered an addition to the lithotripsy procedure (an extracorporeal shock wave). Though malignant ureteral obstacle is treated with a DJ stent, the patency rate has been reduced (Pewowaruk et al., 2020).Double-J or DJ stenting is a clinical technique engaged to repair the drainage of upper urinary tract, in the occurrence of a ureteric difficulty or obstacle (Shilo et al., 2021). After DJ stent implant, stents provide an instant relief the pain by reducing the pressure in the renal pelvis area (Figure 4c).

Figure 4. (a) Double-J (DJ) stent; (b) Application of DJ stent (d) KUV scan image

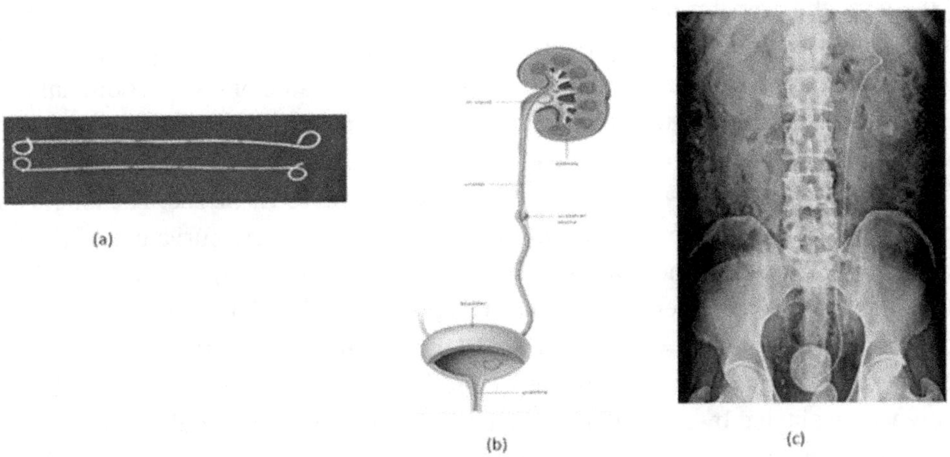

Though, their prolonged usage can reason encrustations and infections, because of bacterial colonization and deposition of crystal on the stent facade, correspondingly. The act of DJ stents and in common of the entire ureteric stents is considered to depend extensively on urine flow area within the ureter stent. Though, veryfew elementary researches about the problem-solved by fluid dynamic factors on stent capabilities have been performed until now. These factors are frequently complicated to evaluate *in-vivo*, involving the completion of difficult and expensive investigational set of rules. The plan of the current study was consequently to build up an artificial model of the ureter (UM) to imitate the fluid dynamic background in a ureter stent. The model of the ureter was planned to reproduce the geometry of pig ureters, and to examine the values of volumetric flow rate (Q), dynamic viscosity (μ), and seriousness of ureteric obstacles that may reason significant pressures in the renal pelvis area. The circulated obstruction obtained by the single stent placing was also counted. Additionally, visualizationof flow experimentations and CFD simulations were done with the intention of additional distinguish the flow area in the ureter model.

Table 1. Three different processes in a 4.6 mm ureter with a 5Fr and 7Fr DJ stent.

Process No.	Side hole number	Gap between the side holes (cm)	Ang. of side hole
1	24	1.0	0^0, 90^0,180^0,270^0
2	24	1.0	0^0, 90^0,180^0,270^0
3	42	0.5	0^0, 90^0,180^0,270^0

MATERIALS METHODS

Here, we have investigated the interaction between urine draining and a DJ-stented ureter. A simulation of the renal pelvis draining through a stented ureter with a simplified geometry has been analyzed with the finite element method and considering the Fluid-Structure Interaction (FSI). The stresses made from the urine and pressures of the intra-abdominal were very small and the strains almost minor, thus, it can be considered that the ureter behaves as a quasi-incompressible and inelastic structure in this pressure area. The finite element model of the ureter was assumed to the given loads: the pressure of intra-abdominal considered on the ureteral wall outer surface by unattached connective tissue adjacent to the ureter, Ureteropelvic junction (UPJ) pressure, applied by urine in the area of the renal pelvis, and Vesicoureteric junction (VUJ) pressure, which is basically the fluid pressure in the bladder. The transient period is 30 sec. The length and diameter of the model were 228 mm and 4.6 mm correspondingly. The geometric curvature of the ureter model was also recognized by estimating the collected clinical data. Assume the (a) Ureteropelvic junction (UPJ) pressure is 652.4 Pa, (b) Vesicoureteric junction (VUJ) pressure is 254.5 Pa, and the (c) Intra-abdominal pressure is 534 Pa.(ii) UPJ pressure = 666.61 Pa.(iii)VUJ pressure = 266.24 Pa.

The current study examined the fluid dynamics phenomenon of urine flow inside the urinary tract. The urine flow was evaluated by computational fluid dynamics (CFD) models. This model was useful to the urinary tract as an entire to generate a non-invasive analytical implement for urinary tract obstacle (due to renal stone) in both the ureter and urethra. The simulations have inferences for urological applications of the analysis of the urinary tract, diagnosis, urine flow properties, and probable urinary disease prevention by initial detection.

Figure 5. (a) Normal ureter simulation model (urine flow); (b) internal flow phenomenon; (b) external wall

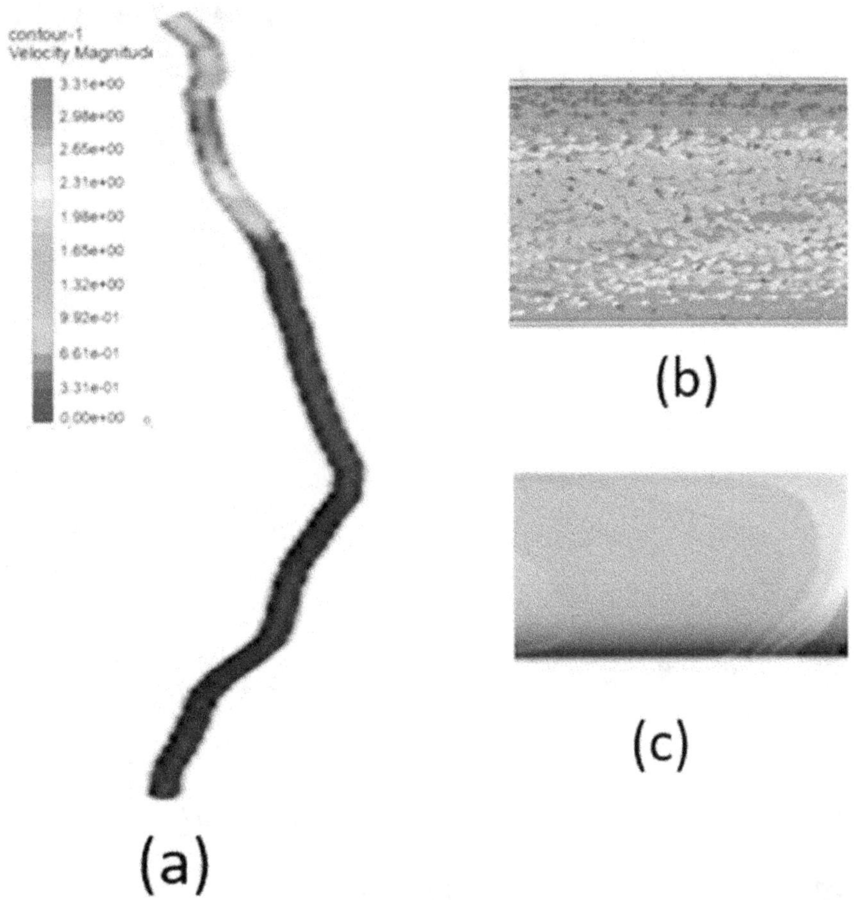

RESULTS AND DISCUSSIONS

An essential restriction of the investigation is that the ureter is basically a fibred material, so it is considered isotropic. Though, most of the stress-strain curvatures presented an additional or a smaller amount of isotropic manners, the intention of this overview is acceptable to a positive scope. Furthermore, to reflect the anisotropy in aspect, the three layers of the ureteral wall and regional variances along the ureter should be measured as stated by Sokolis et al., 2012.

Figure 5 shows the (a) velocity outlines noted in the normal ureteral model, (b) enlarged at the inlet and (c) outlet model. Velocity improved progressively during the tract from the Ureteropelvic junction (UPJ) to the Vesicoureteric junction (VUJ). Moreover, this does not appear useful in light of the equal of stresses and strains comprehended in the ureter. The consequence of submergence on the urine flow around the DJ stent should be evaluated in advance detail. Figure 6 shows (a) the urine flow outlines noted in the stone (ureteral stone) ureteral model, (b) inflamed at the internal ureter wall and (c) outlet, and also seen (d) the maximum pressure occurs during the urine drainage through the blockage ureter.

Figure 6. (a) Urine flow ureter simulation stone ureteral model (Ureteral obstruction); (b) inflamed at the internal ureter; (b) disturbed the urine flow (hematuria occurs); (d) Maximum pressure profile.

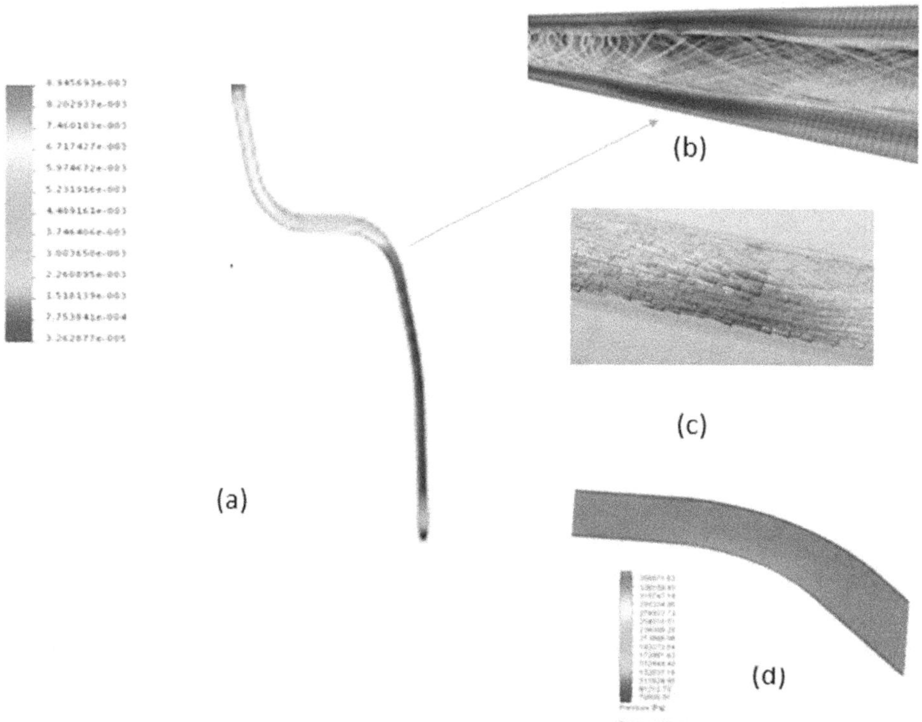

Figure 7. SEM image of DJ stent (a) side wall hole; (b) internal channel

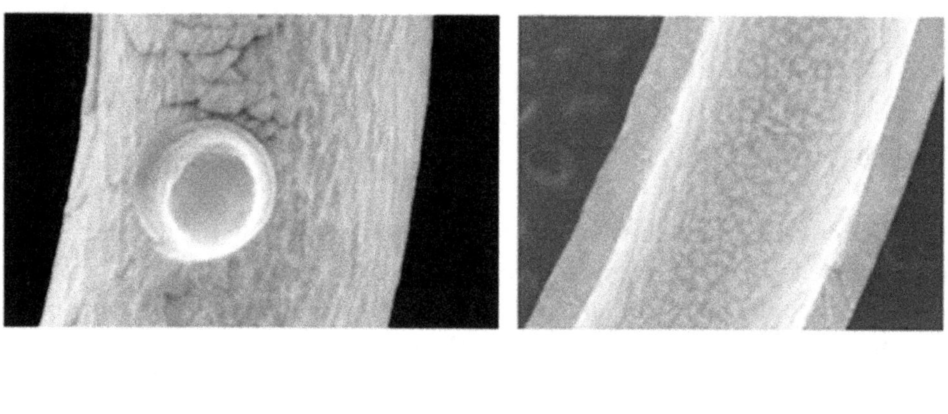

Here, to analyze the performance of different sizes of double-J (DJ) stents in the ureter. The total flow rates (TFR) in the double-J (DJ) stented ureters were estimated with an in-vitro analysis. Figure 7 shows the SEM image of DJ stent (a) side wall hole and (b) internal passage. The total flow rates (TFR) in the 0.67mm 5Fr double-J (DJ) stent were greater than those in the other sizes of double-J (DJ) stents. Figure 8 shows the total flow rate (TFR) in the simulations of the 4.6 mm two different type ureters (5Fr and 7Fr). The results show that the contact between the urine and double-J (DJ) stented ureter is minor. The total flow rate decreased while the size of double-J (DJ) stent increased. Computational fluid dynamics was also used to validate the results. It was presented that the results were well-fixed with the numerical results. The total flow rates are greater in straight ureters compared to curved ureters in addition to in tubular ureters compared to swelled ureters. Though, there has no fixed correlation between the total flow rate and the presence of or the number of side holes.

Figure 8. Simulations of the total flow rate (TFR) of two different type ureters (5Fr and 7Fr).

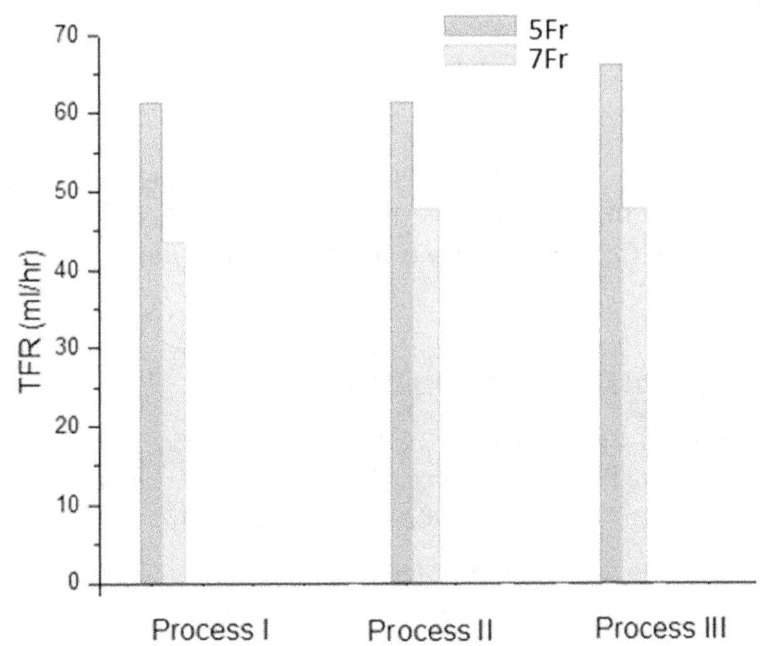

CONCLUSION

A number of material replicas have been chosen to explain the ureter's biomechanical performance and dissimilar positions of constants were experienced to inspect the effect of the general stiffness of the ureter. In proportion to the soft-tissue performance of the ureter, it can be predictable that the relevance of little stresses could make huge deformations. Though, in this study, it is determined that the performance of the ureter during urine flow is almost the similar apart from the model applied and the ureter's stiffness. Even though, the most excellent was the hyper-elastic material model, while it could confine the toe region of the stress-strain result analyzed by Rassoli et al., 2014 more precisely. The stresses formed by the urine and intra-abdominal pressures were extremely little and the strains approximately minor, that why it can be said that the ureter acts as an inflexible structure in this pressure reign. Alternatively, further efforts should be done to gain more practical models of the urinary tract in upcoming works. An ideal cylinder was considered to signify the geometry of the ureter. Further works should employ more practical geometric models offered by MRI or CT images, and, the hypothesis of the characteristics of urine as equivalent to those of water should be accurate by achieving urine characteristics in supplementary tests. Lastly, further experiments of all the features connected to the biomechanical categorization of the natural tissues will be done in upcoming studies.

ACKNOWLEDGMENT

The authors would like to thank IIEST-Shibpur, Centre for Healthcare Science and Technology lab, and Dept. of Urology, R.G. Kar Medical College and Hospital, and thanks to Mrs. Nibedita Bardhan for language proof reading.

REFERENCES

Al, K. F., Denstedt, J. D., Daisley, B. A., Bjazevic, J., Welk, B. K., Pautler, S. E., Gloor, G. B., Reid, G., Razvi, H., & Burton, J. P. (2020). Ureteral Stent Microbiota Is Associated with Patient Comorbidities but Not Antibiotic Exposure. *Cell reports. Medicine*, *1*(6), 100094. doi:10.1016/j.xcrm.2020.100094 PMID:33205072

Beysens, M., & Tailly, T. O. (2018). Ureteral stents in urolithiasis. *Asian Journal of Urology*, *5*(4), 274–286. doi:10.1016/j.ajur.2018.07.002 PMID:30364608

Clavica, F., Zhao, X., ElMahdy, M., Drake, M. J., Zhang, X., & Carugo, D. (2014). Investigating the flow dynamics in the obstructed and stented ureter by means of a biomimetic artificial model. *PLoS One*, *9*(2), e87433. doi:10.1371/journal.pone.0087433 PMID:24498322

Demzik, A., Filippou, P., Chew, C., Deal, A., Mercer, E., Mahajan, S., Wallen, E. M., Tan, H. J., & Smith, A. B. (2021). Gender-Based Differences in Urology Residency Applicant Personal Statements. *Urology*, *150*, 2–8. doi:10.1016/j.urology.2020.08.066 PMID:33035562

Fitts, M. K., Pike, D. B., Anderson, K., &Shiu, Y. T. (2014). Hemodynamic Shear Stress and Endothelial Dysfunction in Hemodialysis Access. *The Open Urology & Nephrology Journal, 7*(Suppl 1 M5), 33–44. doi:10.2174/1874303X01407010033

Gómez-Blanco, J. C., Martínez-Reina, F. J., Cruz, D., Pagador, J. B., Sánchez-Margallo, F. M., & Soria, F. (2016). Fluid Structural Analysis of Urine Flow in a Stented Ureter. *Computational and Mathematical Methods in Medicine*, *5710798*, 1–7. Advance online publication. doi:10.1155/2016/5710798 PMID:27127535

Islamov, M., Sypabekova, M., Kanayeva, D., & Rojas-Solórzano, L. (2017). CFD Modeling of Chamber Filling in a Micro-Biosensor for Protein Detection. *Biosensors (Basel)*, *7*(4), 45. doi:10.3390/bios7040045 PMID:28972568

Kim, K. W., Choi, Y. H., Lee, S. B., Baba, Y., Kim, H. H., & Suh, S. H. (2017). Analysis of Urine Flow in Three Different Ureter Models. *Computational and Mathematical Methods in Medicine*, *5172641*, 1–11. Advance online publication. doi:10.1155/2017/5172641 PMID:28659992

Kinn, A. C., & Lykkeskov-Andersen, H. (2002). Impact on ureteral peristalsis in a stented ureter. An experimental study in the pig. *Urological Research*, *30*(4), 213–218. doi:10.100700240-002-0258-1 PMID:12202937

Leonhäuser, D., Kranz, J., Leidolf, R., Arndt, P., Schwantes, U., Geyer, J., & Grosse, J. O. (2019). Expression of components of the urothelial cholinergic system in bladder and cultivated primary urothelial cells of the pig. *BMC Urology*, *19*(1), 62. doi:10.118612894-019-0495-z PMID:31288793

Matloubieh, J. E., Eghbali, M., & Abraham, N. (2020). Strategies to Encourage Medical Student Interest in Urology. *Current Urology Reports*, *21*(10), 34. doi:10.100711934-020-00984-1 PMID:32767185

Modi, P. K., Kaufman, S. R., Caram, M. V., Ellimoottil, C., Shahinian, V. B., & Hollenbeck, B. K. (2019). Impact of Medicare Office Visit Payment Reform on Urologic Practices. *Urology*, *126*, 83–88. doi:10.1016/j.urology.2019.01.013 PMID:30682462

Oliver, R., Wells, H., Traxer, O., Knoll, T., Aboumarzouk, O., Biyani, C. S., Somani, B. K., & Group, Y. A. U. (2018). Ureteric stents on extraction strings: A systematic review of literature. *Urolithiasis*, *46*(2), 129–136. doi:10.100700240-016-0898-1 PMID:27324264

Park, J. Y., Park, K., & Jeong, S. J. (2020). History of *Investigative and Clinical Urology* and an analysis of published articles. *Investigative and Clinical Urology*, *61*(Suppl 1), S64–S69. doi:10.4111/icu.2020.61.S1.S64 PMID:32055756

Pewowaruk, R., Rutkowski, D., Hernando, D., Kumapayi, B. B., Bushman, W., & Roldán-Alzate, A. (2020). A pilot study of bladder voiding with real-time MRI and computational fluid dynamics. *PLoS One*, *15*(11), e0238404. doi:10.1371/journal.pone.0238404 PMID:33211706

Rassoli, A., Shafigh, M., Seddighi, A., Seddighi, A., Daneshparvar, H., & Fatouraee, N. (2014). Biaxial mechanical properties of human ureter under tension. *Urology Journal*, *11*(3), 1678–1686. PMID:25015616

Shilo, Y., Modai, J., Leibovici, D., Dror, I., & Berkowitz, B. (2021). Comparative study of renal drainage with different ureteral stents subject to extrinsic ureteral obstruction using an in vitro ureter-stent model. *BMC Urology*, *21*(1), 100. doi:10.118612894-021-00865-w PMID:34261481

Sokolis, D. P. (2012). Multiaxial mechanical behaviour of the passive ureteral wall: Experimental study and mathematical characterisation. *Computer Methods in Biomechanics and Biomedical Engineering*, *15*(11), 1145–1156. doi:10.1080/10255842.2011.581237 PMID:21660781

Taguchi, M., Yoshida, K., Sugi, M., Kinoshita, H., & Matsuda, T. (2018). Simplified method using kidney / ureter / bladder x-ray to determine the appropriate length of ureteral stents. *International braz j urol: official journal of the Brazilian Society of Urology, 44*(6), 1224–1233. doi:10.1590/S1677-5538.IBJU.2017.0620

Wheeler, A. P., Morad, S., Buchholz, N., & Knight, M. M. (2012). The shape of the urine stream—From biophysics to diagnostics. *PLoS One*, *7*(10), e47133. doi:10.1371/journal.pone.0047133 PMID:23091609

Yin, F. C., & Fung, Y. C. (1971). Mechanical properties of isolated mammalian ureteral segments. *The American Journal of Physiology*, *221*(5), 1484–1493. doi:10.1152/ajplegacy.1971.221.5.1484 PMID:5124294

Zhang, B., Liu, S., Liu, Y., Wu, B., Zhang, X., Wang, X., Liang, X., Cao, X., Wang, D., & Wu, C. L. (2021). Novel CFD modeling approaches to assessing urine flow in prostatic urethra after transurethral surgery. *Scientific Reports*, *11*(1), 663. doi:10.103841598-020-79505-6 PMID:33436678

Zheng, J., Pan, J., Qin, Y., Huang, J., Luo, Y., Gao, X., & Zhou, X. (2015). Role for intravesical prostatic protrusion in lower urinary tract symptom: A fluid structural interaction analysis study. *BMC Urology*, *15*(1), 86. doi:10.118612894-015-0081-y PMID:26285823

KEY TERMS AND DEFINITIONS

CFD: Computation fluid dynamics (CFD) models effort to simulate the collaboration of gases and liquids where the surfaces are described by boundary situations.

DJ Stent: Ureteral DJ or double-J stents are normally used to release ureteral blockade and commonly as a routine part of the Ureteroscopic processes by urologist surgeons. The placement of DJ stent has the possible side effects for instanceurinary tract infection (UTI), flank pain due to backward urine flow.

Hematuria: Blood in urine is known as a hematuria. The causes of hematuria include inflammation of the kidney, bladder, urethra, kidney or bladder cancer, polycystic kidney disease, and or prostate infection etc.

Kidney Stone: Rigid bonds/deposits formed by salts and minerals inside kidneys, it is also known as renal calculi. Excess body weight, improper diet, various medical disorders, and certain medications and supplements are among the several reasons of kidney stones.

Urology: A clinical and surgical specialty which deals with the treatment of situations concerning the urinary tract of male and female, and the reproductive organs of male.

UTI: An infection of the urinary tract system, also known urinary tract infection. UTI can involve kidneys which is known as pyelonephritis, urethra which is known as urethritis, bladder which is known as cystitis.

APPENDIX: ADDITIONAL INFORMATION

Computational fluid dynamics (CFD) have exemplar shifting possibly in accepting the physiological activities like a flow of fluids in the human body. This interdisciplinary branch of engineering has now prepared a significant clinical influence on the study of cardiovascular and urological disease. Numerous biomechanical engineers and urologist surgeons are presently studying new strategies of ureteral stents to develop the feature of their procedures and the successive recovery and rescue of the patient. With the purpose of benefit during this design procedure, several computational simulations have been established to simulate the performance of dissimilar living tissues and offer a true computational background to assess the stents.

Chapter 13
Computational FEM Application on Percutaneous Nephrolithotomy (PCNL) Minimum Invasive Surgery Through Needle Insertion Process

Ranjit Barua
https://orcid.org/0000-0003-2236-3876
CHST, Indian Institute of Engineering Science and Technology, Shibpur, India

Surajit Das
R. G. Kar Medical College and Hospital, India

Pallab Datta
National Institute of Pharmaceutical Education and Research, Kolkata, India

Amit RoyChowdhury
Indian Institute of Engineering Science and Technology, Shibpur, India

ABSTRACT

PCNL or percutaneous nephrolithotripsy is one of the foremost interventional surgical treatment modalities for big kidney stones, which are more than two centimeters in diameter. With the application of miniaturized procedures, the signs for percutaneous nephrolithotripsy have been increased to smaller renal stones. Particularly for urologist surgeons without an accent on endourology, it is challenging to indicate the developing multitude of existing methods and procedures. Several makers have established different percutaneous nephrolithotripsy methods with changing diameters and different features. The suggestions for the dissimilar methods are intersecting. Reflective studies presented decreased complication rates. This chapter defines the presently available methods for percutaneous stone surgical treatment with their particularities and suggestions and studies the steering of a surgical flexible needle into the kidney to take out the stone by this procedure and finite element model analysis force of surgical needle and deformation of kidney tissue model.

DOI: 10.4018/978-1-7998-9078-2.ch013

INTRODUCTION

Accurate needle insertion into the kidney is an essential and challenging step for successful nephrolithotomy process (Lahme et al., 2001) (Pelit et al., 2017). Numerous plans and surgical performances have been improved to effortlessly attain the appropriate renal approach (Kirac et al., 2013) (Resorlu et al., 2013). Surgical needle placing is very significant in medical surgical area and particularly in 'Urology' as it covers huge area of renal stone surgery (Kim et al., 2020) (Srisubat et al., 2009) (Yuri et al., 2018) (Cabrera et al., 2020). It will help enormously in faultless the puncture in to the kidney for the duration of percutaneous nephrolithotomy process. Apart from applied force, density of the soft tissue, numerous other factors like needle tip shape (whether beveled or diamond), tip sharpness, strength of the shaft of needle and angle of needle in respect to tissue in heterogeneous tissue will impact on velocity of needle (Saussine et al., 2008). Nephrolithiasis is an important universal source of morbidity; comprises a regular urological syndrome which involves between 15% of the world population, with a following clinical reversion rate of around 50%. Modern techniques and surgical progresses have substitute they require for conventional surgery (open) with minimum invasive techniques, for example extracorporeal shockwave lithotripsy and percutaneous nephrolithotomy (Atassi et al., 2020). The choice of the exact renal stone surgical practice typically depends on the stones' shape and size, material composition (Wilhelm et al., 2015), and position of the calculi, the continuation of distal urinary difficulties, and anatomic dissimilarity of the urinary structure (Kumar et al., 2012) (Sabnis et al., 2014). Nowadays, nephrolithotomy is the recognized process for removaing of the *'Staghorn renal stone'* (Desai et al., 2017) (Wang et al., 2020) (Gadzhiey et al., 2020) (Aminsharifi et al., 2016). The surgical process usually involves with three most important phases, initially with the placing of a ureteral catheter to execute a retrospective study to estimate the anatomy of the kidney and to decide whether a renal stone is jamming the urinary tract (Raharia et al., 2019) (Fu et al., 2017). After that, the minimal insertion is done by a surgical needle from the skin en route for the exact calculi position (He et al., 2015). Presently medical science and engineering going hand in hand, different engineering applications are making medical science more advanced, like Virtual reality (VR)/Augmented Reality (AR), image processing, biorobotics etc. [Barua et al., 2022]. Figure 1 shows the percutaneous nephrolithotomy (PCNL) process. The ultimate process concerns the disintegration and extraction renal stone via surgical apparatus, for example forceps, nephroscope, baskets, and lithotripters (Di Grazia et al., 2013) (Giusti et al., 2020) (Choudhury et al., 2021) (Lipsky et al., 2013). The most common problems were bleeding (7.8%), hydrothorax (1.8%), and perforation of renal pelvis (3.4%), blood transfusion (5.7%), and fever (10.5%) (Kallidonis et al., 2016) (De et al., 2015) (Nikić et al., 2014). The success rate and treatment results of the operation are extremely well recognized as being greatly dependent on the exactness and precision of the insertion step, as it have to attain the calculi through an accurate and precise path, which is making this phase the main challenging assignment for surgeons (Ichaoui et al., 2019) (Sourial et al., 2019) (Bozzini et al., 2020).

The perfect renal entrance is one with the purpose of permit total removal of the renal stoner at the same time as reducing the bleeding. Incorrect needle insertions often make difficulties (Barua et al., 2020), for example harms in the kidney and adjacent organs, and in the end discrimination the overall surgical achievement and patient result. Though percutaneous nephrolithotomy (PCNL) is referred as a minimal invasive surgery with numerous related benefits, for example creating small patient opening, minimize the hospitalization time, and improving recovery (Li et al., 2020) (Li et al., 2014). Figure 2 shows the fluoroscopic view of PCNL process during removal of kidney stone. The most familiar difficulties one may get anatomic target limited vision, complexity in managing the surgical tools, preventive mobility in the body, high deftness levels of surgeon hand-eye organization, deflections of the surgical needle, affecting anatomic target, and deformations of the anatomic structure etc. (Barua et al., 2020) (Bhat et al., 2021) (Agarwal et al., 2008) (Izol et al., 2021). In this paper, we have analyzed the performance of the percutaneous nephrolithotomy technique, and also studied FEM analysis force of surgical needle and deformation of kidney tissue model.

Figure 1. (a) Percutaneous nephrolithotomy process; (b) removing the stones from the kidney

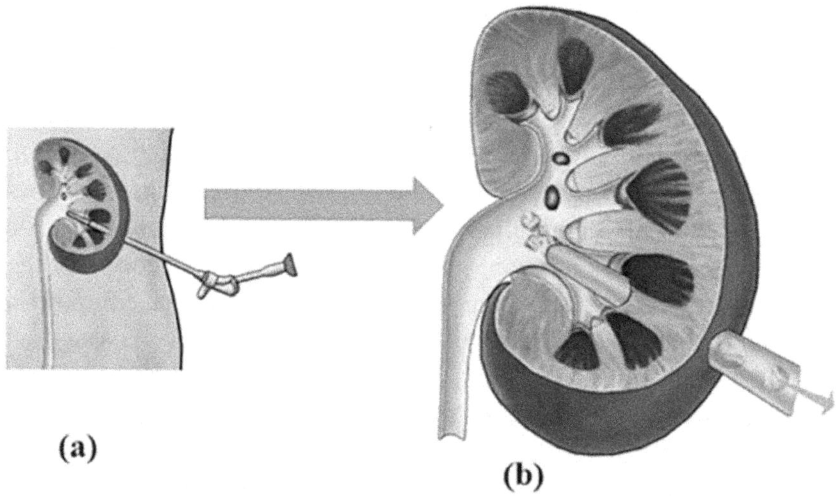

Figure 2. (a) PCNL procedure during removal of kidney stone; (b) Fluoroscopic view of PCNL process

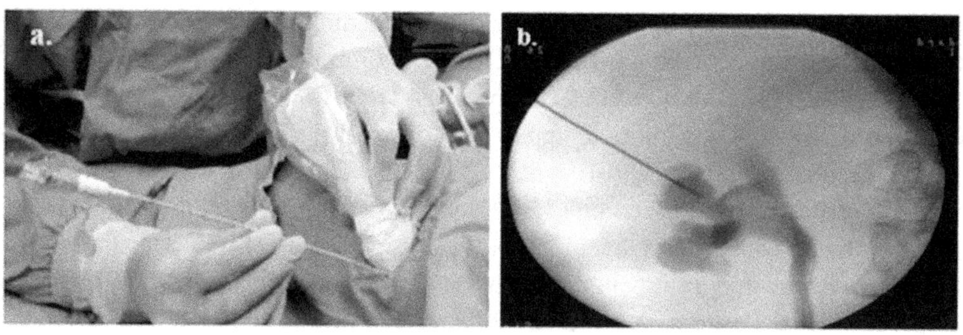

PCNL (PERCUTANEOUS NEPHROLITHOTOMY) PROCEDURE

Renal stones are formed in the kidney or urinary tract because of the crystallization of different chemical compounds or minerals in the urine. Nephrolithotomy is a performance applied to remove renal stones in the kidney or upper ureter area, the cylindrical tube which drains urine from the kidney to the bladder (Hosseini et al., 2019). Percutaneous nephrolithotomy process has been performed on a lot of patients and is established usual of concern for urology patients with renal stones which are big, very hard, or opposed to other shapes of stone healing. It has restored open surgeries for renal stones in the huge majority of renal stone patients (Batagello et al., 2019). Figure 3 shows the nephrolithotomy surgical tools. Usually, the time taken of the surgery is 3 to 4 hours. The surgery is done by creating a small one centimeter cut in the flank area. A tube is positioned throughout the incision inside the kidney by x-ray guidance. A telescope (small) is followed by the passed throughout the tube to facilitate visualize the renal stone, break the stone, and take out it from the patients' body (Gamal et al., 2015). If needed a laser or additional tool named a lithotripter may be helped to break up the renal stone before it can be removed (Prasad et al., 2020). This process has resulted in appreciably reduce post-operative pain, a minimize hospital stay, and earlier come back to the work and daily activities while evaluated to open surgery (Giusti et al., 2020). Percutaneous nephrolithotomy is usually considered an effective safe surgical treatment offering the maximum stone-free rates after the initial treatment as related to the other lithotripsy surgical techniques (minimal invasive) (Cabrera et al., 2020). Serious difficulties even though rare should be anticipated following this nephrolithotomy treatment. The most common complications related with percutaneous nephrolithotomy are being studied aiming on the risk factors, existing supervision, and preventing methods which essential to be taken to decrease their prevalence. Complications for example urine leak from nephrocutaneous fistula, perioperative bleeding, and injury in the pelvicalyceal system (Atassi et al., 2020).

Figure 3. Percutaneous nephrolithotomy surgical tools

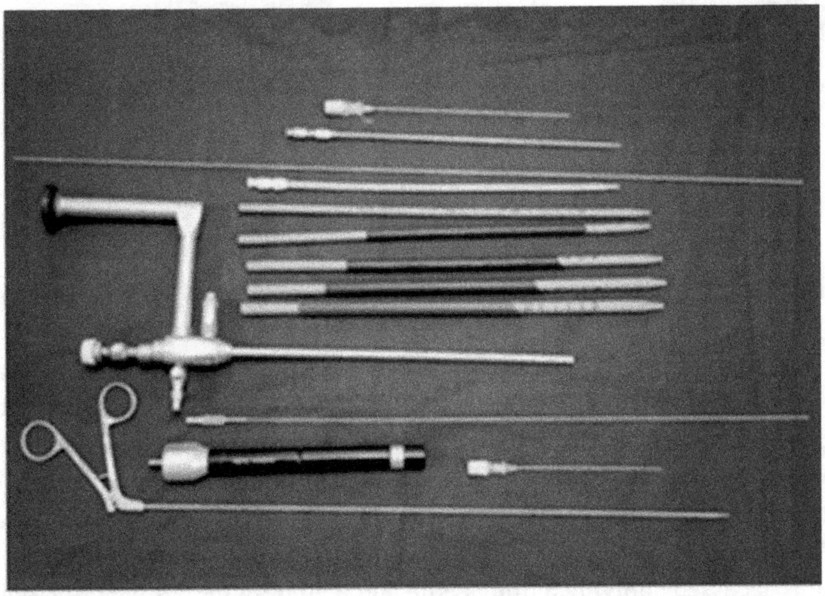

SURGICAL PERFORMANCE OF PCNL (PERCUTANEOUS NEPHROLITHOTOMY)

PCNL (Percutaneous nephrolithotomy) is basically using as a MIS or minimally invasive surgical process for treating complex or large kidney stones (Gadzhiev et al., 2020). From the time when its start, the performance of PCNL has improved by numerous modification (Giusti et al., 2020). The exercise of minimally invasive surgical apparatus for soft biological tissue surgery assures the decreased tissue trauma, minimize process time, and less recovery times for the patient. Percutaneous nephrolithotomy robotics treatment has become an accepted area in modern medical field (Wang et al., 2020). The mobility of organ tissue can make erroneous positional insertions. Mainly in needle steering procedures which use flexible and thin needles, huge deviations can arise between pre-operative data (images) of the patient, from which a process is considered, and the intra-operative prospect, wherever a process is performed (Cabrera et al., 2020). The efficiency of PCNL (Percutaneous nephrolithotomy) has several advantages in particular anatomical places, for example stones at upper urinary tract in pediatric patient (Atassi et al., 2020). To manage this category of present minimally invasive surgery PCNL (Percutaneous nephrolithotomy), the finite element model is very much valuable to calculate the force evaluation for tissue deformation, and secure puncture etc.

Figure 4. (a) Simulation model of needle insertion into the kidney during PCNL surgery; (b) Tissue deformation model of kidney during PCNL surgery

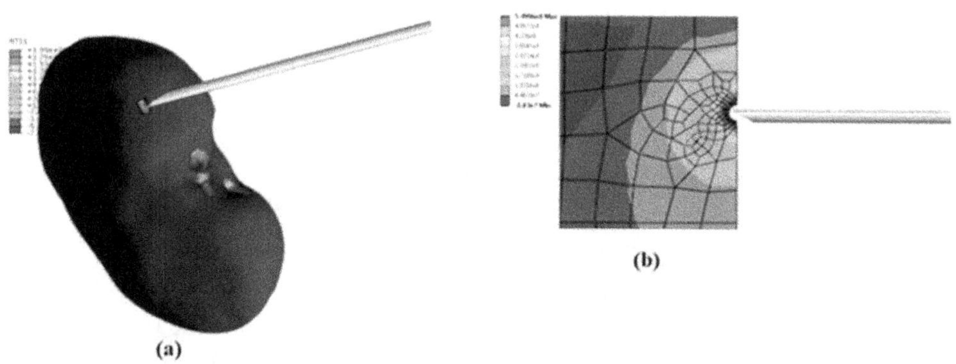

The deformation of the needle during its insertion into the kidney is described in this chapter in the context of intraoperative treatment of urological study based on the integrated imaging-molecular diagnosis. With the help of general anaesthesia, PCNL (Percutaneous nephrolithotomy) is normally performed. Throughout this process, a catheter is positioned in the bladder, the catheter drains urine from the bladder and remains in area with the employ of a balloon. An additional catheter is positioned into the ureter (Wang et al., 2020). A dye or contrast may be infused throughout this catheter to offer a better observation and confirm the exact stone position. This helps access to the kidney while your urinary tract is visualized with the help of ultrasound or x-rays. After stone is to be found, the kidney's collecting scheme is accessed applying a thin needle and a guide-wire is sited (Atassi et al., 2020). The guide-wire allows safe entrée for the nephroscope, a thin tube-like device involved for viewing the inside of the kidney. The entrance route is cautiously opened awaiting access through the nephroscope is feasible. Some stones can straightly be removed by a grasper, which is called nephrolithotomy. Larger stones require to be broken up by a tool like a laser by they can be removed, which is known as nephrolithotripsy (Choudhury et al., 2021). After all stones are removed, the contrast material will use to verify that no stones or small pieces of the broken stone throughout a procedure in the kidney. While the contrast material moves simply during the bladder, which means that there are furthermore any stones inside the ureter. In a number of patients, a temporary small tube known as DJ Stent or Double- J Stent

(6 Fr 10 cm with several side holes) may be positioned to make sure that swelling does not obstruct the drainage of urine. The stent remains in place by curls J-shaped in the kidney and the bladder. At the end of the process 'nephrostomy' (a drainage catheter) may be positioned.

Figure 5. Force model during PCNL surgery (Exp. & Simulation)

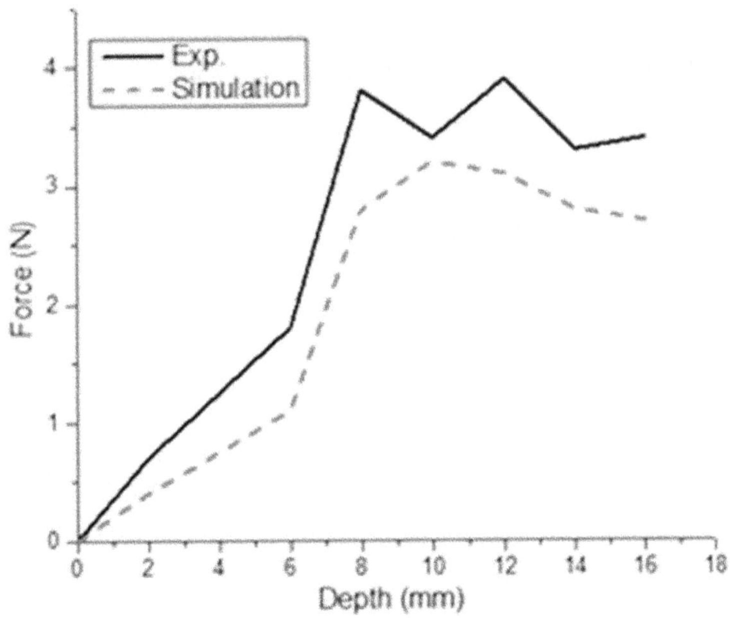

SIMULATION OF PCNL (PERCUTANEOUS NEPHROLITHOTOMY)

A surgical bevel-tip needle (dissimilar of a symmetrical needle tip) was involved to penetrate kidney tissue at an angle (Figure 4a). As the needle inserts at an angle outside the direction, the surgical needle may perhaps bend in the insertion path of the bevel. In this study, the insertion force of the simulation model was applied by the surgical needle on the biological kidney tissue, with the puncturing force at the tip of the surgical bevel needle and the friction forces the length of the needle shaft (Figure 4b). Here, it was assumed that the bending force of the surgical needle was minor considered to the applied forces (elastic) by the kidneys' tissue to the surgical needle. We initially explain model of the kidney tissue and technique for computing deformations of the tissue because of the applied needle force. The model should also consist of the mechanical properties of tissue material and the FEM (finite element mesh) boundary conditions. In our present performance, we estimated soft kidney tissues as homogeneous, linearly elastic, and isotropic materials. For all separated tissue set, the model involves as enter the material properties of the tissue, like the Poisson ratio, Young's modulus, and density. Every element in the mesh may be allocated distinctive properties, which permits for the simulation of numerous tissue types in single mesh. The nodes are significant elements inside tissues are controlled to be set.

Here, it was assumed that the surgical needle was not bending external portion. Once the needle has entered the tissue, it will bend in the direction of the bevel-tip. During the real-time PCNL surgery, the maximum insertion force was noticed at 3.9 N; in case of simulation, the maximum insertion force was 3.1 N (Figure.5). As the needle may be punctured at any position, it is typically essential to adapt the reference mesh in real-time to check that factor boundaries are nearby where the needle tip and frictional force must be used. For analyzing the tip force of the surgical needle, a node is retained at the location of the needle tip throughout insertion process (Figure 6).

Figure 6. Result of the biomechanics force data during PCNL (Exp. and Simulation)

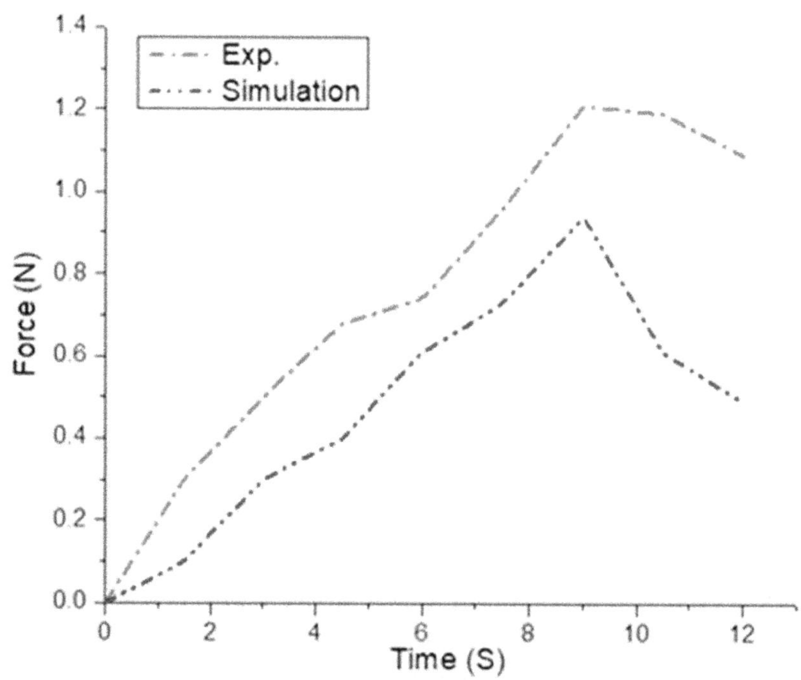

CONCLUSION

The number of PCNL (Percutaneous nephrolithotomy) performances is progressively increasing, even though the reality that the amount of trained urologists involving with this procedure has not kept up with this development factor.11 The spotlight on methodological and technologic progresses that may influence improved and easier nephrolithotomy puncture. Computer-aided surgeries and modern strategies are subject of current research, as they may play a significant role in the upcoming of nephrolithotomy. These methods have the possible to give precious interfaces between surgical tools and anatomic structures, tissue deformation, surgical robots, and improving nephrolithotomy planning and control. So, steering coordination could characterize a step advance to decrease the dependency of medical imaging and connected inadequacy in percutaneous conduction. It is predicted that future enhancements in nephrolithotomy will include the riddance of radiation and decrease of surgical costs and time. Even though the progress and advance of autonomous robots may become wider the employ of PCNL (Percutaneous nephrolithotomy) to urologist surgeons less familiar with minimally invasive surgery, it also increases demanding robustness and protection issues which have to be undertake in the future.

ACKNOWLEDGMENT

The authors would like to thank IIEST-Shibpur, Centre for Healthcare Science and Technology lab, Department of Urology, RG Kar Medical College and Hospital, Kolkata, and thanks to Mrs. Nibedita Bardhan for language proof reading.

REFERENCES

Agrawal, M. S., Agrawal, M., Gupta, A., Bansal, S., Yadav, A., & Goyal, J. (2008). A randomized comparison of tubeless and standard percutaneous nephrolithotomy. *Journal of Endourology*, *22*(3), 439–442. doi:10.1089/end.2007.0118 PMID:18257738

Aminsharifi, A., Irani, D., Masoumi, M., Goshtasbi, B., Aminsharifi, A., & Mohamadian, R. (2016). The management of large staghorn renal stones by percutaneous versus laparoscopic versus open nephrolithotomy: A comparative analysis of clinical efficacy and functional outcome. *Urolithiasis*, *44*(6), 551–557. doi:10.100700240-016-0877-6 PMID:27032961

Atassi, N., & Knoll, T. (2020). Future of kidney stone management: surgical intervention miniaturization of PCNL: where is the limit? *Current Opinion in Urology*, *30*(2), 107–112. doi:10.1097/MOU.0000000000000713 PMID:31895077

Barua, R., & Das, S. (2022). Improvements of Virtual and Augmented Reality for Advanced Treatments in Urology. In L. Coelho, R. Queirós, & S. Reis (Eds.), *Emerging Advancements for Virtual and Augmented Reality in Healthcare* (pp. 117–131). IGI Global. doi:10.4018/978-1-7998-8371-5.ch008

Barua, R., Datta, S., Datta, P., & Roy Chowdhury, A. (2020). Experimental analysis the tissue deformation of needle insertion process in tissue engineering. *IEEE 1st International Conference for Convergence in Engineering (ICCE)*, 83-85. 10.1109/ICCE50343.2020.9290598

Barua, R., Giria, H., Datta, S., Roy Chowdhury, A., & Datta, P. (2020). Force modeling to develop a novel method for fabrication of hollow channels inside a gel structure. *Proceedings of the Institution of Mechanical Engineers. Part H, Journal of Engineering in Medicine*, *234*(2), 223–231. doi:10.1177/0954411919891654 PMID:31774361

Batagello, C. A., Barone Dos Santos, H. D., Nguyen, A. H., Alshara, L., Li, J., Marchini, G. S., Vicentini, F. C., Torricelli, F., Danilovic, A., Pereira, J. G., Rose, E., Srougi, M., Nahas, W. C., Mazzucchi, E., & Monga, M. (2019). Endoscopic guided PCNL in the prone split-leg position versus supine PCNL: A comparative analysis stratified by Guy's stone score. *The Canadian Journal of Urology*, *26*(1), 9664–9674. PMID:30797250

Bhat, A., Katz, J. E., Smith, N., & Shah, H. N. (2021). Nephropleural fistula after supracostal approach for PCNL: Report of two cases with review of literature. *BMJ Case Reports*, *14*(4), e241360. doi:10.1136/bcr-2020-241360 PMID:33795284

Bozzini, G., Aydogan, T. B., Müller, A., Sighinolfi, M. C., Besana, U., Calori, A., Lorenzo, B., Govorov, A., Pushkar, D. Y., Pini, G., Pastore, A. L., Romero-Otero, J., Rocco, B., & Buizza, C. (2020). A comparison among PCNL, Miniperc and Ultraminiperc for lower calyceal stones between 1 and 2 cm: A prospective, comparative, multicenter and randomised study. *BMC Urology*, *20*(1), 67. doi:10.118612894-020-00636-z PMID:32522171

Cabrera, J. D., Manzo, B. O., Torres, J. E., Vicentini, F. C., Sánchez, H. M., Rojas, E. A., & Lozada, E. (2020). Mini-percutaneous nephrolithotomy versus retrograde intrarenal surgery for the treatment of 10-20 mm lower pole renal stones: A systematic review and meta-analysis. *World Journal of Urology*, *38*(10), 2621–2628. doi:10.100700345-019-03043-8 PMID:31813026

Choudhury, S., Talukdar, P., Mandal, T. K., & Majhi, T. K. (2021). Supine versus prone PCNL in lower calyceal stone: Comparative study in a tertiary care center. *Urologia*, *88*(2), 148–152. doi:10.1177/0391560320962404 PMID:33028166

De, S., Autorino, R., Kim, F. J., Zargar, H., Laydner, H., Balsamo, R., Torricelli, F. C., Di Palma, C., Molina, W. R., Monga, M., & De Sio, M. (2015). Percutaneous nephrolithotomy versus retrograde intrarenal surgery: A systematic review and meta-analysis. *European Urology*, *67*(1), 125–137. doi:10.1016/j.eururo.2014.07.003 PMID:25064687

Desai, M., Sun, Y., Buchholz, N., Fuller, A., Matsuda, T., Matlaga, B., Miller, N., Bolton, D., Alomar, M., & Ganpule, A. (2017). Treatment selection for urolithiasis: Percutaneous nephrolithomy, ureteroscopy, shock wave lithotripsy, and active monitoring. *World Journal of Urology*, *35*(9), 1395–1399. doi:10.100700345-017-2030-8 PMID:28303335

Di Grazia, E., & La Rosa, P. (2013). Split-leg percutaneous nephrolithotomy: a safe and versatile technique. *Archivio italiano di urologia, andrologia: organo ufficiale (di) Societa italiana di ecografia urologica e nefrologica, 85*(2), 82–85.

Fu, W., Yang, Z., Xie, Z., & Yan, H. (2017). Intravenous misplacement of the nephrostomy catheter following percutaneous nephrostolithotomy: Two case reports and literature review. *BMC Urology*, *17*(1), 43. doi:10.118612894-017-0233-3 PMID:28615052

Gadzhiev, N., Malkhasyan, V., Akopyan, G., Petrov, S., Jefferson, F., & Okhunov, Z. (2020). Percutaneous nephrolithotomy for staghorn calculi: Troubleshooting and managing complications. *Asian Journal of Urology*, *7*(2), 139–148. doi:10.1016/j.ajur.2019.10.004 PMID:32257807

Gamal, W., Moursy, E., Hussein, M., Mmdouh, A., Hammady, A., & Aldahshoury, M. (2015). Supine pediatric percutaneous nephrolithotomy (PCNL). *Journal of Pediatric Urology*, *11*(2), 78.e1–78.e785. doi:10.1016/j.jpurol.2014.10.012 PMID:25819602

Giusti, G., & De Lisa, A. (2020). PCNL in the prone position VS PCNL in the modified supine Double-S position: Is there a better position? A prospective randomized trial. *Urolithiasis*, *48*(1), 63–69. doi:10.100700240-018-1088-0 PMID:30456414

He, X., Xie, D., Du, C., Zhu, W., Li, W., Wang, K., Li, Y., Lu, H., & Guo, F. (2015). Improved nephrostomy tube can reduce percutaneous nephrolithotomy postoperative bleeding. *International Journal of Clinical and Experimental Medicine*, *8*(3), 4243–4249. PMID:26064336

Hosseini, S. R., Mohseni, M. G., Aghamir, S., & Rezaei, H. (2019). Effect of Irrigation Solution Temperature on Complication of Percutaneous Nephrolithotomy: A Randomized Clinical Trial. *Urology Journal*, *16*(6), 525–529. PMID:30882166

Ichaoui, H., Samet, A., Ben Hadjalouane, H., Hermi, A., Hedhli, H., Bakir, M. A., Khiari, R., & Ghozzi, S. (2019). Percutaneous Nephrolithotomy (PCNL): Standard Technique Versus Tubeless - 125 Procedures. *Cureus*, *11*(3), e4251. doi:10.7759/cureus.4251 PMID:31131174

Izol, V., Deger, M., Akdogan, N., Ok, F., Bayazit, Y., & Aridogan, I. A. (2021). The Effect of Percutaneous Nephrolithotomy on the Estimated Glomerular Filtration Rate in Patients with Chronic Kidney Disease. *Journal of Endourology*, *35*(5), 583–588. doi:10.1089/end.2020.0512 PMID:33054416

Kallidonis, P., Panagopoulos, V., Kyriazis, I., & Liatsikos, E. (2016). Complications of percutaneous nephrolithotomy: Classification, management, and prevention. *Current Opinion in Urology*, *26*(1), 88–94. doi:10.1097/MOU.0000000000000232 PMID:26555687

Kim, C. H., Chung, D. Y., Rha, K. H., Lee, J. Y., & Lee, S. H. (2020). Effectiveness of Percutaneous Nephrolithotomy, Retrograde Intrarenal Surgery, and Extracorporeal Shock Wave Lithotripsy for Treatment of Renal Stones: A Systematic Review and Meta-Analysis. *Medicina (Kaunas, Lithuania)*, *57*(1), 26. doi:10.3390/medicina57010026 PMID:33396839

Kirac, M., Bozkurt, Ö. F., Tunc, L., Guneri, C., Unsal, A., & Biri, H. (2013). *Comparison of retrograde intrarenal surgery and mini-percutaneous nephrolithotomy in management of lower-pole renal stones with a diameter of smaller than 15 mm*. Academic Press.

Kumar, P., Bach, C., Kachrilas, S., Papatsoris, A. G., Buchholz, N., & Masood, J. (2012). Supine percutaneous nephrolithotomy (PCNL): 'in vogue' but in which position? *BJU International*, *110*(11c, 11 Pt C), E1018–E1021. doi:10.1111/j.1464-410X.2012.11188.x PMID:22564784

Lahme, S., Bichler, K. H., Strohmaier, W. L., & Götz, T. (2001). Minimally invasive PCNL in patients with renal pelvic and calyceal stones. *European Urology*, *40*(6), 619–624. doi:10.1159/000049847 PMID:11805407

Li, Q., Wan, L., Liu, S., Li, M., Chen, L., Hou, Z., & Zhang, W. (2020). Clinical efficacy of enhanced recovery after surgery in percutaneous nephrolithotripsy: A randomized controlled trial. *BMC Urology*, *20*(1), 162. doi:10.118612894-020-00728-w PMID:33081762

Li, X., Long, Q., Chen, X., He, D., & He, H. (2014). Real-time ultrasound-guided PCNL using a novel SonixGPS needle tracking system. *Urolithiasis*, *42*(4), 341–346. doi:10.100700240-014-0671-2 PMID:24965272

Lipsky, M. J., Shapiro, E. Y., Cha, D. Y., & Gupta, M. (2013). Modified-PCNL without modified instruments: A description of technique. *Journal of Endourology*, *27*(6), 684–687. doi:10.1089/end.2012.0604 PMID:23268559

Nikić, P., Durutović, O., Kajmaković, B., Nale, D., Bumbaširević, U., Radovanović, M., Milenković-Petronić, D., & Džamić, Z. (2014). Complications associated with percutaneous nephrolitholapaxy (PCNL)—Our experience and literature review. *Acta Chirurgica Iugoslavica, 61*(1), 51–56. doi:10.2298/ACI1401051N PMID:25782226

Pelit, E. S., Kati, B., Çanakci, C., Sağir, S., & Çiftçi, H. (2017). Outcomes of miniaturized percutaneous nephrolitotomy in infants: Single centre experience. *International Braz J Urol: Official Journal of the Brazilian Society of Urology, 43*(5), 932–938.

Prasad, M. K., Varshney, R. K., Jain, P., Choudhary, A. K., Khare, A., & Jheetay, G. S. (2020). Postoperative analgesic efficacy of fluoroscopy-guided erector spinae plane block after percutaneous nephrolithotomy (PCNL): A randomized controlled study. *Saudi Journal of Anaesthesia, 14*(4), 480–486. doi:10.4103ja.SJA_26_20 PMID:33447190

Raharja, P., Atmoko, W., Rasyid, N., & Birowo, P. (2019). Safety and Effectiveness of Externalized Ureteral Catheter in Tubeless Percutaneous Nephrolithotomy. *Urology Journal, 17*(5), 456–461. PMID:31422576

Resorlu, B., Unsal, A., Ziypak, T., Diri, A., Atis, G., Guven, S., Sancaktutar, A. A., Tepeler, A., Bozkurt, O. F., & Oztuna, D. (2013). Comparison of retrograde intrarenal surgery, shockwave lithotripsy, and percutaneous nephrolithotomy for treatment of medium-sized radiolucent renal stones. *World Journal of Urology, 31*(6), 1581–1586. doi:10.100700345-012-0991-1 PMID:23179732

Sabnis, R. B., Chhabra, J. S., Ganpule, A. P., Abrol, S., & Desai, M. R. (2014). Current role of PCNL in pediatric urolithiasis. *Current Urology Reports, 15*(7), 423. doi:10.100711934-014-0423-4 PMID:24898187

Saussine, C., Lechevallier, E., & Traxer, O. (2008). Les variantes techniques de la NLPC (PCNL: technical variations). *Progres en Urologie: Journal de l'Association Francaise d'Urologie et de la Societe Francaise d'Urologie, 18*(12), 897–900.

Sourial, M. W., Francois, N., Box, G. N., & Knudsen, B. E. (2019). Supracostal access tubeless percutaneous nephrolithotomy: Minimizing complications. *World Journal of Urology, 37*(7), 1429–1433. doi:10.100700345-018-2518-x PMID:30302593

Srisubat, A., Potisat, S., Lojanapiwat, B., Setthawong, V., & Laopaiboon, M. (2009). Extracorporeal shock wave lithotripsy (ESWL) versus percutaneous nephrolithotomy (PCNL) or retrograde intrarenal surgery (RIRS) for kidney stones. *Cochrane Database of Systematic Reviews*, (4), CD007044. PMID:19821393

Wang, J., Bai, Y., Yin, S., Cui, J., Tang, Y., Wang, Z., Chen, B., Li, H., Wei, W., & Wang, J. (2020). Risk factors for deterioration of renal function after percutaneous nephrolithotomy in solitary kidney patients with staghorn calculi. *Translational Andrology and Urology, 9*(5), 2022–2030. doi:10.21037/tau-20-916 PMID:33209666

Wilhelm, K., Fritsche, H. M., & Netsch, C. (2015). Perkutane Steintherapie heute: Standard-, Mini-, Micro-, Ultramini-PCNL (Percutaneous Stone Treatment Today: Standard-, Mini-, Micro-, Ultramini-PCNL). *Aktuelle Urologie, 46*(4), 297–302. doi:10.1055-0035-1555863 PMID:26227130

Yuri, P., Hariwibowo, R., Soeroharjo, I., Danarto, R., Hendri, A. Z., Brodjonegoro, S. R., Rasyid, N., Birowo, P., & Widyahening, I. S. (2018). Meta-analysis of Optimal Management of Lower Pole Stone of 10 - 20 mm: Flexible Ureteroscopy (FURS) versus Extracorporeal Shock Wave Lithotripsy (ESWL) versus Percutaneus Nephrolithotomy (PCNL). *Acta Medica Indonesiana*, *50*(1), 18–25. PMID:29686172

KEY TERMS AND DEFINITIONS

Bevel Needle: The bevel needle is such type of clinical needle used in the medicinal or clinical procedures.

DJ Stent: The DJ stent is known as JJ or double J stent is a ureteral stenting process to place a flexible thin tube (polymer/plastic) which is temporarily in the ureter to relieve urine drains from the kidney into the bladder in the incident of an obstruction.

FEM: The finite element method or FEM is an extensively used technique for numerically explaining difference equations rising in mathematical, engineering, and biomedical modeling.

MIS: Minimally invasive surgery or MIS, surgeons apply a variety of performances to surgery with a smaller amount of damage to the patient's body than with conventional surgery. Normally, MIS is held by less pain, reduces the hospital stay, and less difficulties.

PCNL: Percutaneous nephrolithotomy or PCNL is a performance applied to take away renal calculi or stones in the kidney or upper ureter which are too large for other procedures of stone treatment for example shock wave lithotripsy or ureteroscopy.

Renal Stone: A kidney or renal stone is basically a hard deposit of minerals and salts, in concentrated urine, it sticks to each together. While passing or moving through the urinary tract, it can be painful. The most common sign is unadorned pain, typically in the side of the abdomen and frequently related to nausea.

APPENDIX: ADDITIONAL INFORMATION

Kidney or renal stone or calculi is a urological complaint that is considered by a high reappearance rate. The movement of the kidney stone causes renal colic or pain and the obstruction or block by renal calculi could affect the damage of renal function. To remove the obstacle, urologists select different treatments for dissimilar size renal calculi (0.3 cm < dia. < 0.6 cm). As the guidelines mention PCNL (percutaneous nephrolithotomy), of which normal access tracts are 24–30 Fr, which is suggested management of patients with ureteral or renal stones more than 20 mm or and for lesser stones like 10–20 mm of the lower-pole stones when anatomic factors make ESWL (extracorporeal shockwave lithotripsy) unfavorable.

Chapter 14
Biomechanics of an Avian Flight:
Exploration of the Complex Interactions Between the Musculoskeletal, Sensory, and Neural Systems

Yugesh Ramdhun
National Institute of Technology, Rourkela, India

Priyobroto Basu
https://orcid.org/0000-0002-9271-4611
National Institute of Technology, Rourkela, India

Ravi Kant A.
https://orcid.org/0000-0001-5586-746X
National Institute of Technology, Rourkela, India

Thirugnanam A.
National Institute of Technology, Rourkela, India

ABSTRACT

Birds have always fascinated scientists and opened their eyes to new areas of flight mechanisms through biomimicry of these flyers. These flyers can sustain and control the flight through clever maneuvering by flapping their wings. It involves an intricate aerodynamic force production to generate sufficient lift force to overcome the drag and perform useful maneuvers. Development of the forces differs widely among the species due to natural selection of the flyers. As far as the flapping mechanism is considered, it is very efficient. The differences in flight mechanisms may be explored by determining the kinematics, kinetics, and aerodynamics of the flyers. Wing kinematics determines the aerodynamic forces, which vary with flapping speed. The maneuverability and stability are regulated by complex muscle action and neural control allowing the flyer to perform specific tasks. Computational models have emerged as powerful tools to predict the flow around the flyers with potential exploration of the complex interactions between the skeletal system, sensory system, and neural control of the flight.

DOI: 10.4018/978-1-7998-9078-2.ch014

INTRODUCTION

Humans have been fascinated by the beauty and intricacies of an avian flight for ages, preferably with the desire of performing the same locomotion; the flapping of wings. In dreams, science fiction, and legends, human beings have always aspired to fly using wings. Despite the state-of-the-art research in the area, resulting in the development of flight models such as airplanes, fighter planes, and helicopters, it appears that it would require a little more time to master it. Birds have already have mastered these flying techniques since their appearance on the earth, and it seems like they are just flaunting their ability on us.

The evolution of flight is distinctive among animal adaptations, especially in how the evolution has transformed the species from runners to flyers. According to the arboreal theory, there had been a pressure to evolve to fly (or glide) among the species on trees. However, the cursorial theory states no intermediate gliding stage for ground running bipeds that evolved into active fliers (Ostrom, 1979). Since the late Cretaceous period, birds have evolved from theropods and are considered "*living dinosaurs*" (Chiappe, 2009). Our current understanding of avian flight evolution has strengthened with evidence from morphological studies to molecular data. There is nearly a perfect harmony between the form and the flight performance of these species. From take-off to landing, the wingbeat kinematics is very particular to the desired motion. Understanding the wing motions during level flight, including forward flapping, gliding, soaring, stooping, slow flight, fast flight, and hovering, gives a deep insight into the manoeuvre. The profound admiration of the biomechanics of flight comes from arising importance of fluid mechanics and the structural effects on design problems of air vehicles. The The anatomical features of nature's flyers can be described as advanced manufacturing devices which are capable of aerodynamic control, aeroelasticity, and sensing (Heers et al., 2018). Deep appreciation and recognition of flight biomechanics with the application of computer simulations have led to the development of bio-inspired flight systems such as micro air vehicles (Han et al., 2021).

In this chapter, we review the nature of forces acting on the bird during the flight and explore as to how it controls the flight and performs the manoeuvre. We further look at the aerodynamics of avian flight from the computation perspective. Following this, the problem of avian flight has been addressed from a biophysics perspective by introducing the wing anatomy and correlating the physiology to the flight. The different control strategies have been described to understand the consequence of geometrical changes during different flight activities. Since decision-making for the choice of manoeuvre is an important part of flight control, we address the issues of the sensory-motor control of the bird also. The biomimetic design of a flapping wing has been discussed as an overview for bio-inspired design. We then conclude with the chapter overview including the future scopes and limitations of the current understanding of flight biomechanics.

AERODYNAMICS OF THE FLIGHT

Wing kinematics brings about the motion of air around the wings, essentially leading to the generation of the lift and thrust, depending on the nature of the flow. To fly, the bird has to generate sufficient force (lift and propulsion) as shown in figure 1 to overcome the gravitational pull and drag before gaining momentum for the flight manoeuvre.

Lift is the force generated to overcome the body's weight and enable flight. As the body propels through a fluid, it experiences resistance to the motion, known as drag. Of the drag, the friction between the solid surface of the bird and air (skin friction) is prominent and depends on the surface property of the body (feathers) and the air. The smooth feathers produce less drag. Another source of drag depends on the shape of the species and is called the form drag. The drag component dependent on the generation of lift is called the induced drag. The direction of flight is well coordinated by the vortex's induced flow, giving rise to a downward aerodynamic force. This force is the induced drag as it is formed by the vortices' action and depends on the magnitude of lift. Thrust is the force that the bird generates to move through the air that overcomes drag and weight. The magnitudes of lift (L), drag (D), and thrust (T), thus generated varies with the wing shape, factors affecting are wing area (A), density of fluid (ρ), kinematic viscosity (v), flow stream velocity relative to the wing (V), and area between the direction of air flow and surface of the wing (α).

The above factors are related as follows

$$L = \frac{1}{2}\rho V^2 A C_L . \tag{1}$$

$$D = \frac{1}{2}\rho V^2 A C_D . \tag{2}$$

$$T = \frac{1}{2}\rho V^2 A C_T . \tag{3}$$

Where, C_L, C_D, and C_T are lift coefficient, drag coefficient and thrust coefficient which can be determined experimentally. The drag coefficient which depends on the Reynolds number (R_e), density (ρ) and viscosity (μ) of the fluid is given by,

$$R_e = Vc \cdot \frac{\rho}{\mu} . \tag{4}$$

The flight power requirement can be determined in the form of the metabolic power input which is equal to the sum of mechanical power output and thermal power losses acting on the skeletal system for flapping, stooping and other kinematic behaviours. Muscular power output is the sum of the inertial power required for the wing strokes and aerodynamic power (P_{aero}) for the flight. P_{aero} is given as the sum the parasite power P_D and induced power $P_{i.}$

$$P_{aero} = P_i + P_D \tag{5}$$

Where, P_i is equal to the kinetic energy in the downward direction per unit time and can be expressed in terms of lift.

$$P_i = \frac{1}{2}\frac{L^2}{\pi\left(\left(\frac{1}{2}\right)b\right)^2 \rho V}. \tag{6}$$

Parasite power as,

$$P_D = DV = \frac{1}{2}\rho V^3 A C_D. \tag{7}$$

Where, b is the diameter equal to the wingspan.

Figure 1. Representation of the lift (L), drag (D), thrust (T), and weight (W) acting on the body of the avian species during flight. β. the angle of attack, θ. the stroke plane angle, α. the pronation angle relative to the body,

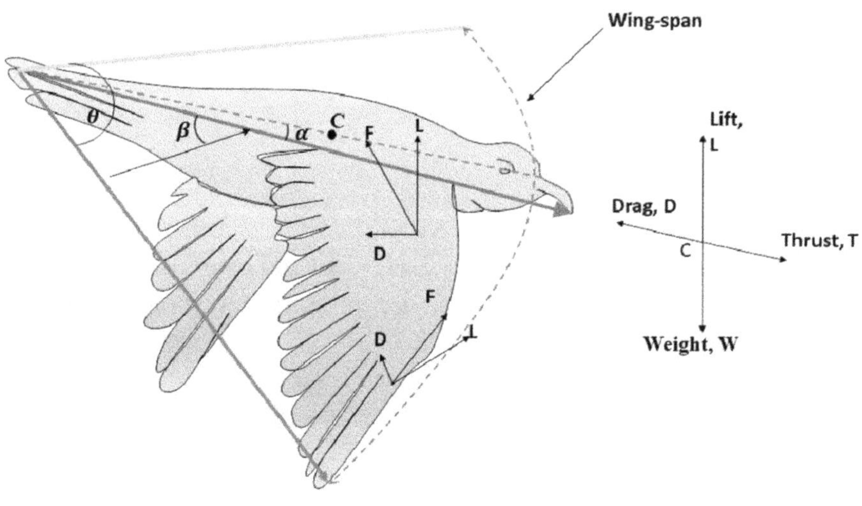

COMPUTER SIMULATIONS OF THE FLIGHT

The underlying principles of the bird flight may be understood by considering flow over the NACA (National Advisory Committee for Aeronautics) aerofoils, involving wind tunnel test (NACA, 1917). Considering that the quest for aeroplane design was inspired by the birds in its earliest development, we now go back to the aerofoil design as an inspiration for exploring the bird flight, especially the flow of air around the bird wing. The experiments performed in laboratory limits the investigation of flow over the wings owing to the dynamic events of the wing flapping along the direction of bird flight. Computer simulation of the aerofoil employing numerical methods eliminates such limitation for flow exploration. Based on the classical studies, flow over the wing surface may be described in the following way.

Significance of the Vortices

Vortices are very much influential in setting up pressure over the wing surface. These are the circulating current of air which is produced with the movement of an aeroplane. Depending on the nature of circulation (clockwise or anti-clockwise), the aerodynamic lift produce may be positive or negative.

Let us consider an example of a fixed wing motion by referring to figure 2. With onset of motion of the wing in forward direction, the air over the top surface tends to move towards the trailing edge and along the downward direction (downwash). However the air from the bottom surface also arrives at the trailing edge to meet the air traversing along the top surface, causing interaction and development of circulation in the flow. It could be noted that the flow arises from bottom surface to the top surface due to difference in the pressure at top and bottom proximal to the tailing edge. At the leading edge, the air flow is upward and positive (upwash). We now establish that the upward motion of air at the leading edge and downward motion of air at the trailing edge leads to circulation around the wing (bound vortex). Local to the surface, the local vortex circulate around and stay attached to the top surface of the wing. The bound vortices are found to be strongest at mid-region of wing span and weaken as one recede towards the wing tip. Based on the information of the circulation of the fluid (or strength of the local vortex) across the span, one may evaluate the lift across the span of the wing by applying the Kutta–Joukowski theorem. This gives us a well-known Lanchester–Prandtl theory, which emphasizes that the lift generation is not uniform but assumes a distribution (such as an elliptical load) along the span (Anderson Jr, 2010).

Along the span of the wing, the trailing edge receives the flow of air from both the top and bottom surface as illustrated in figure 3. Since air at the bottom surface arrives at higher pressure and encounters the downward direction of flow at higher momentum from the top surface, there is essentially any scope available for the air to move along the steam wise direction to expend the excess energy resulting from the pressure difference. This excess energy may be diverted cross-stream and towards the wing tip. As a result of cross-flow, wing-tip vortices are formed as the fluid from the lower wing spills up over and at the wing tips to accommodate the low pressure region (wing tip vortex). The bound vortices continue with the wing tip vortices and combines with the downwash to form a fast-spinning vortex which continue trailing down the wing tip (wake vortex). Considering the principle of mass and momentum conservation in qualitative sense for a compressible fluid, the air has the liberty to travel along the span wise direction (due to least resistance offered to fluid flow) to the wing tip and enter the top surface to fill up the low pressure region. In continuation, there is an occurrence of lateral movement of the fluid towards the wing tip at the bottom surface and away from the wing tip at the top surface. It may be reasoned that the fluid at the higher pressure on the bottom surface has the tendency to spread laterally and the fluid at the top surface has the tendency to shrink along the stream wise direction to accommodate fluid rolling off at the wing tip. This essentially contributes to spiralling of the wing tip vortices as the move down the stream. Trailing vortices are known to increase drag since a part of the energy goes into the production of turbulence; leading to an inducted drag and reduction in the effect lift force. According to the Helmholtz theory, the vortex must end up in a closed loop forming a horseshoe vortex. The trailing vortices on the either side of the wings are connected by a vortex over some distance downstream (starting vortex). During the start of the flow over the wing, there is a large difference of the fluid velocity at the trailing edge, which eventually rolls up forming the starting vortex. With increasing momentum, the starting vortex becomes stronger to shed. However, for steady flows the circulation of the starting vortex equals (but counter clockwise rotation) to the circulation around the aerofoil. At this point we notice that the vortex no longer increases in strength with no more addition of vortices from leading edge; satisfies the Kutta-Joukowski condition. The flow assumes a steady flow.

Figure 3. Comparison of aerofoil and bird wing kinematics (showing trace of the centre of gravity or CG of the cross-section of the aerofoil and wing) along the direction of propulsion and formation of vortices around the wing.

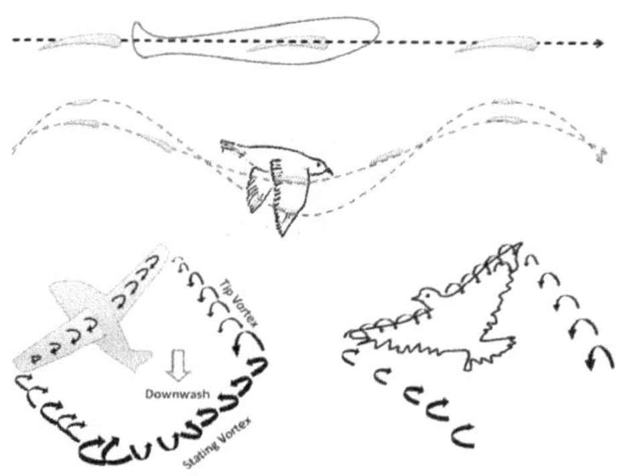

Movement of the Stagnation Point

A stagnation point is a point where the flow ceases to zero velocity. Laminar flow around the wing leads to stagnation of flow at both the edges. Depending on the circulation (Kutta condition) and the AoA, the stagnation point may drift along the surface of the wing. For circulation satisfying Kutta condition, the rear stagnation point positions at the trailing edge of the wing. At zero and lower circulation the rear stagnation point appears on the upper surface and near the trailing edge, while at higher circulation the rear stagnation point appears on the lower surface proximal to the trailing edge. As the AoA increase, the flow cease to turbulence at the trailing edge. Flow along the stream wise direction ceases and reverses at some point on the top surface, also referred to as the flow separation point. The viscous nature of the fluid tends to rolls off the fluid to form eddies in the boundary layer. If the eddies are small and confined to the surface, it will roll and reattach at a point close to the trailing edge (flow reattachment point). With increasing AoA the separation point traverses along the leading edge. The turbulence wake behind the wing does not favour the flight, rather contributes to drag. The drag increase significantly with increasing AoA where beyond critical AoA (where lift is maximum), any further increase in the AoA will abruptly reduce the lift force causing stall.

Aerodynamics of the Flapping Wing

Unsteady flow around the aerofoil is a problem of interest to aerospace engineers; especially when dealing of flapping wing for possible applications in micro air vehicle (MAV) as discussed in the next section. Had there been no flapping, the bird flight can be compared to an aeroplane with fixed wing. Following numerous studies, the scope of flight capabilities of an aerofoil with fixed wing (without flapping) would range from low speed to high speed as determined by the shape of the wing section; for example – a low camber thin wing design offers low drag and high speed; a thick/ thin and deep camber generates higher lift forces at low speed; and a reflex trailing edge provides low lift and high drag offering greater stability by limiting CoP (centre of pressure) movement (Abbott 2012).

In previous section, the flow around the wing has considered for a fixed wing with inflow at a constant speed of U m/s. If we assume no inflow in the flow domain (zero inflow) then the flow around the wing would be resulting from wing motion at U m/s (for e.g. albatross gliding). From ground frame of reference, it would appear as the bird travelling a speed of U m/s without flapping. However if wing movement were also considered along the vertical direction then the resulting flow will have an effect of flapping.

The flight of a bird is driven by the nature of flapping, which leads to the generation of forces and vortices that determine how the flow interacts with the wing and the flight trajectory. Over one cycle of the wing flap, the trajectory followed by the wing's cross-section at the mid region and those near the tip are different. Trace of the wing close to the tip region has long amplitude with wide AoA while at the mid-span the trace has small amplitude with narrow range of AoA.

With stroke of a wing (downward direction), the fluid tend to move around the wings causing circulation around the tips; vortex at the leading edge are referred to as the leading edge vortex (LEV). LEVs contribute to generation of a lift force to overcome the weight of the bird during unsteady flight. If by some means or by specific manoeuvre of the wing, the LEVs are allowed to stay attached on to the wing or delayed then the duration of lift force generation may be prolonged (Eldredge & Jones, 2019). If the LEVs detach then there is the sudden stall of the flight. The bird adopts rapid manoeuvres to take advantage of the peak lift forces generation above the critical AoA while ensuring attachment of the vortices to the surface. However, if the vortex shed prematurely before the end of the flapping cycle then there the no enhancement in lift. The condition of stall in flapping (dynamic stall) is observed well beyond the critical AoA of the fixed wing (Yu et al., 2017). Understand the dynamics of vortex formation during flapping would be very useful to explore the science of avian flight (Lin-Lin et al., 2016). It is still a subject of investigation as to how the vortices stay close to the wing for longer duration of time during complex wing motions (Nan & Crowther, 2017).

WING STRUCTURE

Avian wings vary in size and shape among the species. The wing serves to develop the lift and propulsive forces through various wing movements such as the upstroke and downstroke. The wingspan, generally measured as the wing area, is the critical factor determining aerodynamic forces. Spreadable feathers, intrinsic muscles and skeletal joints allow wing morphing (transient shape changes during flight). The circular area swept by the wings is given by $\pi b2/4$ where b is the wingspan, the perpendicular distance between the wingtips. The wing architecture is measured as wing loading (ratio of average body mass, M, to wing area, A*)* and wing aspect ratio (squared wingspan to area of the wing). The aspect ratio is given as b2/A and the wing loading is given as *M/A.* The bird wings can be distinguished in different forms based on the specific adaptation of the avian species. The elliptical wing is short, broad and have moderate wing loading and typifies birds like jays and woodpeckers. The broad soaring wings are typical of vultures and eagles and have a long and broad shape. High-speed wings have a higher aspect ratio and wing loading, which helps falcons and hummingbirds. The long soaring wings are confined to birds to oceanic soaring birds like albatrosses as they are long, pointed and slender with a high aspect ratio and wing loading (Dhawan, 1991). The wings have lightweight structures with feathers that contribute to aerodynamic forces required for flight.

A thorough study of skeletal component and muscular component in the wing structure is required to understand the actuation of flight. The avian wings have the usual arm structure of mammals and reptiles. The upper arm comprises the humerus and the lower arm is made up of the radius and the ulna which is connected to the carpometacarpals connected to the metacarpals, refer to illustration in figure 4. The digits are modified particularly for flight. The pectoral girdle consists of the scapula, the clavicle and the coracoid. The coracoid is the strongest bone of the pectoral girdle and moves ventrally and posteriorly to synchronise with the sternum. The coracoid length determines the distance between the sternum and shoulder joint. The pair of clavicles is fused to form the furcula. The shoulder joint connects the wings to the body through the proximal condyle of the humerus to the scapula and the coracoid. The primaries are long flight feathers connected through connective tissue to the phalanges and metacarpal. The secondaries are the long flight feathers growing from the ulna. Both the primaries and secondaries are the profile of the wing. The feathers are connected by tendons along the upper side of the patagium. As shown in figure 4, feathers fill in the gaps between the primaries and the secondaries. The alula is the first digit in the birds' wings which move freely. All these bones have specific composition for flight function ability. The cranial, femoral and humeral are densest and hence the bone stiffness and strength increases. The bones are thinner than other terrestrial animals and are hollow. The small relative volume of birds' wings reduces the muscle size and energy requirement.

The distribution and size of muscles vary among the bird species. The flight muscles can be classified into three types: pectoral, accessory, and tensor. Pectoral muscles contract to allow for the upstroke and downstroke movement of the wing. Accessory muscles aid in elevating and depressing the wings during flight (Biewener, 2011). Tensor muscles keep the propatagium fully stretched for the wing extension during flight. The pectoralis major is the largest muscle of the wing with a triangular shape. The former covers the whole breast of the bird, taking up to 8-11% of the total body mass, and is known as *"Breast muscle"* (Cao & Jin, 2020). It attaches to the sternum and narrows to an attachment area on the humerus. The pectoralis minor lies on the dorsal side of the pectoralis major. The supracoracoideus muscle is the second in mass and powers the upstroke during flight and hovering. The former attaches broadly to the sternum and circuitously attaches to the humerus.

FROM MUSCULAR CONTRACTION TO KINEMATICS

Contraction of the skeletal muscles brings about the movement of the wing and defines as to how the wing traces its path in the three-dimensional space (wing kinematics). The pectorialis major is the primary pronator and depressor of the wing and they rotate the humerus around the shoulder joint of the wing. The pectoral muscles have an essential role of producing necessary mechanical force for downstrokes during flight and hovering. The pectoralis major is activated halfway during upstroke and therefore slows and reverses the upstroke. Activation of the muscle is actively the pectorialis minor helps birds in swift take off as well as controlling the upstroke and backstroke. The supracoracoideus muscle is the primary supinator and elevator of the wing and therefore contributes to crucial angular variations reducing counterproductive upstroke forces. The upstroke generates the aerodynamic lift. The supracoracoideus muscle plays a critical role in hummingbirds during hovering flight by generating 25% of the necessary vertical force providing significant aerodynamic power (Cheng et al., 2016). The supracoracoideus muscles stores elastic energy to reduce the required power output of flight muscles. The muscles at the upper surface of the wing membrane are connected to the feathers through tendon that extend the bunch of feathers as the elbow and metacarpals extend. The secondaries are curved downwards by the tendons to increase the twist angle. The torque acting at the primaries translates through the digits to the elbow, and the resultant torque produced for the motion of the secondaries translates from the humerus to the shoulder. This torque at these joints is generated by the contraction of muscles which direct the motion of different bones. Stretch and twist is generated by the wrist and elbow joint and flapping and swinging is performed by the shoulder joint which form the "*flapping–twist–swing*" motion (Li et al., 2020).

Figure 4. Musculoskeletal diagram of the bird of relevance to flight

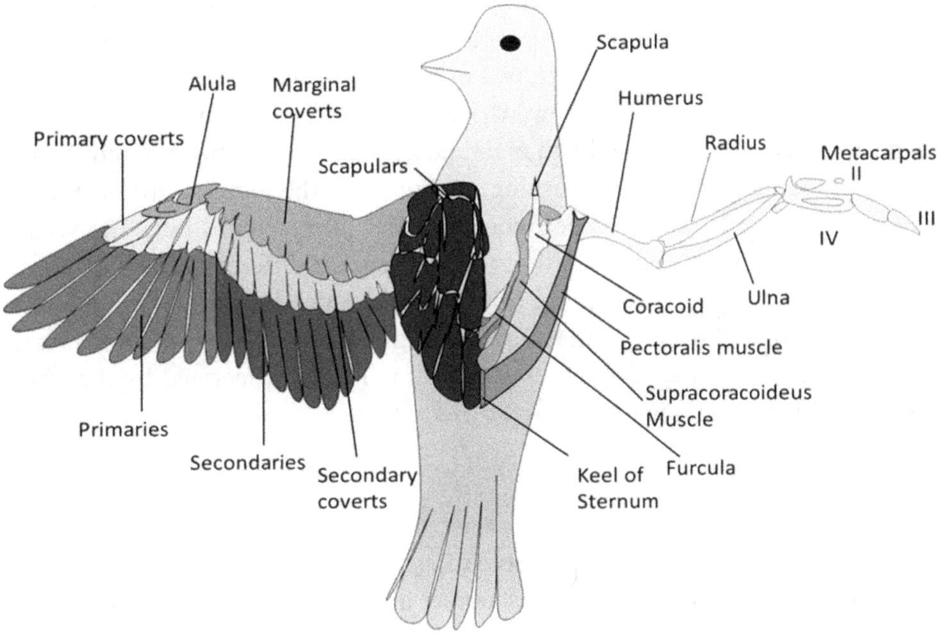

FLIGHT CONTROL STRATEGIES

Understanding the control strategies involved in the manoeuvring of a successful flight gives an insight on how the avian species have adapted to such an advanced locomotion. In control sense, it is desirable to study the wingbeat kinematics so as to understand the variations for different tasks (Marey, 1980). Wingbeat mechanism varies among species as much as there are differences in walking styles of human beings. Scholars continue to understand and investigate the wingbeat mechanics of different species (Videler, 2006).

Wing and Tail Coordination

Understanding the joint kinematics for the upstroke and downstroke is essential. During upstroke, the axis rotation of the metacarpus bone leads the motion (refer to section 2 on wing structure). The wrist joint has major contribution to the vortex formation. Supination of the wrist joint occurs during downstroke and pronation occurs during upstroke causing the wingtip reversal. With the joint coordinate system, the net torque at the joint can be computed through inverse dynamics method. The level of extension and flexion varies for different species. The shape of the wings varies for different birds according to the constraint to flexion and extension at the radius and ulna. The shoulder joint supinates during upstroke at the same time pronation of metacarpus begins. The humerus remains heavily supinated for the transition between upstroke and downstroke. The shoulder joint is most responsible for the motion of the wing tip and the elbow joint causes most of the dorso-ventral motion of the wing. The wing joints mostly have the same motion however the timing, the torque and the excursion angles vary for different external wing kinematics.

The tail forms an essential part of bird's flight. The tails also vary among the avian species not only for the purpose of sexual selection but also for suiting the aerodynamic requirements. After expanding its tail, the bird lowers it in the direction of the angle of attack during slow flight. While, it accelerates, the bird reduces the angle of the tail relative to airflow. For quicker motion, it furls the tail. Before landing, a bird increases the tail spread and raises it upwards. This gives a control moment to the body to keep it in the vertical position. Birds use their tails to maintain aerodynamic stability, control angle of attack of wings by balancing the moment, control turning, and produce the required lift during acceleration, slow flight, and turning (Thomas, 1997. Experimental studies have demonstrated the variations of the tail position of Harris hawk during glide in a tilting wind tunnel at different speeds (Tucker, 1992). As the wings flex and sweep during high speed, the tails furl in order to reduce the drag. Similarly, during slow flight, the wings are spread widely with the centre of gravity moved forward and hence the Harris hawk generates lift using the tail for balancing the moment. The lift generated by the tail depends on the acceleration and the downward movement of the airflow.

Landing and Take off

During take-off, the bird needs to increase its forward speed by a quick flapping motion to generate enough lift to overcome gravity. The wings must generate the required lift to balance weight and accelerate the forward motion with the speed still low. The wing spreads wide to adjust the flapping motion. Some birds jump to facilitate take-off and generate the lift quicker. They beat their wings at a high angle of attack and most of the acceleration is provided by downstroke lift. Parrotlets increase lift and drag during take-off to maximise the resultant force vector (Deetjen et al., 2017). They limit power required to fly short distances with effective take-off angles. Pigeons continue to accelerate after leaving the perch and their kinematic parameters rotate through larger angles during take-off. Their stroke plane is inclined steeply downward to accelerate the take-off (Berg & Biewener, 2010).

To land, the bird must reduce its speed, reduce the lift and gradually come to a stop. The bird tilts backwards, raising its angle of attack and increasing the drag to slow the flight. The large angle of attack disrupts the airflow on the wings creating turbulent eddies that cause stalling. However, the bird still needs some lift and activates the alula, located on the edge of the wing near the wrist joint. Activating the alula elevates it and creates space between the wing and the quills (Dvořák, 2016). The speed before the touchdown is too high and the bird tilts back more with nearly vertical wings beating forward to apply the brakes for the landing. The landing varies among species and depends mostly on body mass, shape and wing size.

Forward Flapping Flight and Intermittent Flight

Flapping flight is considered the *"power-on"* flight since energy is continuously supplied to flap the wings and generate adequate lift and thrust forces. The main aim is to propel the body and overcome gravity. Here, the birds vary the stroke plane angle (θ) and pronation angle of the wing (α) to change the direction of the forces accordingly. The pronation angle is the angle of rotation of the wing with respect to the axis of the body. The bird adducts its wings during upstroke especially during slow flight as there is no necessary lift production. Most definitive data of power consumed during forward flapping are obtained from the kinematic study of birds flying through wind tunnels. In these studies, the upstroke and downstroke kinematics are monitored and the kinematic variables are observed including body stroke angle, wingbeat frequency, body angle and wrist path. There are considerable differences of wingtip path and wrist path among species. In pigeons, the wingtips show a "figure of eight" pattern at 6 ms-1 to 8 ms-1 and an elliptical pattern at 10 ms-1 to 20 ms-1 whereas in magpies, the wingtip shows an elliptical path throughout (Tobalske and Dial, 1996).

Energy can be saved during flapping flights by intermittent flight by reducing the wingbeat frequency. Flap bounding birds perform rhythmic flight with an up and down sinusoidal path. Birds flap their wings during low tracks, increase their speed, and then fold their wings close to the body while accelerating along the upper path. Drag is reduced with their wings folded, and this is advantageous during quick flight as drag force is proportional to velocity squared. The power required can be reduced by 35% for chaffinch this way (Rayner et al., 2001). Tobalske (2007) monitored the behaviour of flap bounding flight for six different avian species by varying their body mass by a factor of 10 (Tobalske, 1996). The flapping frequency was highly variable during the flight. At higher speeds, the stroke plane angles and tail spread decreased. With high speed, the birds flap glide and they reduced the percentage of flap bounds. Pigeons folded their wings closer to their body during glides at higher speed.

Gliding and Soaring

Gliding can be analysed as a horizontal powered flight occurring in still air, rising air and slope current. The power is supplied by gravity and the body moving in a straight line inclined downwards with outstretched wings. During gliding, the bird's weight overcome the air resistance in the forward motion; therefore, large birds can glide effectively regularly. The bird gliding at an angle with respect to the horizontal loses height at a specific rate known as the 'sinking speed'. The weight can be resolved into two components here to form the drag and lift equation. For a steady glide, the total drag is equal to $W\sin\alpha$ and the lift is equal to $W\cos\alpha$. The ability of a bird to glide effectively at high speed for long distances with a minimum sinking speed and the aerodynamic conditions under which the bird sustains itself in the air for the longest duration have been studied with interest for biomimetic purpose.

During soaring, the bird increases or maintains the altitude of flight without any relative wing flapping. It does so by taking advantage of rising air (updrafts), the upward velocity component maintains the flight height. These updrafts are called thermals, which occur when airstream hits the cliffs, building or hills caused by uneven heating of air near the earth's surface. Warmer air is less dense and rises, but cools down at high altitudes and sink. Birds use these thermals to fly in circular motion. Argentavis magnificiens was one of the world's most giant known flying birds from 6 million years ago and weighed approximately 70 kg with a wingspan of 7m. The former used mainly slope soaring in the windward slopes of the Andes, Argentina, and mainly depended on thermals for its flight to hunt (Chatterjee et al., 2007). Albatrosses take advantage of the velocity gradients for dynamic soaring, which is a complex fight maneuver in which the bird uses energy from the horizontal component of the airstream enabling the bird to fly continuously without flapping (Sachs, 2005).

Hovering Mechanism

Hovering flight requires more power than other flight motions. A higher lift requirement for the hovering flight is accomplished by increasing the frequency of wing beating and increasing the horizontal area swept by the wings. Hovering at one spot is quite common for kestrels and others, but opposing the wind motion is difficult as well. Hummingbirds do change their positions sideways or backward while hovering for juice suction. During hovering, the resultant lift is vertical and equals to the body weight. In hummingbirds, the movement of the wings is almost sinusoidal with respect to angular displacement (Weis-Fogh et al., 1972). In free-flying hummingbirds, the lift production is asymmetric as the downstroke produces 75% of the weight support (Nan et al., 2019). Hummingbirds achieve their aerobatic beats with a fully extended wing. Wing inversion occurs during the transition from the upstroke and downstroke of hummingbirds' wings which is directed by the motion of the shoulder joint.

The power equation that has been provided in section 3.2 cannot be applied for hovering as V is zero. The power is calculated through the vertical momentum change that supports the weight (Dhawan, 1991). Here we assume uniform induced speed and minimal friction drag experienced by the flapping wings and the body. The induced velocity, u_t is obtained from the change in momentum in the vertical direction supporting the weight, W, and is given by

$$u_i = \left(\frac{2}{\rho \pi} \frac{W}{b^2} \right)^{\frac{1}{2}}. \tag{8}$$

The power required for hovering is then given by

$$P_h = W u_i. \tag{9}$$

The ratio of the power required for hovering to the power required for horizontal flight increases with the size of the birds and therefore smaller birds can effectively hover.

Turning Flight

As a bird moves in a circular motion, its wings are banked in the direction of the center of the circle and the body experiences a centripetal force. Turning flight is essential to understand to know how birds are able to change the direction of flight. The birds' wings are widespread during turning flight, and the primary feathers' tips extend wide. The bird has to generate more lift on one wing than the other to control the roll angle and adjust the wingtip. Then to control the pitch angle, the bird changes its direction of attack. To go up, the bird body makes an upward angle relative to the airstream and to go down, the bird makes a downward angle. To control yaw angle, the birds twist the wingtips in the desired direction.

Stooping

Stooping is a high-speed attack dive. Eagles often use the stooping technique to catch their prey. After identifying their target prey, the bird dives straight down by folding its wings to reduce drag and accelerate the flight to increase the attack force. The peregrine falcon is known to have one of the fastest speeds while stooping. It generates higher acceleration at the end of its attack and has a very steep descent angle. The falcons' stoops from high altitude have a higher catch success than low stoops since high stoops give them better control. High altitude stoops have a higher velocity and generate higher aerodynamic forces (Mills et al., 2018).

NEURAL CONTROL OF FLIGHT MANOEUVRES

The wingbeat kinematics results from the coordination between the sensory and neuromuscular system as shown in figure 5. The nervous system controls this coordination. In section 6, we have understood how the bird navigates as a pilot and the different kinematic executions. The nervous system is responsible for executing these actions. The motor control system is a complex organization of the sensors of the bird body, the nervous system, the motor output, and the actuators. The different sensory systems provide the necessary and relevant nervous input on different temporal scales resulting in an organized behavioral control (Altshuler et al., 2015).

Figure 5. A system's viewpoint of a bird flight from the kinematics of wing motion to generation of aerodynamic forces through use of the sensory and neuromuscular system for feedback regulation of the flight.

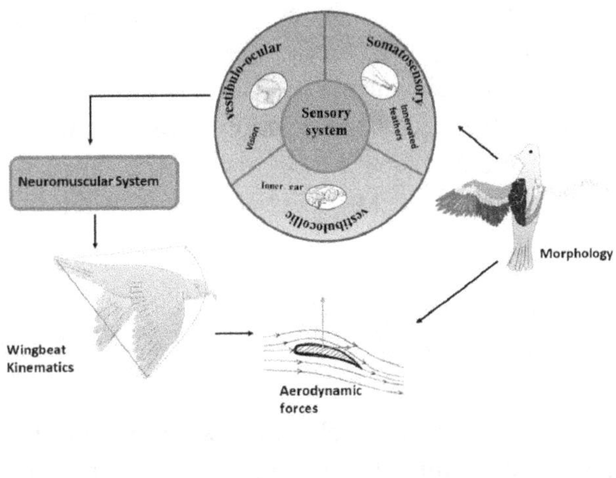

The somatosensory system detects the mechanical forces acting within the or on the bird's body by a rapid technique called mechanosensation which allows fast responses to invisible stimuli. This explains the rapid maneuvering of flight. The somatosensory system can be categorized into the trigeminal system (supplied to the beak) and the spinal system (supplied to the body surface, wings and legs). The system includes cutaneous receptors which are specialized to activation through mechanical, thermal or noxious stimuli. Pressure variations of a wider range of frequencies occur due to the low viscosity of air may have caused physiological specialisations in avian somatosensation. Mechanosensors are situated under feather follicles and any force experienced by all body feathers triggers a force feedback system (Saxod, 1996). The translation of these forces to the monitoring of the forces at the wings is still unknown. Mechanosensory system reduces the reaction time if uncoupled with the visual system. Flight stabilization may result from a reflex loop, the complex circuitry from vision to detection and then to the motor control. Birds are one of the most visually dependent species. The acuity of many birds surpasses that of other living beings especially those for birds of prey and quick flight (Draper, 1966). The size of the birds' eyes advocates for their importance as they are the largest relative to the body size of other species. The incoming light passes the lens, the vitreous chamber and then to the retina which absorbs the incoming light for sensing and then integrates information in it which is sent to the brain. The retinas in avian species have double cones which are responsible for processing achromatic motion and luminance cues and color information is integrated by four different classes of single cones (Hunt et al., 2009). The density of the receptor cells in the bird's eye retina allows them to see in dim light. Birds can track a moving object easily due to their double fovea. Polarized light for dim light at sunrise and sunset helps the birds navigate and hence sensitive light polarization is one of the key features for maneuvering flight (Muheim, 2011).

The vestibular system relays sensory information on the head movement to the brain and it contributes to stabilizing the head during flight. The vestibular stimuli determine flight coordination extensively and are responsible for the variations in flight style and abilities (Walsh et al., 2013). For proper image stabilization in the high-speed motion, the movement of the eyes should be in a definite range. The visual and vestibular signals drive the head stabilization reflexes for these specific actions Birds have to keep their heads fixed relative to the horizontal despite variations in the body axis rotation to develop stable platform for visualization. When collars restricting the neck from executing any angular motion were put on pigeons, they could not control the flight as they were unable to maintain a fixed head position with respect to the horizontal (Warrick et al., 2002). The eye motion triggered by the vestibular system for adjustments is called the vestibulo-ocular (VOR) reflex (Wallman et al., 1982) and the vestibulocollic (VCR) reflex triggers the head position adjustments (Goldberg & Cullen, 2011). The VOR makes inadequate adjustments for instability in the fixed head position, compensating when the VCR is activated to contribute in the fixed head position (Haque & Dickman, 2005). Vestibular reflexes stabilize gaze and aid in posture control and stabilizing head, tail, and body positions relative to the direction of flight.

BIOMIMETICS: FROM BIOMECHANICS OF AVIAN FLIGHT TO BIOINSPIRED DESIGN

Biomimetics is the combination of biology and technology where newer discoveries are inspired from nature. By biomimetic design of a micro air vehicle or MAV, we refer to the mimicking all the kinds of maneuvers the bird would exhibit. Designs by nature are more efficient e.g., the albatross with hinged wing tips is efficient in countering the effects of gusts and turbulence. By biomimicry, the nature's untold mysteries of flight can now be fused with technology artificial birds in application perspective.

Mechanical Hummingbird Project or MHP is one such design. To mimic the Hummingbird motion a single actuator system is utilized to perform flapping about biaxial direction by employing cam based system using a four-bar mechanism (McIntosh et al., 2006). The structure comprises of motors, gear wheels, connecting rods, follower and guide, springs, wing carriers and wing spars. The wing carrier attaches with the wing spar and body of the mechanical Hummingbird by a pin. There are three directions of movement to enable movement about the pin joint, feathering axis and addition motion involving deflection. The flapping motion is executed by the four-bar arrangement using motor, two gear wheels and wing carrier. The right gear is driven directly by the motor but the left gear indirectly by an idler gear. The gear wheels move the connecting rods which in turn move the wing carriers and wing spar causing flapping of the wings. The cam system drives the feathering motion. Torsion spring connects wing spar and wing carrier at its two ends. The torsion spring undergoes compression during wing spar motion generating moment about the wing spar axis. Bending spring is used to allow slight misalignment while transfer of torque from follower to wing spar. Designs were made to provide the liberty drive various kinds of movement. The sequence of motion is described, the follower rotates about its' feathering axis thus promoting rotation of the wing, after this, the follower slides along the back side of the guide during its downward motion. During this process, the torsion spring remaining compressed supplies the required moment and the bending spring prevents the follower getting separated during the sliding motion. As the follower reaches the bottom of the guide the wing takes into fully pronated form. Next, the follower from the bottom edge moves to the guide's top edge, sliding across its front side. Since no flange is present in the follower's back side, the wings remain in previous pronated form only.

The project's inspiration (MHP) was to explore the possibility of hovering with a single actuator. Experiments were performed on the mechanical Hummingbird to measure forces developed during upstroke and downstroke at various flap rates. To facilitate flight control, authors collected the data using dSPACE data acquisition board and processed using MATLAB. A force-torque sensor was used to regulate the linear and rotational motions of the MAV. Forces were measure in the Cartesian coordinate system for the six complete flap cycles. Two peaks were observed in both upstroke and downstroke motions in the force vs. time plot. The first negative peak occurred before upstroke due to rotational forces caused by the wing's feathering motion and the second one during its translational motion. The positive peaks were observed at first when the MHP feathers were prior to downstroke and second as the wing translate during downstroke. Lift force vs. flapping Rate plot showed that the experimental results obtained are higher than the prescribed theoretical model.

Besides mere replication of the wing kinematics, new concepts in biomimicry such as the *"morphing"* has also evolved to address various flight challenges. Developing specific mission designed MAVs is a challenge. It would be more economical to construct a MAV that can adapt via morphing for better performances such as during take-off, hover, and cruise (Li et al., 2018). Morphing, besides allowing for folding and unfolding of the wings, deformation ability also improves the flight performance. The lift forces generated on both the wings can be vectorially resolved into lateral and vertical components. Though the vertical components cancel out due to gravity, the lateral components generate a torque on the MAV which causes the MAV to roll. During steady flow, the rolling torque can be controlled by autopilots. But, when the MAV launches its wings from folded state quickly, the autopilots become ineffective. Constructing both the wings on same horizontal plane can solve this challenge (Zheng, 2019). MAV can be designed by keeping both the wings on same horizontal plane by proper synchronization of wing position and change of sweep angle. Such a design can be developed using an axle, two wing brackets and two racks; where, the two wing brackets and two wings are joined with each other through bolts. These two wings are bolted with a mount which is spring or motor driven. The spring performs wing positioning so that both wings are on same axis, while motor adjusts the sweeping angle.

CONCLUSION

Understanding flight mechanism provides an immense opportunity to mimic the physics of avian flight and design a bio-inspired model for possible application in navigation. From wing structure to kinematics to aerodynamics, we can appreciate the level of technicality that goes into the design of a bio-inspired model of a bird. Understanding the concept of bionic aerodynamics presents new challenges to developing new autonomous and efficient air vehicles. This chapter provides a deep insight on the aerodynamic changes during bird flight and the effective flight manoeuvre for different species. The use of computational methods is useful to address the issues due to structural design and maneuverability. While the biomimicry of the bird kinematics imitates the range of movement observable in a bird during the take-off, on-flight and landing, controllability (level of sophistication required to perform a specific manoeuvre) equally plays a key role in actuating the mechanical bird to follow through the specific manoeuvre. We have a vague idea of the functioning of the important flight muscles. Hence, there is a need of effective simulation models to investigate muscle functions and the interactions between bones, feathers, and muscles to accomplish bird flight. The understanding of flapping flight is still inadequate. Potential exploration of the complex interactions between the skeletal system, sensory system, and neural flight control would pave for further understanding of the mechanics involved in the avian flight

REFERENCES

Abbott, I. H., & Von Doenhoff, A. E. (2012). *Theory of wing sections: Including a summary of airfoil data*. Courier Corporation.

Ahmed, R. I. (2016). *Theoretical Foundation and Numerical Analysis on Coandă Effect for Micro-Air Vehicle* (Doctoral dissertation). Universiti Putra Malaysia.

Ahmed, R. I., Djojodihardjo, H., Talib, A. R. A., & Rafie, A. M. (2017). Review on Progress and Application of Active Flow Control Devices-Coandă Effect on Unmanned Aerial Vehicles. *Pertanika Journal of Scholarly Research Reviews*, *3*(1).

Altshuler, D. L., Bahlman, J. W., Dakin, R., Gaede, A. H., Goller, B., Lentink, D., Segar, P. S., & Skandalis, D. A. (2015). The biophysics of bird flight: Functional relationships integrate aerodynamics, morphology, kinematics, muscles, and sensors. *Canadian Journal of Zoology*, *93*(12), 961–975. doi:10.1139/cjz-2015-0103

Anderson, J. D. Jr. (2010). *Fundamentals of aerodynamics*. Tata McGraw-Hill Education.

Batchelor, C. K., & Batchelor, G. K. (2000). *An introduction to fluid dynamics*. Cambridge university press. doi:10.1017/CBO9780511800955

Berg, A. M., & Biewener, A. A. (2010). Wing and body kinematics of takeoff and landing flight in the pigeon (Columba livia). *The Journal of Experimental Biology*, *213*(10), 1651–1658. doi:10.1242/jeb.038109 PMID:20435815

Biewener, A. A. (2011). Muscle function in avian flight: Achieving power and control. *Philosophical Transactions of the Royal Society of London. Series B, Biological Sciences*, *366*(1570), 1496–1506. doi:10.1098/rstb.2010.0353 PMID:21502121

Cao, T., & Jin, J. P. (2020). Evolution of flight muscle contractility and energetic efficiency. *Frontiers in Physiology*, *11*, 1038. doi:10.3389/fphys.2020.01038 PMID:33162892

Chatterjee, S., Templin, R. J., & Campbell, K. E. (2007). The aerodynamics of Argentavis, the world's largest flying bird from the Miocene of Argentina. *Proceedings of the National Academy of Sciences of the United States of America*, *104*(30), 12398–12403. doi:10.1073/pnas.0702040104 PMID:17609382

Cheng, B., Tobalske, B. W., Powers, D. R., Hedrick, T. L., Wethington, S. M., Chiu, G. T., & Deng, X. (2016). Flight mechanics and control of escape manoeuvres in hummingbirds. I. Flight kinematics. *The Journal of Experimental Biology*, *219*(22), 3518–3531. doi:10.1242/jeb.137539 PMID:27595850

Chiappe, L. M. (2009). Downsized Dinosaurs: The Evolutionary Transition to Modern Birds. *Evolution (New York)*, *2*(2), 248–256. doi:10.100712052-009-0133-4

Deetjen, M. E., Biewener, A. A., & Lentink, D. (2017). High-speed surface reconstruction of a flying bird using structured light. *The Journal of Experimental Biology*, *220*(11), 1956–1961. doi:10.1242/jeb.149708 PMID:28348041

Dhawan, S. (1991). Bird flight. Sadhana. *Academy Proceedings in Engineering Sciences*, *16*(4), 275–352.

Draper, M. H. (1966). Avian Physiology. *Quarterly Journal of Experimental Physiology and Cognate Medical Sciences: Translation and Integration*, *51*(4), 381–382. doi:10.1113/expphysiol.1966.sp001873

Dvořák, R. (2016). Aerodynamics of bird flight. *EPJ Web of Conferences*, *114*(01001).

Eldredge, J. D., & Jones, A. R. (2019). Leading-edge vortices: Mechanics and modeling. *Annual Review of Fluid Mechanics*, *51*(1), 75–104. doi:10.1146/annurev-fluid-010518-040334

Goldberg, J. M., & Cullen, K. E. (2011). Vestibular control of the head: Possible functions of the vestibulocollic reflex. *Experimental Brain Research*, *210*(3), 331–345. doi:10.100700221-011-2611-5 PMID:21442224

Han, J., Hui, Z., Tian, F., & Chen, G. (2021). Review on bio-inspired flight systems and bionic aerodynamics. *Chinese Journal of Aeronautics*, *34*(7), 170–186. doi:10.1016/j.cja.2020.03.036

Haque, A., & Dickman, J. D. (2005). Vestibular gaze stabilization: Different behavioral strategies for arboreal and terrestrial avians. *Journal ofNeurophysiology*, *93*(3), 1165–1173. doi:10.1152/jn.00966.2004 PMID:15525803

Heers, A. M., Rankin, J. W., & Hutchinson, J. R. (2018). Building a Bird: Musculoskeletal Modeling and Simulation of Wing-Assisted Incline Running During Avian Ontogeny. *Frontiers in Bioengineering and Biotechnology*, *6*, 140. Advance online publication. doi:10.3389/fbioe.2018.00140 PMID:30406089

Hunt, D. M., Carvalho, L. S., Cowing, J. A., & Davies, W. L. (2009). Evolution and spectral tuning of visual pigments in birds and mammals. *Philosophical Transactions of the Royal Society of London. Series B, Biological Sciences*, *364*(1531), 2941–2955. doi:10.1098/rstb.2009.0044 PMID:19720655

Li, B., Zhu, Q., Guo, S., Yang, F., Li, Y., Zhu, Z., Chen, S., Song, R., & Li, Y. (2020). Design and experiment of a bionic flapping wing mechanism with flapping–twist–swing motion based on a single rotation. *AIP Advances*, *10*(6), 065018. doi:10.1063/5.0008792

Li, D., Zhao, S., Da Ronch, A., Xiang, J., Drofelnik, J., Li, Y., Zhang, L., Wu, Y., Kintscher, M., Monner, H. P., Rudenko, A., Guo, S., Yin, W., Kirn, J., Storm, S., & De Breuker, R. (2018). A review of modelling and analysis of morphing wings. *Progress in Aerospace Sciences*, *100*, 46–62. doi:10.1016/j.paerosci.2018.06.002

Lin-Lin, Z., Hui, G., & Chui-Jie, W. (2016). Three-dimensional numerical simulation of a bird model in unsteady flight. *Computational Mechanics*, *58*(1), 1-11.

Marey, E. J. (1890). *Physiologie du mouvement: vol des oiseaux, 207. G.* Masson. doi:10.5962/bhl.title.115247

McIntosh, S. H., Agrawal, S. K., & Khan, Z. (2006). Design of a mechanism for biaxial rotation of a wing for a hovering vehicle. *IEEE/ASME Transactions on Mechatronics*, *11*(2), 145–153. doi:10.1109/TMECH.2006.871089

Mills, R., Hildenbrandt, H., Taylor, G. K., & Hemelrijk, C. K. (2018). Physics-based simulations of aerial attacks by peregrine falcons reveal that stooping at high speed maximizes catch success against agile prey. *PLoS Computational Biology*, *14*(4), e1006044. doi:10.1371/journal.pcbi.1006044 PMID:29649207

Muheim, R. (2011). Behavioural and physiological mechanisms of polarized light sensitivity in birds. *Philosophical Transactions of the Royal Society of London. Series B, Biological Sciences*, *366*(1565), 763–771. doi:10.1098/rstb.2010.0196 PMID:21282180

Nabawy, M. R., & Crowther, W. J. (2017). The role of the leading edge vortex in lift augmentation of steadily revolving wings: A change in perspective. *Journal of the Royal Society, Interface*, *14*(132), 159. doi:10.1098/rsif.2017.0159 PMID:28747395

Nan, Y., Peng, B., Chen, Y., & McGlinchey, D. (2019). From studying real hummingbirds to designing hummingbird-like robots—A literature review. *IEEE Access: Practical Innovations, Open Solutions*, *7*, 131785–131804. doi:10.1109/ACCESS.2019.2939491

Ostrom, J. H. (1979). Bird Flight: How Did It Begin? Did birds begin to fly" from the trees down" or" from the ground up"? Reexamination of Archaeopteryx adds plausibility to an" up from the ground" origin of avian flight. *American Scientist*, *67*(1), 46–56. PMID:434589

Rayner, J. M., Viscardi, P. W., Ward, S., & Speakman, J. R. (2001). Aerodynamics and energetics of intermittent flight in birds. *American Zoologist*, *41*(2), 188–204. doi:10.1093/icb/41.2.188

Reba, I. (1966). Applications of the Coanda effect. *Scientific American*, *214*(6), 84–93. doi:10.1038scientificamerican0666-84

Sachs, G. (2005). Minimum shear wind strength required for dynamic soaring of albatrosses. *The Ibis*, *147*(1), 1–10. doi:10.1111/j.1474-919x.2004.00295.x

Saxod, R. (1996). Ontogeny of the cutaneous sensory organs. *Microscopy Research and Technique*, *34*(4), 313–333. doi:10.1002/(SICI)1097-0029(19960701)34:4<313::AID-JEMT4>3.0.CO;2-P PMID:8807616

Schlichting, H., & Gersten, K. (2016). *Boundary-layer theory*. Springer.

Smith, N. F. (1972). Bernoulli and Newton in fluid mechanics. *The Physics Teacher*, *10*(8), 451–455. doi:10.1119/1.2352317

Thomas, A. L. (1997). On the tails of birds. *Bioscience*, 215–225.

Tobalske, B. W. (2007). Biomechanics of bird flight. *The Journal of Experimental Biology*, *210*(18), 3135–3146. doi:10.1242/jeb.000273 PMID:17766290

Tobalske, B. W., & Dial, K. (1996). Flight kinematics of black-billed magpies and pigeons over a wide range of speeds. *The Journal of Experimental Biology*, *199*(2), 263–280. doi:10.1242/jeb.199.2.263 PMID:9317775

Tucker, V. A. (1992). Pitching equilibrium, wing span and tail span in a gliding Harris' hawk, *Parabuteo unicinctus*. *The Journal of Experimental Biology*, *165*(1), 21–41. doi:10.1242/jeb.165.1.21

Videler, J. J. (2006). *Avian flight*. Oxford University Press. doi:10.1093/acprof:oso/9780199299928.001.0001

Wallman, J. O. S. H., Velez, J. O. S. E., Weinstein, B., & Green, A. E. (1982). Avian vestibuloocular reflex: Adaptive plasticity and developmental changes. *Journal of Neurophysiology*, *48*(4), 952–967. doi:10.1152/jn.1982.48.4.952 PMID:6982960

Walsh, S. A., Iwaniuk, A. N., Knoll, M. A., Bourdon, E., Barrett, P. M., Milner, A. C., Nudds, R. L., Abel, R. L., & Dello Sterpaio, P. (2013). Avian cerebellar floccular fossa size is not a proxy for flying ability in birds. *PLoS One*, *8*(6), e67176. doi:10.1371/journal.pone.0067176 PMID:23825638

Warrick, D. R., Bundle, M. W., & Dial, K. P. (2002). Bird maneuvering flight: Blurred bodies, clear heads. *Integrative and Comparative Biology*, *42*(1), 141–148. doi:10.1093/icb/42.1.141 PMID:21708703

Weis-Fogh, T. (1972). Energetics of hovering flight in hummingbirds and in Drosophila. *The Journal of Experimental Biology*, *56*(1), 79–104. doi:10.1242/jeb.56.1.79

Yu, J. M., Leu, T. S., & Miau, J. J. (2017). Investigation of reduced frequency and freestream turbulence effects on dynamic stall of a pitching airfoil. *Journal of Visualization / the Visualization Society of Japan*, *20*(1), 31–44. doi:10.100712650-016-0366-6

Chapter 15
Viromechanics of Fluid: A Novel Computational Method to Model the COVID-19 Transmission

Nima Norouzi
https://orcid.org/0000-0002-2546-4288
Chongqing Medical University, China

ABSTRACT

This study deals with estimating the trajectory of COVID-19 coronavirus adhering to horizontally projected respiratory droplets, considering the geographic altitude. The size of the viruses and respiratory droplets are the factors that determine the trajectory of the microparticles in a viscous medium such as air. For this purpose, a graphic comparison of the diameters and masses of the microparticles that are produced in respiratory activity has been made. The estimation of the vertical movement of the microparticles through the air is based on Stokes' law, and it was determined that respiratory droplets smaller than 10µm in diameter have very small terminal velocities; in practice, they are floating for brief seconds before evaporating in the air. Regarding the horizontal displacement of respiratory droplets, frames from Beggs determine its scope.

INTRODUCTION

Physics allows us to make an important approximation of the dynamics of microparticles, such as the COVID-19 coronaviruses, whose average diameter is 0.1 µm or 100 nm (Beggs, 2020). Droplets resulting from speech with diameters between 3.5 to 5 µm (Abuhegazy et al., 2020), an intermediate-range is respiratory droplets with diameters between 5 to 10 µm (Diwan et al., 2020), and Flügge droplets with diameters equal to or greater than 100 µm. CO_2, O, N, and water vapor are emitted; in sneezing or coughing, microparticles of water of different diameters are emitted. Transmission by COVID-19 is due in part to the fact that these viruses adhere to respiratory droplets of 5 -10 µm that result from respiratory activity (Feng et al., 2020), there are also Flügge droplets with diameters equal to or greater than 100 µm, with a higher viral load. The difference in diameters between a coronavirus and a respiratory droplet

DOI: 10.4018/978-1-7998-9078-2.ch015

is around 50 to 100 times; that is, the virus is much smaller in size and mass than a respiratory droplet. Therefore, the virus cannot be considered "heavy" concerning means of transport which are respiratory droplets; In the case of making a comparison between the new coronavirus concerning other viruses of the same family, it can be observed that their masses do not differ substantially, this difference in masses and diameters does not influence their displacement concerning distances around 1 m.

As part of the methodology, a graphic comparison of the different microparticles produced in exhalation and the sneeze or cough of a person with respiratory symptoms has been initially made. Stokes's law has been used, assuming that air is a viscous fluid with a laminar regime of low Reynolds number, this case is close to an isolated environment in which there is no ventilation or movement of people; if micro air flows are considered due to variations in pressure, temperatures and the movement of people; the analysis is complex due to stochastic variables that are framed in non-linear dynamics. Air, in principle, is a homogeneous medium where physicochemical processes take place; in this viscous medium, there are mass and heat transport mechanisms, as a particular case the transport or displacement of respiratory droplets from one point to another, the other components of respiratory activity are aerosols that are vehicles for airborne virus transmission due to their diameter that is 35 to 50 times greater than that of a COVID-19 coronavirus. A 10-micron respiratory droplet that falls into the air experiences the viscosity of the medium in such a way that its terminal velocity would be approximately 3mm-1; that is, it is floating from the perspective of an observer at rest. According to the WHO, these respiratory droplets can be transferred between people when they are in close contact (within 1 m) with other people with respiratory symptoms (Fauci et al., 2020). It has been indicated that the virus is "heavy." However, the results obtained in the estimation show the opposite; its size and mass do not influence the kinematics of the respiratory droplets they adhere to; they do not describe parabolas in the air because the viscosity restricts its movement (Stokes Law). The experimental frames of the research carried out by Scharfman (Scharfman et al., 2016) have made it possible to estimate the movement of respiratory droplets.

METHODS AND MATERIALS

The procedure consisted in graphically visualizing the sizes and diameters of the different particles that are produced in the exhalation, sneeze, or cough of a person with respiratory symptoms; a calculation of the masses of the microparticles understudy has been carried out; This procedure has made it possible to different sizes and masses of the 0.1 μm diameter coronaviruses and the transport media: droplets resulting from the speech of 3.5 to 5 μm from respiratory droplets of 5 to 10 μm and a 100 μm de Flügge drop diameter. As demonstrated in the present study, the movement of a 0.1 μm virus is not equal to a 100 μm drop in a viscous medium.

The objective has been to estimate the trajectory of COVID-19 coronavirus adhered to respiratory droplets and projected horizontally, considering the geographical altitude. To achieve this purpose, the laws of Newtonian dynamics have been used (Liu et al., 2021), Stokes' law (Lotfi et al.,2020), and Scharfman's experimental frames. The analysis of the movement in two dimensions XY has been carried out; in each axis, the equations of the movement and their respective solutions have been obtained. Subsequently, the respective constants have been determined. The equations describe the movement of the respiratory droplets considering the immobility of the air and its viscosity, assuming a laminar regime of low Reynolds number. The graphs of the kinematic equations for the cases studied to describe the estimation of motion.

To estimate the dynamics of the microparticles at a different geographical point at sea level, the viscosity value at altitude is required, which varies because the amount of mass per unit volume is less concerning sea level, for; for this purpose, a calculator called the 1976 Standard Atmosphere Calculator - Digital Dutch has been used, in this computer program only the altitude of the place and the temperature are entered, as a reference an altitude of 3000 meters above sea level has been chosen and a temperature of 20 °C; the cities or towns on Earth are between 0 and 5000 m.a.s.l., hence the chosen altitude is representative(Le et al., 2020).

ESTIMATION OF VERTICAL MOVEMENT

A vertically descending spherical particle is exposed to three forces: the viscous force of air (F_v), the thrust force (E), and the weight of the particle (w). Stokes's law allows measuring the friction force experienced by a small spherical particle in a viscous fluid and a laminar regime (Mutuku et al., 2020).

Figure 1. Forces acting on a particle in a viscous fluid

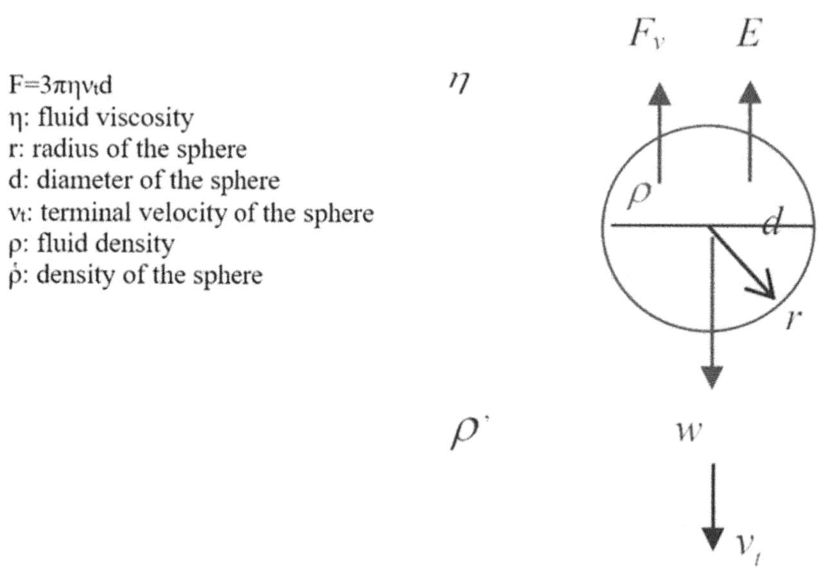

The equation of motion in the y-axis (vertical direction) is given by Newton's Second Law, considering the thrust and the viscous force in the positive vertical direction and the weight of the particle (respiratory droplet, COVID-19 coronavirus) in the opposite direction from above:

$$m\frac{d^2y}{dt^2} = \frac{1}{6}\pi d^3 \rho' g + 3\pi\eta d v_y - mg \tag{1}$$

The terminal velocity of respiratory droplets and COVID-19 coronavirus in the air occurs when the microparticles achieve an invariable rate of descent, then the acceleration in the y-axis is zero(Mittal& Seo, 2020):

$$v_t = \frac{1}{18}\frac{d^2 g}{\eta}(\rho - \rho')t \tag{2}$$

The vertical (down) offset is:

$$y = \frac{1}{18}\frac{d^2 g}{\eta}(\rho - \rho')t \tag{3}$$

RESULTS AND DISCUSSION

Mass and Diameters

The movement of bodies is a physical phenomenon in principle, which can be measured from general movement laws; however, the size factor influences its movement. Classical mechanics allows evaluating the movement of bodies or particles in certain ranges of lengths or sizes and cannot be applied at the quantum or atomic level in the same way, even in the range - around the micrometer - that the analysis of the movement of the droplets of respiratory activity, geometric magnitudes must be considered; the nano and micrometric size influence the movement of the particles in a viscous medium such as air, for this reason, the diameter of the different particles resulting from respiratory activity has been graphically compared. In the graph made, the diameter of the particles and the differences in each case are considered; Regarding the mass, a table has been prepared considering their diameters and densities for each particle. The following scale graph shows the difference in diameters between the COVID-19 coronavirus (Ndaïrou et al., 2020), the droplets resulting from speech (Nouri et al., 2021), the respiratory droplets (Peng et al., 2020), and the Flügge drop.

When measuring the thickness of a standard sheet of paper with a micrometer, we will see that it is approximately equal to 0.1 mm or 100 μm, that is, the largest drop seen in figure 2 is equal to the thickness of a standard sheet of paper, we give an idea of the magnitude ranges of the microdroplets; a drop of Flügge can be perceived by a human being depending on age and visual capacity. Therefore, respiratory droplets of 5 or 10 μm could not be visualized individually but jointly. Below is a table that allows distinguishing the masses of different particles: coronavirus COVID-19, droplet resulting from speech, respiratory droplets, and Flügge drop:

The diameter ratio is different from the mass ratio, in the case of a COVID-19 coronavirus and a respiratory droplet, the diameter ratio is:

$$\frac{d_{\text{droplet}}}{d_{virus}} = 50 \tag{4}$$

Comparing their masses:

$$\frac{m_{\text{droplet}}}{m_{virus}} = \frac{\rho_{aqua}V_{droplet}}{\rho_{virus}V_{virus}} = \frac{d^3_{\text{droplet}}}{d^3_{virus}} = 125000 \quad (5)$$

Figure 2. Graphical comparison of the diameters of a COVID-19 coronavirus, droplet resulting from speech, respiratory droplets, and a Flügge droplet

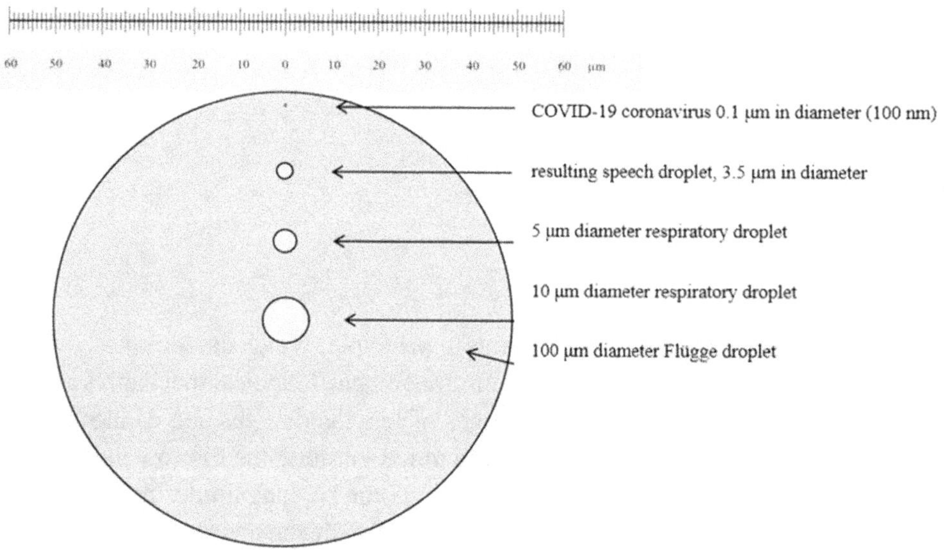

Table 1. Masses of particles product of the respiratory activity of a human being

Particle	Mean particle diameter (µm)	Density (gk/m³)	Mass (kg)
COVID-19	0.1	≈10e+3	5.24e-19
Droplet resulting from speech	3.5-5	10e+3	22.45e-15-65.45e-15
Respiratory droplet	5	10e+3	65.45e-15
Respiratory droplet	10	10e+3	5.23e-13
Flügge drop	100	10e+3	5.23e-10

The mass of a respiratory droplet of 5 µm in diameter is approximately 1.25×10^5 times greater (125,000 times) the mass of a COVID-19 coronavirus 0.1 µm in diameter, this comparison allows us to affirm that a virus or several that have this diameter that adheres to the surface of a respiratory droplet of 5 or 10 µm in diameter, they would not influence the total mass (respiratory droplet and virus). From this, we can conclude that the COVID-19 coronavirus "is not heavy" concerning droplets resulting from speech (aerosol range) or respiratory droplets or Flügge droplets. It could be considered slightly heavier than other or similar viruses, but it would not influence the terminal velocity of the transport droplet; Due to the thermodynamic conditions of the air, a 10 μm water droplet can evaporate at a distance of 2.1 cm in fall (Shao et al.,2021), this distance traveled in the fall corresponds to 7 seconds on average, agrees With our daily observation when a film of water is on any surface, it evaporates in a few seconds.

Estimation of the Range, Speed, and Acceleration

A horizontally projected respiratory droplet movement has been assumed, considering air resistance and without random microflows. The resultant force experienced by the microdroplet is four, assuming that the air is immobile and that in the vertical direction, they achieve terminal velocity:

Figure 3. This figure shows the forces that act on the dynamics of a microdroplet of water

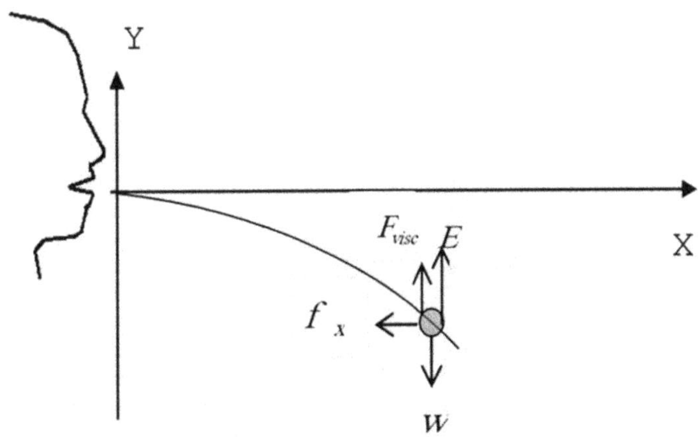

The respective equations in the XY plane are:

$$m\frac{d^2x}{dt^2}=-f_x \tag{6}$$

$$m\frac{d^2y}{dt^2}=E+F_{visc}-w \tag{7}$$

On the x-axis, the differential equation of motion is:

$$\frac{d^2x}{dt^2}=\frac{-f_x}{m} \tag{8}$$

The friction force f_x is proportional to the speed of the object in the x-direction: $f_x=kv_x$ (Setti et al., 2021), where k is a constant, then:

$$\frac{d^2x}{dt^2} = -\frac{k}{m}v_x \tag{9}$$

For analysis, we have a new constant: a=k/m, then:

$$\frac{d^2x}{dt^2} = -\alpha v_x \tag{10}$$

From here, we have the three kinematic variables:

$$x = \frac{v_0}{\alpha}[1-e^{-\alpha t}] \tag{11}$$

$$v_x = v_0 e^{-\alpha t} \tag{12}$$

$$\alpha_x = -\alpha v_0 e^{-\alpha t} \tag{13}$$

Where v_o is the initial velocity in the x-direction, α is the attenuation constant of the medium related to the viscosity and the mass of the microparticle in motion; knowing the initial velocity, the constant α can be determined. From the Scharfman frames (Shafaghi et al., 2020), the initial velocity has been estimated considering the frame where a distance of 0.03 m is observed and the time of 0.005 s, hence the initial velocity is vo = 6 ms^{-1}; substituting these values in equation (11) and through an iterative process the value of α = 3.5 is obtained. The equations are now:

$$x = \frac{6.0}{3.5}[1-e^{-3.5t}] \tag{14}$$

$$v_x = 6.0e^{-3.5t} \tag{15}$$

$$\alpha_x = -21e^{-3.5t} \tag{16}$$

Extrapolating for 2 seconds after the expulsion of the respiratory droplets and applying equations (14), (15), and (16), the following graphs are obtained:

The displacement does not change substantially in the range of 1 to 2 s; velocity and acceleration decrease exponentially;they tend to zero in the interval of 1 and 2 s.

Figure 4. The displacement, velocity, and horizontal acceleration in (a), (b), and (c) respectively vary exponentially in the interval of 0 and 0.150 s (Velavan& Meyer, 2020)

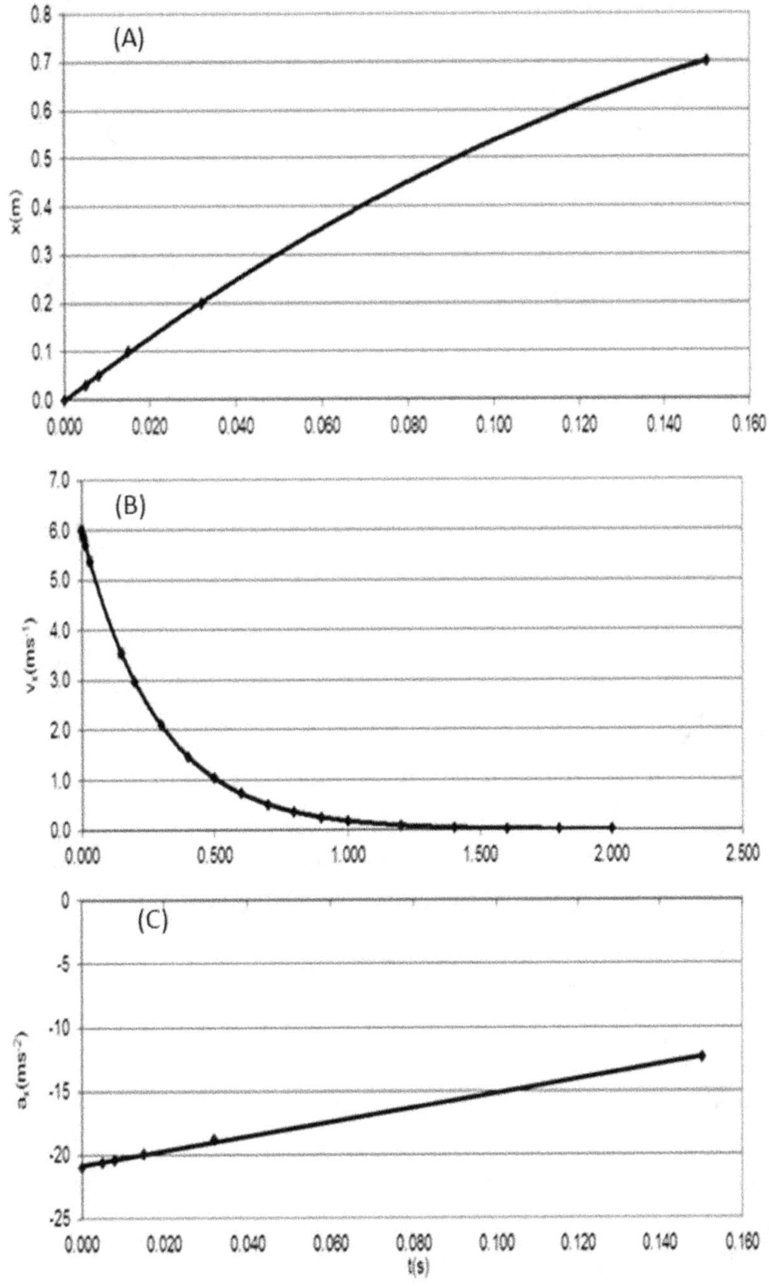

Figure 5. (a) The displacement of the respiratory droplet reaches 1.66 m in 1 s, (b) the velocity and acceleration decrease exponentially concerning time, (c) Respiratory droplet acceleration, extrapolation for 2 s

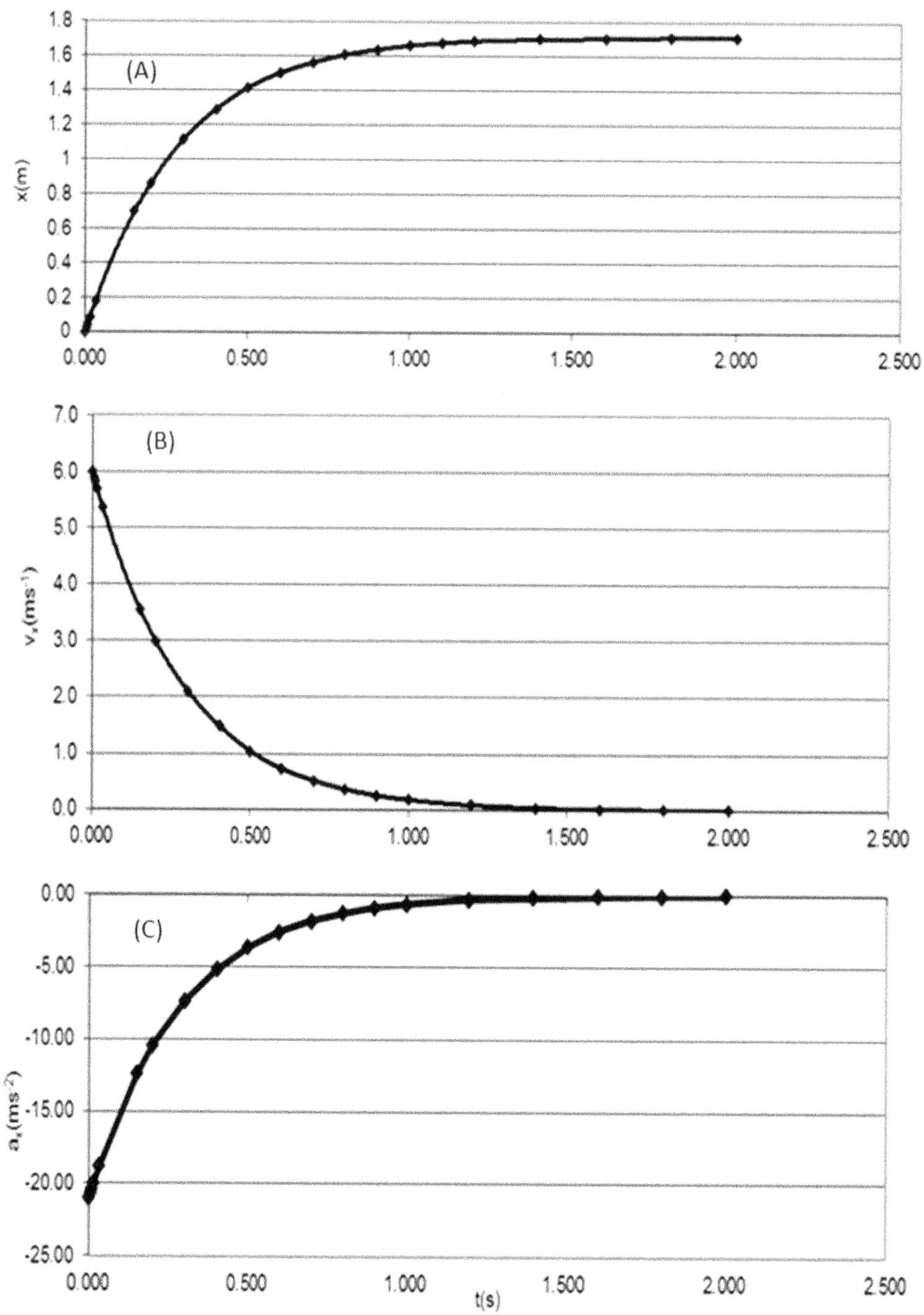

Estimation of the Range, Speed, and Acceleration

In a geographic location at 0 m.a.s.l .; the value of the average acceleration of gravity is $g=9.80m/s^2$, the viscosity of air for a temperature of 20°C (at 1 atm pressure) is $\eta=1.83x10^5N.s/m^2$ (Velavan & Meyer,2020), the density of water $\rho=10^3kg/m^3$ and the air is $\rho'=1.20kg/m^3$, on the other hand, the approximate density of a COVID-19 coronavirus is 1 g/cm^3 (Zuo et al., 2020), obtained through an indirect analysis.

Table 2. The estimate of the terminal velocity of a coronavirus, a droplet resulting from speech, respiratory droplets, and a Flügge drop in the air at 0 m.a.s.l. is shown

Particle	Mean particle diameter (μm)	Density (gk/m^3)	Terminal velocity (m/s)
COVID-19	0.1	≈ 10e+3	2.97e-7
Droplet resulting from speech	3.5-5	10e+3	9.10e-5 – 7.43e-4
Respiratory droplet	5	10e+3	7.43e-4
Respiratory droplet	10	10e+3	2.97e-3
Flügge drop	100	10e+3	1.19

The terminal velocity tends to be zero; that is, it could be suspended in the air if it is free of contact with other microparticles; the resulting droplets of speech have very low terminal velocities, for example, 0.7 mms^{-1}, their descent could not be visually distinguished. The 10 μm respiratory droplet has a terminal velocity of 2.97 mms^{-1}, approximately 3 millimeters per second, descending very slowly. Assuming the adherence of a COVID-19 coronavirus in respiratory droplets, the terminal velocity will not change in practice; therefore, the buoyancy of respiratory droplets in air predominates. The only masses that fall to the ground are the Flügge droplets with diameters greater than or equal to 100 μm. The rate of falling of water droplets and a COVID-19 coronavirus at an altitude of 3000 m.a.s.l. has been estimated. At an atmospheric pressure of 70108 Pa and a temperature of 20oC, the atmospheric density at this height is $\rho'=0.85kg/m^3$, the viscosity of air is $\eta=1.81x10^5N.s/m^2$ (1976 Standard Atmosphere Calculator - Digital Dutch).

The terminal velocities obtained at an altitude of 3000 m.a.s.l. are approximately equal to the speeds found at sea level; these results are due to the viscosity of the air in both places do not differ substantially.

Table 3. The terminal velocities depend on the diameter or size of the microdroplets of water, in the same way, the terminal velocity of a virus that would correspond to the aerosol range

Particle	Mean particle diameter (μm)	Density (gk/m^3)	Terminal velocity (m/s)
COVID-19	0.1	≈ 10e+3	3e-7
Droplet resulting from speech	3.5-5	10e+3	9.20e-5–7.51e-4
Respiratory droplet	5	10e+3	7.51e-4
Respiratory droplet	10	10e+3	3.01e-3
Flügge drop	100	10e+3	1.20

Path of a Respiratory Droplet in the XY Plane

With the results previously obtained regarding the kinematic variables in the x-axes, and a new table has been completed as a function of time: x (t), y (t); in this way, it has been possible to obtain the graphs of y (x) for 2 and 5 s. The trajectory of a respiratory droplet in an interval of 2 s is practically a horizontal line as seen in figure 6, the range is 1.71 m in 2 s, at this instant, its speed and acceleration are approximately zero, the respiratory droplet would be immobile and floating, the viscosity would prevent it from moving even with constant speed, that is, the particle would be at rest for the assumed conditions. If 5 seconds is considered, the respiratory droplet will remain floating without the possibility of advancing in the x-direction; this coincides with our daily perception; that is, the exhaled air does not travel more than 1.71 m of continuously.

The following graph shows the trajectory of the respiratory droplet of 5 μm, which describes a polynomial curve in a vertical length of 1.5 mm and in a horizontal distance of 1.71 m for a time of 2. If observed from a few meters away, the trajectory can be observed as a line in the ranges described above.

Figure 6. (a) Path of a respiratory droplet in the XY plane, the scale on the y axis is only millimeters; (b) the trajectory is a line if viewed at normal scale

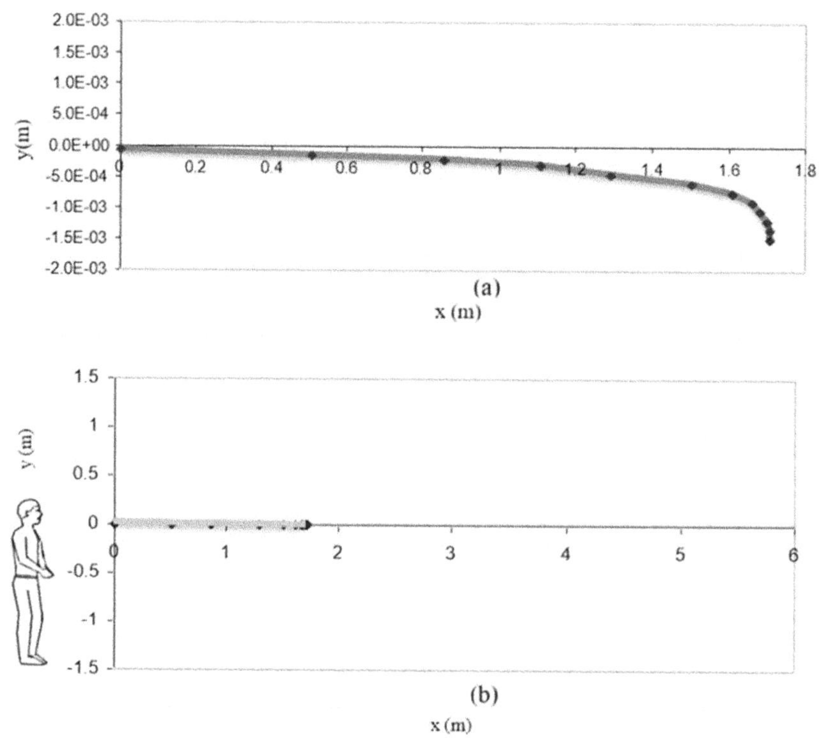

Using the equations of motion for the displacement in the XY plane, we have the graph that allows us to see the trajectories of a 5 µm respiratory droplet (blue color) and a 100 µm Flügge drop (red color) for 5 seconds; It can be seen that the trajectory of a respiratory droplet is a line segment and remains floating, while the Flügge drop describes a polynomial curve, the drop hits the Earth's surface in 5 seconds.

If we consider air flows from 2 s, these flows would contribute to the spread of respiratory droplets and aerosols to an even greater distance than 1.71 m and allow them to reach different heights.

Figure 7. The trajectory in the XY plane of a respiratory droplet (blue color curve) and a Flügge droplet (red color), the size factor determines the trajectory in each case

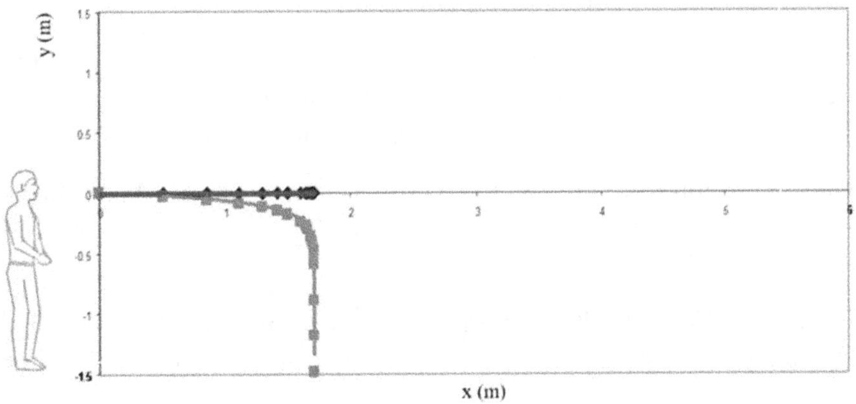

CONCLUSION

Respiratory droplets of 5-10 µm have been determined to float in the air in the vicinity of a person with respiratory symptoms, the cough or sneeze of a human being would lead to these droplets reaching 1.65 m distance in 1 s slowing down rapidly to reach 1.71m in 2 seconds; for 5 seconds, the respiratory droplets of 5 -10 µm would be found floating due to their terminal velocity that is between $7.43x10^{-4}ms^{-1}$ to $2.97x10^{-3}$ ms^{-1}, this range of speeds is too small to be detected by the human being. By action of the thermodynamic conditions of the air, these microdroplets evaporate around 7 seconds. The terminal droplet velocity resulting from the speech of 3.5 µm in diameter (aerosol) is still much lower, $9.10x10^{-5}$ ms^{-1}; these droplets can also carry COVID-19 coronavirus. The only drops of water that fall about 1.71 m away from the patient with respiratory symptoms due to COVID-19 are the drops of Flügge that manage to reach the ground or another object in the surrounding environment; therefore, a COVID-19 can displace in the air adhering to respiratory droplets in the environment of the person with respiratory symptoms. Geographic altitude minimally influences the movement of respiratory droplets.

REFERENCES

Abuhegazy, M., Talaat, K., Anderoglu, O., & Poroseva, S. V. (2020). Numerical investigation of aerosol transport in a classroom with relevance to COVID-19. *Physics of Fluids*, *32*(10), 103311. doi:10.1063/5.0029118 PMID:33100808

Beggs, C. B. (2020). Is there an airborne component to the transmission of COVID-19?: a quantitative analysis study. MedRXiv. doi:10.1101/2020.05.22.20109991

Diwan, S. S., Ravichandran, S., Govindarajan, R., & Narasimha, R. (2020). Understanding transmission dynamics of COVID-19-type infections by direct numerical simulations of cough/sneeze flows. *Transactions of the Indian National Academy of Engineering*, *5*(2), 255–261. doi:10.100741403-020-00106-w

Fauci, A. S., Lane, H. C., & Redfield, R. R. (2020). *Covid-19—navigating the uncharted*. Academic Press.

Feng, Y., Marchal, T., Sperry, T., & Yi, H. (2020). Influence of wind and relative humidity on the social distancing effectiveness to prevent COVID-19 airborne transmission: A numerical study. *Journal of Aerosol Science*, *147*, 105585. doi:10.1016/j.jaerosci.2020.105585 PMID:32427227

Le, T. T., Andreadakis, Z., Kumar, A., Román, R. G., Tollefsen, S., Saville, M., & Mayhew, S. (2020). The COVID-19 vaccine development landscape. *Nature Reviews. Drug Discovery*, *19*(5), 305–306. doi:10.1038/d41573-020-00073-5 PMID:32273591

Liu, H., He, S., Shen, L., & Hong, J. (2021). Simulation-based study of COVID-19 outbreak associated with air-conditioning in a restaurant. *Physics of Fluids*, *33*(2), 023301. doi:10.1063/5.0040188 PMID:33746488

Lotfi, M., Hamblin, M. R., & Rezaei, N. (2020). COVID-19: Transmission, prevention, and potential therapeutic opportunities. *Clinica Chimica Acta*, *508*, 254–266. doi:10.1016/j.cca.2020.05.044 PMID:32474009

Mittal, R., Ni, R., & Seo, J. H. (2020). The flow physics of COVID-19. *Journal of Fluid Mechanics*, 894.

Mutuku, J. K., Hou, W. C., & Chen, W. H. (2020). An overview of experiments and numerical simulations on airflow and aerosols deposition in human airways and the role of bioaerosol motion in COVID-19 transmission. *Aerosol and Air Quality Research*, *20*(6), 1172–1196. doi:10.4209/aaqr.2020.04.0185

Ndaïrou, F., Area, I., Nieto, J. J., & Torres, D. F. (2020). Mathematical modeling of COVID-19 transmission dynamics with a case study of Wuhan. *Chaos, Solitons, and Fractals*, *135*, 109846. doi:10.1016/j.chaos.2020.109846 PMID:32341628

Nouri, Z., Norouzi, N., Norouzi, N., Ataei, E., & Azizi, S. (2021). Virologic microparticle fluid mechanics simulation: COVID-19 transmission inside an elevator space. *International Journal of Computational Materials Science and Engineering*, *10*(02), 2150007. doi:10.1142/S204768412150007X

Peng, S., Chen, Q., & Liu, E. (2020). The role of computational fluid dynamics tools on investigation of pathogen transmission: Prevention and control. *The Science of the Total Environment*, *746*, 142090. doi:10.1016/j.scitotenv.2020.142090 PMID:33027870

Setti, L., Passarini, F., De Gennaro, G., Barbieri, P., Perrone, M. G., Borelli, M., ... Miani, A. (2020). *Airborne transmission route of COVID-19: why 2 meters/6 feet of inter-personal distance could not be enough*. Academic Press.

Shafaghi, A. H., Rokhsar Talabazar, F., Koşar, A., & Ghorbani, M. (2020). On the effect of the respiratory droplet generation condition on COVID-19 transmission. *Fluids*, *5*(3), 113. doi:10.3390/fluids5030113

Shao, S., Zhou, D., He, R., Li, J., Zou, S., Mallery, K., Kumar, S., Yang, S., & Hong, J. (2021). Risk assessment of airborne transmission of COVID-19 by asymptomatic individuals under different practical settings. *Journal of Aerosol Science*, *151*, 105661. doi:10.1016/j.jaerosci.2020.105661 PMID:32968325

Velavan, T. P., & Meyer, C. G. (2020). The COVID-19 epidemic. *Tropical Medicine & International Health*, *25*(3), 278–280. doi:10.1111/tmi.13383 PMID:32052514

Zuo, Y. Y., Uspal, W. E., & Wei, T. (2020). Airborne transmission of COVID-19: Aerosol dispersion, lung deposition, and virus-receptor interactions. *ACS Nano*, *14*(12), 16502–16524. doi:10.1021/acsnano.0c08484 PMID:33236896

ADDITIONAL READING

Cao, X. (2020). COVID-19: Immunopathology and its implications for therapy. *Nature Reviews. Immunology*, *20*(5), 269–270. doi:10.103841577-020-0308-3 PMID:32273594

Daniel, J. (2020). Education and the COVID-19 pandemic. *Prospects*, *49*(1), 91–96. doi:10.100711125-020-09464-3 PMID:32313309

Fauci, A. S., Lane, H. C., & Redfield, R. R. (2020). Covid-19—navigating the uncharted.

Le, T. T., Andreadakis, Z., Kumar, A., Román, R. G., Tollefsen, S., Saville, M., & Mayhew, S. (2020). The COVID-19 vaccine development landscape. *Nature Reviews. Drug Discovery*, *19*(5), 305–306. doi:10.1038/d41573-020-00073-5 PMID:32273591

Liu, H., He, S., Shen, L., & Hong, J. (2021). Simulation-based study of COVID-19 outbreak associated with air-conditioning in a restaurant. *Physics of Fluids*, *33*(2), 023301. doi:10.1063/5.0040188 PMID:33746488

Lotfi, M., Hamblin, M. R., & Rezaei, N. (2020). COVID-19: Transmission, prevention, and potential therapeutic opportunities. *Clinica Chimica Acta*, *508*, 254–266. doi:10.1016/j.cca.2020.05.044 PMID:32474009

Solomon, I. H., Normandin, E., Bhattacharyya, S., Mukerji, S. S., Keller, K., Ali, A. S., Adams, G., Hornick, J. L., Padera, R. F. Jr, & Sabeti, P. (2020). Neuropathological features of Covid-19. *The New England Journal of Medicine*, *383*(10), 989–992. doi:10.1056/NEJMc2019373 PMID:32530583

KEY TERMS AND DEFINITIONS

Antibody Tests: The body responds to viral infection by producing antibodies that help neutralize the virus. Blood tests (also called serology tests or serology immunoassays) can detect the presence of such antibodies.[65] Antibody tests can be used to assess what fraction of a population has once been infected, which can then be used to calculate the disease's mortality rate. They can also be used to determine how much antibody is contained in a unit of convalescent plasma for COVID-19 treatment or to verify if a given vaccine generates an adequate immune response.

Antigen Tests: An antigen is the part of a pathogen that elicits an immune response. Antigen tests look for antigen proteins from the viral surface. In the case of a coronavirus, these are usually proteins from the surface spikes. SARS-CoV-2 antigens can be detected before the onset of COVID-19 symptoms (as soon as SARS-CoV-2 virus particles) with more rapid test results but with less sensitivity than PCR tests for the virus.

COVID-19 Vaccines: A COVID-19 vaccine is a vaccine intended to provide acquired immunity against severe acute respiratory syndrome coronavirus 2 (SARS-CoV-2), the virus that causes coronavirus disease 2019 (COVID-19). Before the COVID-19 pandemic, an established body of knowledge existed about the structure and function of coronaviruses causing diseases like severe acute respiratory syndrome (SARS) and the Middle East respiratory syndrome (MERS). This knowledge accelerated the development of various vaccine technologies in early 2020. On 10 January 2020, the SARS-CoV-2 genetic sequence data was shared through GISAID, and by 19 March, the global pharmaceutical industry announced a major commitment to addressing COVID-19. The COVID-19 vaccines are widely credited for their role in reducing the spread, severity, and death caused by COVID-19.

Imaging Test: Typical visible features on CT initially include bilateral multilobar ground-glass opacities with a peripheral or posterior distribution. COVID-19 can be identified with higher precision using CT than with RT-PCR. Subpleural dominance, crazy paving, and consolidation may develop as the disease evolves. Chest CT scans and chest x-rays are not recommended for diagnosing COVID-19. Radiologic findings in COVID-19 lack specificity.

Personal Protective Equipment(PPE): Personal protective equipment is protective clothing, or as sometimes called EPI are helmets, goggles, or other garments or equipment designed to protect the wearer's body from injury or infection. The hazards addressed by protective equipment include physical, electrical, heat, chemicals, biohazards, and airborne particulate matter.

Polymerase Chain Reaction (PCR): Polymerase chain reaction (PCR) is a process that amplifies (replicates) a small, well-defined segment of DNA many hundreds of thousands of times, creating enough of it for analysis. Test samples are treated with certain chemicals that allow DNA to be extracted. Reverse transcription converts RNA into DNA. Reverse transcription-polymerase chain reaction (RT-PCR) first uses reverse transcription to obtain DNA, followed by PCR to amplify that DNA, creating enough to be analyzed. RT-PCR can thereby detect SARS-CoV-2, which contains the only RNA. The RT-PCR process generally requires a few hours. These tests are also referred to as molecular or genetic assays.

Rapid Diagnostic Test (RDT): RDTs typically use a small, portable, positive/negative lateral flow assay that can be executed at the point of care. RDTs may process blood samples, saliva samples, or nasal swab fluids. RDTs produce colored lines to indicate positive or negative results.

Sniff Tests: Sudden loss of smell can be used to screen people daily for COVID-19. The National Institutes of Health study showed that those infected with SARS-CoV-2 could not smell a 25% mixture of ethanol and water. Because various conditions can lead to the loss of smell, a sniff test would not be definitive but indicate the need for a PCR test. Because the loss of the sense of smell shows up before other symptoms, there has been a call for widespread sniff testing. Health care bureaucracies have generally ignored sniff tests even though they are quick, easy, and capable of being self-administered daily. This has led some medical journals to write editorials supporting the adoption of sniff testing.

Compilation of References

Abbott, I. H., & Von Doenhoff, A. E. (2012). *Theory of wing sections: Including a summary of airfoil data*. Courier Corporation.

Abhilash, A. S., Baker, B. M., Trappmann, B., Chen, C. S., & Shenoy, V. B. (2014). Remodeling of fibrous extracellular matrices by contractile cells: Predictions from discrete fiber network simulations. *Biophysical Journal*, *107*(8), 1829–1840. doi:10.1016/j.bpj.2014.08.029 PMID:25418164

Abu-Faraj, Z. O., Harris, G. F., Smith, P. A., & Hassani, S. (2015). Human gait and clinical movement analysis. Wiley Encyclopedia of Electrical and Electronics Engineering, 1-34. doi:10.1002/047134608X.W6606.pub2

Abuhegazy, M., Talaat, K., Anderoglu, O., & Poroseva, S. V. (2020). Numerical investigation of aerosol transport in a classroom with relevance to COVID-19. *Physics of Fluids*, *32*(10), 103311. doi:10.1063/5.0029118 PMID:33100808

Additive manufacturing staircase. (2022). https://resources.renishaw.com/en/details/additive-manufacturing-staircase--78019

Adham, M., Gournier, J.-P., Favre, J.-P., De La Roche, E., Ducerf, C., Baulieux, J., Barral, X., & Pouyet, M. (1996). Mechanical Characteristics of Fresh and Frozen Human Descending Thoracic Aorta. *The Journal of Surgical Research*, *64*(1), 32–34. doi:10.1006/jsre.1996.0302 PMID:8806470

Agozzino, L., de Vivo, F., Falco, A., de Luca, L., Schinosa, T., & Cotrufo, M. (1994). Surgical pathology of the aortic valve: Gross and histological findings in 1120 excised valves. *Cardiovascular Pathology*, *3*(3), 155–161. doi:10.1016/1054-8807(94)90024-8 PMID:25990991

Agrawal, M. S., Agrawal, M., Gupta, A., Bansal, S., Yadav, A., & Goyal, J. (2008). A randomized comparison of tubeless and standard percutaneous nephrolithotomy. *Journal of Endourology*, *22*(3), 439–442. doi:10.1089/end.2007.0118 PMID:18257738

Ahmadian, M., Wang, Y., & Sul, H. S. (2010). Lipolysis in adipocytes. *The International Journal of Biochemistry & Cell Biology*, *42*(5), 555–559. doi:10.1016/j.biocel.2009.12.009 PMID:20025992

Ahmed, R. I. (2016). *Theoretical Foundation and Numerical Analysis on Coandă Effect for Micro-Air Vehicle* (Doctoral dissertation). Universiti Putra Malaysia.

Ahmed, R. I., Djojodihardjo, H., Talib, A. R. A., & Rafie, A. M. (2017). Review on Progress and Application of Active Flow Control Devices-Coandă Effect on Unmanned Aerial Vehicles. *Pertanika Journal of Scholarly Research Reviews*, *3*(1).

Ahmed, S. A., & Giddens, D. P. (1984). Pulsatile poststenotic flow studies with laser Doppler anemometry. *Journal of Biomechanics*, *17*(9), 695–705. doi:10.1016/0021-9290(84)90123-4 PMID:6238968

Akhtaruzzaman, M., Shafie, A. A., & Khan, M. R. (2016). Gait Analysis: Systems, Technologies, and Importance. *Journal of Mechanics in Medicine and Biology*, *16*(7), 1630003. doi:10.1142/S0219519416300039

Alexander, R. M. (2003). Modelling approaches in biomechanics. *Philosophical Transactions of the Royal Society of London. Series B, Biological Sciences*, *358*(1437), 1429–1435. doi:10.1098/rstb.2003.1336 PMID:14561333

Al, K. F., Denstedt, J. D., Daisley, B. A., Bjazevic, J., Welk, B. K., Pautler, S. E., Gloor, G. B., Reid, G., Razvi, H., & Burton, J. P. (2020). Ureteral Stent Microbiota Is Associated with Patient Comorbidities but Not Antibiotic Exposure. *Cell reports. Medicine*, *1*(6), 100094. doi:10.1016/j.xcrm.2020.100094 PMID:33205072

Allen, A. A., Caldwell, G. L., & Fu, F. H. (1995). Functional anatomy and biomechanics of the meniscus. *Operative Techniques in Sports Medicine*, *2*(3), 152–163.

Altshuler, D. L., Bahlman, J. W., Dakin, R., Gaede, A. H., Goller, B., Lentink, D., Segar, P. S., & Skandalis, D. A. (2015). The biophysics of bird flight: Functional relationships integrate aerodynamics, morphology, kinematics, muscles, and sensors. *Canadian Journal of Zoology*, *93*(12), 961–975. doi:10.1139/cjz-2015-0103

Aminsharifi, A., Irani, D., Masoumi, M., Goshtasbi, B., Aminsharifi, A., & Mohamadian, R. (2016). The management of large staghorn renal stones by percutaneous versus laparoscopic versus open nephrolithotomy: A comparative analysis of clinical efficacy and functional outcome. *Urolithiasis*, *44*(6), 551–557. doi:10.100700240-016-0877-6 PMID:27032961

Amis, A. A. (2004). The Biomechanics of Ligaments. *Biomechanics and Biomaterials in Orthopedics*, 550–563.

Anderson, D. E., & Madigan, M. L. (2013). Effects of age-related differences in femoral loading and bone mineral density on strains in the proximal femur during controlled walking. *Journal of Applied Biomechanics*, *29*(5), 505–516. doi:10.1123/jab.29.5.505 PMID:23185080

Anderson, J. D. Jr. (2010). *Fundamentals of aerodynamics*. Tata McGraw-Hill Education.

Anderson, K., Strickland, S. M., & Warren, R. (2001). Hip and groin injuries in athletes. *The American Journal of Sports Medicine*, *29*(4), 521–533. doi:10.1177/03635465010290042501 PMID:11476397

Anderson, L. M., Bonanno, D. R., Hart, H. F., & Barton, C. J. (2020). What are the benefits and risks associated with changing foot strike pattern during running? A systematic review and meta-analysis of injury, running economy, and biomechanics. *Sports Medicine (Auckland, N.Z.)*, *50*(5), 885–917. doi:10.100740279-019-01238-y PMID:31823338

Andersson, H. I., Halden, R., & Glomsaker, T. (2000). Effects of surface irregularities on flow resistance in differently shaped arterial stenosis. *Journal of Biomechanics*, *33*(10), 1257–1262. doi:10.1016/S0021-9290(00)00088-9 PMID:10899335

Ansari, S., & Yamaoka, Y. (2017). Survival of Helicobacter pylori in gastric acidic territory. *Helicobacter*, *22*(4), e12386. doi:10.1111/hel.12386 PMID:28402047

Anton, S. R., & Sodano, H. A. (2007). A review of power harvesting using piezoelectric materials (2003-2006). *Smart Materials and Structures*, *16*(3), R1–R21. doi:10.1088/0964-1726/16/3/R01

Arab, A. Z., Merdji, A., Benaissa, A., Roy, S., Bouiadjra, B. A., Layadi, K., Ouddane, A., & Mukdadi, O. M. (2020). Finite-Element analysis of a lateral femoro-tibial impact on the total knee arthroplasty. *Computer Methods and Programs in Biomedicine*, *1*(192).

Argento, G., Simonet, M., Oomens, C. W. J., & Baaijens, F. P. T. (2012). Multi-scale mechanical characterization of scaffolds for heart valve tissue engineering. *Journal of Biomechanics*, *45*(16), 2893–2898. doi:10.1016/j.jbiomech.2012.07.037 PMID:22999107

Ashkavand, Z., Malekinejad, H., & Vishwanath, B. S. (2013). The pathophysiology of osteoarthritis. *Journal of Pharmacy Research*, *7*(1), 132–138. doi:10.1016/j.jopr.2013.01.008

Atassi, N., & Knoll, T. (2020). Future of kidney stone management: surgical intervention miniaturization of PCNL: where is the limit? *Current Opinion in Urology*, *30*(2), 107–112. doi:10.1097/MOU.0000000000000713 PMID:31895077

Aversa, R., Petrescu, F.I., Petrescu, R.V., & Apicella, A. (2016). *Biofidel FEA modeling of customized hybrid biological hip joint design Part II: Flexible stem trabecular prostheses*. Academic Press.

Avvari, R. K. (2015). *Bio-mechanics of the distal stomach and duodenum: An insight into mechanisms of duodenogastric reflux and duodenal mixing*. Department of Biological Sciences and Bioengineering.

Avvari, R. K. (2020). Enteric and Central Nervous System Mediated Control of Digestive Processes in the Small Intestine: A Coprocessor-Processor Paradigm. *Food Science and Engineering*, 39-44.

Avvari, R. K. (2019a). Managing motility disorders of the antro-pyloro-duodenal segment: A biomedical engineering perspective. *United J Biochem Biotechnol*, *1*, 1–20.

Avvari, R. K. (2019b). Effect of local longitudinal shortening on the transport of luminal contents through small intestine. *Lixue Xuebao*, *35*(1), 45–60. doi:10.100710409-018-0809-5

Avvari, R. K. (2019c). Biomechanics of the small intestinal contractions. In *Digestive system- Recent advances*. IntechOpen.

Avvari, R. K. (2021a). Effects of stationary contraction of the small intestine on digestion. *Journal of Mechanics in Medicine and Biology*, *21*(02), 2150021. doi:10.1142/S0219519421500214

Avvari, R. K. (2021b). Is Rheology a Concern in GI Physiology? *Evolutions in Mechanical Engineering*, *3*(3), 1–4.

Avvari, R. K. (2021c). Theoretical modeling of the resistance to gastric emptying and duodenogastric reflux due to pyloric motility alone, presuming antral and duodenal quiescence. *Journal of Theoretical Biology*, *508*, 110460. doi:10.1016/j.jtbi.2020.110460 PMID:32891592

Avvari, R. K. (2021d). A novel two-indenter based micro-pump for lab-on-a-chip application: Modeling and characterizing flows for a non-Newtonian fluid. *Meccanica*, *56*(3), 569–583. doi:10.100711012-020-01303-1

Avvari, R. K. (2021e). Managing motility disorders of the gastrointestinal segment and obesity through electrical stimulation. *Health and Technology*, *11*(6), 1175–1189. doi:10.100712553-021-00590-2

Baas, E., & Kuiper, J. H. (2008). A numerical model of heterogeneous surface strains in polymer scaffolds. *Journal of Biomechanics*, *41*(6), 1374–1378. doi:10.1016/j.jbiomech.2008.01.018 PMID:18353333

Babu, A. N., Meng, X., Zou, N., Yang, X., Wang, M., Song, Y., Cleveland, J. C., Weyant, M., Banerjee, A., & Fullerton, D. A. (2008). Lipopolysaccharide Stimulation of Human Aortic Valve Interstitial Cells Activates Inflammation and Osteogenesis. *The Annals of Thoracic Surgery*, *86*(1), 71–76. doi:10.1016/j.athoracsur.2008.03.008 PMID:18573401

Baca, A., Dabnichki, P., Heller, M., & Kornfeind, P. (2009). Ubiquitous computing in sports: A review and analysis. *Journal of Sports Sciences*, *27*(12), 1335–1346. doi:10.1080/02640410903277427 PMID:19764000

Back, L. H., & Banerjee, R. K. (2000). Estimated flow resistance increase in a spiral human coronary artery segment. *Journal of Biomechanical Engineering*, *122*(6), 675–677. doi:10.1115/1.1319661 PMID:11192391

Bahr, R. (2009). No injuries, but plenty of pain? On the methodology for recording overuse symptoms in sports. *British Journal of Sports Medicine*, *43*(13), 966–972. doi:10.1136/bjsm.2009.066936 PMID:19945978

Baker, R. (2006). Gait analysis methods in rehabilitation. *Journal of Neuroengineering and Rehabilitation, 3*(1), 1–10. doi:10.1186/1743-0003-3-4 PMID:16512912

Baker, R. (2007). The history of gait analysis before the advent of modern computers. *Gait & Posture, 26*(3), 331–342. doi:10.1016/j.gaitpost.2006.10.014 PMID:17306979

Baleani, M., Fognani, R., & Toni, A. (2001). Initial stability of a cementless acetabular cup design: Experimental investigation on the effect of adding fins to the rim of the cup. *Artificial Organs, 25*(8), 664–669. doi:10.1046/j.1525-1594.2001.025008664.x PMID:11531719

Banaszkiewicz, P. A., & Kader, D. F. (2014). Classic papers in orthopaedics. *Classic Papers in Orthopaedics, 10*(15), 1–624.

Barbone, P., &Oberai, A. (2010). *A Review of the Mathematical and Computational Foundations of Biomechanical Imaging*. . doi:10.1007/978-90-481-3575-2_13

Bartlett, R. (1999). *Sports Biomechanics : Reducing injury and improving performance*. E & FN Spon.

Bartlett, R. (2014). *Introduction to sports biomechanics: Analysing human movement patterns*. Routledge. doi:10.4324/9781315889504

Barua, R., Datta, S., Datta, P., & Roy Chowdhury, A. (2020). Experimental analysis the tissue deformation of needle insertion process in tissue engineering. *IEEE 1st International Conference for Convergence in Engineering (ICCE)*, 83-85. 10.1109/ICCE50343.2020.9290598

Barua, R., & Das, S. (2022). Improvements of Virtual and Augmented Reality for Advanced Treatments in Urology. In L. Coelho, R. Queirós, & S. Reis (Eds.), *Emerging Advancements for Virtual and Augmented Reality in Healthcare* (pp. 117–131). IGI Global. doi:10.4018/978-1-7998-8371-5.ch008

Barua, R., Giria, H., Datta, S., Roy Chowdhury, A., & Datta, P. (2020). Force modeling to develop a novel method for fabrication of hollow channels inside a gel structure. *Proceedings of the Institution of Mechanical Engineers. Part H, Journal of Engineering in Medicine, 234*(2), 223–231. doi:10.1177/0954411919891654 PMID:31774361

Batagello, C. A., Barone Dos Santos, H. D., Nguyen, A. H., Alshara, L., Li, J., Marchini, G. S., Vicentini, F. C., Torricelli, F., Danilovic, A., Pereira, J. G., Rose, E., Srougi, M., Nahas, W. C., Mazzucchi, E., & Monga, M. (2019). Endoscopic guided PCNL in the prone split-leg position versus supine PCNL: A comparative analysis stratified by Guy's stone score. *The Canadian Journal of Urology, 26*(1), 9664–9674. PMID:30797250

Batchelor, C. K., & Batchelor, G. K. (2000). *An introduction to fluid dynamics*. Cambridge university press. doi:10.1017/CBO9780511800955

Beckenbaugh, R. D., & Ilstrup, D. M. (1978). Total hip arthroplasty. *The Journal of Bone and Joint Surgery, 60*(3), 306–313. doi:10.2106/00004623-197860030-00005 PMID:649633

Beggs, C. B. (2020). Is there an airborne component to the transmission of COVID-19?: a quantitative analysis study. MedRXiv. doi:10.1101/2020.05.22.20109991

Berg, A. M., & Biewener, A. A. (2010). Wing and body kinematics of takeoff and landing flight in the pigeon (Columba livia). *The Journal of Experimental Biology, 213*(10), 1651–1658. doi:10.1242/jeb.038109 PMID:20435815

Bermejo, J., Postigo, A., & Baumgartner, H. (2021). The year in cardiovascular medicine 2020: Valvular heart disease. *European Heart Journal, 42*(6), 647–656. doi:10.1093/eurheartj/ehaa1060 PMID:33388778

Beysens, M., & Tailly, T. O. (2018). Ureteral stents in urolithiasis. *Asian Journal of Urology*, *5*(4), 274–286. doi:10.1016/j.ajur.2018.07.002 PMID:30364608

Bhat, A., Katz, J. E., Smith, N., & Shah, H. N. (2021). Nephropleural fistula after supracostal approach for PCNL: Report of two cases with review of literature. *BMJ Case Reports*, *14*(4), e241360. doi:10.1136/bcr-2020-241360 PMID:33795284

Bhatia, A., & Vesely, I. (2005). *The Effect of Glycosaminoglycans and Hydration on the Viscoelastic Properties of Aortic Valve Cusps*. Academic Press.

Bhattacharya, S., Roy, S., Rana, M., Banerjee, S., Karmakar, S. K., & Biswas, J. K. (2019). Biomechanical performance of a modified design of dynamic cervical implant compared to conventional ball and socket design of an artificial intervertebral disc implant: A finite element study. *Journal of Mechanics in Medicine and Biology*, *19*(04), 1950017. doi:10.1142/S0219519419500179

Bhosale, A. M., & Richardson, J. B. (2008). Articular cartilage: Structure, injuries and review of management. *British Medical Bulletin*, *87*(1), 77–95. doi:10.1093/bmb/ldn025 PMID:18676397

Biewener, A. A. (2011). Muscle function in avian flight: Achieving power and control. *Philosophical Transactions of the Royal Society of London. Series B, Biological Sciences*, *366*(1570), 1496–1506. doi:10.1098/rstb.2010.0353 PMID:21502121

Billiar, K. L., & Sacks, M. S. (2000). Biaxial mechanical properties of the natural and glutaraldehyde treated aortic valve cusp - Part I: Experimental results. *Journal of Biomechanical Engineering*, *122*(1), 23–30. doi:10.1115/1.429624 PMID:10790826

Birmingham, E., Grogan, J. A., Niebur, G. L., McNamara, L. M., & McHugh, P. E. (2013, April). Computational modelling of the mechanics of trabecular bone and marrow using fluid structure interaction techniques. *Annals of Biomedical Engineering*, *41*(4), 814–826. doi:10.100710439-012-0714-1 PMID:23519534

Biswas, J. K., Roy, S., Majumder, S., Karmakar, S. K., Saha, S., & Roychowdhury, A. (2018). Artificial Intervertebral Disc Replacement to Provide Dynamic Stability to the Lumbar Spine: A Finite Element Study. *Journal of Long-Term Effects of Medical Implants*, *28*(2), 101–109. doi:10.1615/JLongTermEffMedImplants.2018025397 PMID:30317959

Biswas, J. K., Roy, S., Pradhan, R., Rana, M., & Majumdar, S. (2019). Effects of cervical disc replacement and anterior fusion for different bone conditions: a finite element study. *International Journal for Multiscale Computational Engineering, 17*(4). doi:10.1615/IntJMultCompEng.2019030212

Biswas, N., Manna, N. K., Datta, A., Mandal, D. K., & Benim, A. C. (2020). Role of aspiration to enhance MHD convection in protruded heater cavity. *Progress in Computational Fluid Dynamics*, *20*(6), 363–378. doi:10.1504/PCFD.2020.111408

Biswas, N., Manna, N. K., Mandal, D. K., & Chamkha Ali, J. (2021b). Effect of surface waviness on MHD thermo-gravitational convection of Cu−Al2O3−water hybrid nanofluid in a porous oblique enclosure. *Physica Scripta*, *96*(10), 105002. doi:10.1088/1402-4896/ac0f94

Biswas, N., Manna, N. K., Mandal, D. K., & Gorla, R. S. R. (2021a). Magnetohydrodynamic mixed bioconvection of oxytactic microorganisms in a nanofluid-saturated porous cavity heated with a bell-shaped curved bottom. *International Journal of Numerical Methods for Heat & Fluid Flow*, *31*(12), 3722–3751. doi:10.1108/HFF-10-2020-0668

Bodhak, S., Nath, S., & Basu, B. (2009). Friction and Wear Properties of Novel HDPE—HAp—Al2O3 Biocomposites against Alumina Counterface. *Journal of Biomaterials Applications*, *23*(5), 407–433. doi:10.1177/0885328208090012 PMID:18667457

Bogdanova, M., Zabirnyk, A., Malashicheva, A., Enayati, K. Z., Karlsen, T. A., Kaljusto, M. L., Kvitting, J. P. E., Dissen, E., Sullivan, G. J., Kostareva, A., Stensløkken, K. O., Rutkovskiy, A., & Vaage, J. (2019). Interstitial cells in calcified aortic valves have reduced differentiation potential and stem cell-like properties. *Scientific Reports*, *9*(1), 1–13. doi:10.103841598-019-49016-0 PMID:31506459

Bonnel, F., Toullec, E., Mabit, C., & Tourné, Y. (2010). Chronic ankle instability: Biomechanics and pathomechanics of ligaments injury and associated lesions. *Orthopaedics & Traumatology: Surgery & Research*, *96*(4), 424–432. doi:10.1016/j.otsr.2010.04.003 PMID:20493797

Boobalan, V., & Shankar, S. (2013). Investigation on Various Proximal Femoral Stem Shapes for Human Hip Prosthesis Using Finite Element Concepts. *Trends in Biomaterials & Artificial Organs*, 27.

Borghi, A., New, S. E. P., Chester, A. H., Taylor, P. M., & Yacoub, M. H. (2013). Acta Biomaterialia Time-dependent mechanical properties of aortic valve cusps : Effect of glycosaminoglycan depletion. *Acta Biomaterialia*, *9*(1), 4645–4652. doi:10.1016/j.actbio.2012.09.001 PMID:22963848

Bozzini, G., Aydogan, T. B., Müller, A., Sighinolfi, M. C., Besana, U., Calori, A., Lorenzo, B., Govorov, A., Pushkar, D. Y., Pini, G., Pastore, A. L., Romero-Otero, J., Rocco, B., & Buizza, C. (2020). A comparison among PCNL, Miniperc and Ultraminiperc for lower calyceal stones between 1 and 2 cm: A prospective, comparative, multicenter and randomised study. *BMC Urology*, *20*(1), 67. doi:10.118612894-020-00636-z PMID:32522171

Brazile, B., Wang, B., Wang, G., Bertucci, R., Prabhu, R., Patnaik, S. S., Butler, J. R., Claude, A., Brinkman-Ferguson, E., Williams, L. N., & Liao, J. (2015). On the Bending Properties of Porcine Mitral, Tricuspid, Aortic, and Pulmonary Valve Leaflets. *Journal of Long-Term Effects of Medical Implants*, *25*(1–2), 41–53. doi:10.1615/JLongTermEffMedImplants.2015011741 PMID:25955006

Breuls, R. G., Sengers, B. G., Oomens, C. W., Bouten, C. V., & Baaijens, F. P. (2002). Predicting local cell deformations in engineered tissue constructs: A multilevel finite element approach. *Journal of Biomechanical Engineering*, *124*(2), 198–207. doi:10.1115/1.1449492 PMID:12002129

Bridgett, L. A., & Linthorne, N. P. (2007). Changes in long jump take-off technique with increasingrun-up speed. *Journal of Sports Sciences*, 37–41.

Broedersz, C. P., Mao, X., Lubensky, T. C., & MacKintosh, F. C. (2011). Criticality and isostaticity in fibre networks. *Nature Physics*, *7*(12), 983–988. doi:10.1038/nphys2127

Buchanan, J. R. Jr, Kleinstreuer, C., & Comer, J. K. (2000, April). (20000). Rheological effects on pulsatile hemodynamics in a stenosed tube. *Computers & Fluids*, *29*(6), 695–724. doi:10.1016/S0045-7930(99)00019-5

Buckwalter, J. A. (1998). Articular cartilage: Injuries and potential for healing. *The Journal of Orthopaedic and Sports Physical Therapy*, *28*(4), 192–202. doi:10.2519/jospt.1998.28.4.192 PMID:9785255

Bullough, P., & Goodfellow, J. (1968). The significance of the fine structure of articular cartilage. *The Journal of Bone and Joint Surgery. British Volume*, *50*(4), 852–857. doi:10.1302/0301-620X.50B4.852 PMID:5706888

Bures, J., Cyrany, J., Kohoutova, D., Förstl, M., Rejchrt, S., Kvetina, J., & Kopacova, M. (2010). Small intestinal bacterial overgrowth syndrome. *World Journal of Gastroenterology*, *16*(24), 2978. doi:10.3748/wjg.v16.i24.2978 PMID:20572300

Butcher, J. T., Mahler, G. J., & Hockaday, L. A. (2011). Aortic valve disease and treatment: The need for naturally engineered solutions. *Advanced Drug Delivery Reviews*, *63*(4), 242–268. doi:10.1016/j.addr.2011.01.008 PMID:21281685

Butcher, J. T., & Nerem, R. M. (2006). Valvular endothelial cells regulate the phenotype of interstitial cells in co-culture: Effects of steady shear stress. *Tissue Engineering*, *12*(4), 905–915. doi:10.1089/ten.2006.12.905 PMID:16674302

Cabrera, J. D., Manzo, B. O., Torres, J. E., Vicentini, F. C., Sánchez, H. M., Rojas, E. A., & Lozada, E. (2020). Mini-percutaneous nephrolithotomy versus retrograde intrarenal surgery for the treatment of 10-20 mm lower pole renal stones: A systematic review and meta-analysis. *World Journal of Urology*, *38*(10), 2621–2628. doi:10.100700345-019-03043-8 PMID:31813026

Camilleri & Choi. (1997). Irritable bowel syndrome. *Alimentary Pharmacology & Therapeutics*, *11*(1), 3-15.

Cao, T., & Jin, J. P. (2020). Evolution of flight muscle contractility and energetic efficiency. *Frontiers in Physiology*, *11*, 1038. doi:10.3389/fphys.2020.01038 PMID:33162892

Carey, M. C., Small, D. M., & Bliss, C. M. (1983). Lipid digestion and absorption. *Annual Review of Physiology*, *45*(1), 651–677.

Carlisle, C. R., Coulais, C., & Guthold, M. (2010). The mechanical stressâ€"strain properties of single electrospun collagen type I nanofibers. *Acta Biomaterialia*, *6*(8), 2997–3003. doi:10.1016/j.actbio.2010.02.050 PMID:20197123

Carlisle, C. R., Coulais, C., Namboothiry, M., Carroll, D. L., Hantgan, R. R., & Guthold, M. (2009). The mechanical properties of individual, electrospun fibrinogen fibers. *Biomaterials*, *30*(6), 1205–1213. doi:10.1016/j.biomaterials.2008.11.006 PMID:19058845

Caro, C. G., Pedley, T. J., Schroter, R. C., & Seed, W. A. (1978). The mechanics of the circulation. Oxford Medical.

Caro, C. G., Fitz-Gerald, J. M., & Schroter, R. C. (1971). Atheroma and arterial wall shear: Observation, correlation, and proposal for a shear dependent mass transfer mechanism for atherogenesis. *Proceedings of the Royal Society of London. Series B, Biological Sciences*, *117*, 109–159. PMID:4396262

Cassanova, R. A., & Giddens, D. P. (1978). Disorder distal to modeledstenoses in steady and pulsatile flow. *Journal of Biomechanics*, *11*(10-12), 441–453. doi:10.1016/0021-9290(78)90056-8 PMID:730759

Chambers, H. G., & Sutherland, D. H. (2002). A practical guide to gait analysis. *The Journal of the American Academy of Orthopaedic Surgeons*, *10*(3), 222–231. doi:10.5435/00124635-200205000-00009 PMID:12041944

Chaparro, D., Dargam, V., Alvarez, P., Yeung, J., Saytashev, I., Bustillo, J., Loganathan, A., Ramella-Roman, J., Agarwal, A., & Hutcheson, J. D. (2020). A Method to Quantify Tensile Biaxial Properties of Mouse Aortic Valve Leaflets. *Journal of Biomechanical Engineering*, *142*(10), 1–7. doi:10.1115/1.4046921 PMID:32291440

Chatterjee, S., Ghosh, U.B., & Majumder, S., RoyChowdhury, A., & Pal, S. (2013). Customization of Design for Hip Implants. *Indian Journal of Biomechanics*, *4*, 58–62.

Chatterjee, S., Kobylinski, S., & Basu, B. (2018). Finite Element Analysis to Probe the Influence of Acetabular Shell Design, Liner Material, and Subject Parameters on Biomechanical Response in Periprosthetic Bone. *Journal of Biomechanical Engineering*, *140*(10), 101014. doi:10.1115/1.4040249 PMID:30029239

Chatterjee, S., Roy, S., Majumder, S., & RoyChowdhury, A. (2020). Biomechanical Analysis to Probe Role of Bone Condition and Subject Weight in Stiffness Customization of Femoral Stem for Improved Periprosthetic Biomechanical Response. *Journal of Biomechanical Engineering*, *142*(10), 101002. doi:10.1115/1.4046973 PMID:32320044

Chatterjee, S., Sarkar, S., Kalidindi, S. R., & Basu, B. (2019). Periprosthetic biomechanical response towards dental implants, with functional gradation, for single/multiple dental loss. *Journal of the Mechanical Behavior of Biomedical Materials*, *94*, 249–258. doi:10.1016/j.jmbbm.2019.03.001 PMID:30928669

Chatterjee, S., Templin, R. J., & Campbell, K. E. (2007). The aerodynamics of Argentavis, the world's largest flying bird from the Miocene of Argentina. *Proceedings of the National Academy of Sciences of the United States of America*, *104*(30), 12398–12403. doi:10.1073/pnas.0702040104 PMID:17609382

Chen, H., & Kassab, G. (2010). *Computational Modeling of Coronary Stents.* . doi:10.1007/978-90-481-3575-2_6

Chen, D. W., Lin, C. L., Hu, C. C., Wu, J. W., & Lee, M. S. (2012). Finite element analysis of different repair methods of Vancouver B1 periprosthetic fractures after total hip arthroplasty. *Injury, 43*(7), 1061–1065. doi:10.1016/j.injury.2012.01.015 PMID:22336128

Cheney, J. A., Konow, N., Bearnot, A., & Swartz, S. M. (2015). A wrinkle in flight: The role of elastin fibres in the mechanical behaviour of bat wing membranes. *Journal of the Royal Society, Interface, 12*(106), 20141286. doi:10.1098/rsif.2014.1286 PMID:25833238

Chen, F., Peng, X., Li, T., Chen, S., Wu, X. F., Reneker, D. H., & Hou, H. (2008). Mechanical characterization of single high-strength electrospun polyimide nanofibres. *Journal of Physics. D, Applied Physics, 41*(2), 025308. doi:10.1088/0022-3727/41/2/025308

Cheng, B., Tobalske, B. W., Powers, D. R., Hedrick, T. L., Wethington, S. M., Chiu, G. T., & Deng, X. (2016). Flight mechanics and control of escape manoeuvres in hummingbirds. I. Flight kinematics. *The Journal of Experimental Biology, 219*(22), 3518–3531. doi:10.1242/jeb.137539 PMID:27595850

Chen, J., Kim, J., Shao, W., Schlecht, S. H., Baek, S. Y., Jones, A. K., Ahn, T., Ashton-Miller, J. A., Banaszak Holl, M. M., & Wojtys, E. M. (2019). An Anterior Cruciate Ligament Failure Mechanism. *The American Journal of Sports Medicine, 47*(9), 2067–2076. doi:10.1177/0363546519854450 PMID:31307223

Chen, J., Ying, C., Yip, Y., Sone, E. D., & Simmons, C. A. (2009). Identification and Characterization of Aortic Valve Mesenchymal Progenitor Cells with Robust Osteogenic Calcification Potential. *American Journal of Pathology, 174*(3), 1109–1119. doi:10.2353/ajpath.2009.080750 PMID:19218344

Chen, X.-Y., Zhang, C.-Y., Nie, E.-M., & Zhang, M. (2012). Treatment planning of implants when 3 mandibular posterior teeth are missing: A 3-dimensional finite element analysis. *Implant Dentistry, 21*(4), 340–343. doi:10.1097/ID.0b013e31825cbc67 PMID:22814561

Chiappe, L. M. (2009). Downsized Dinosaurs: The Evolutionary Transition to Modern Birds. *Evolution (New York), 2*(2), 248–256. doi:10.100712052-009-0133-4

Choi, H. W., & Barakat, A. (2010). *Computational Modeling of ATP/ADP Concentration at the Vascular Surface.* . doi:10.1007/978-90-481-3575-2_2

Choudhury, S., Talukdar, P., Mandal, T. K., & Majhi, T. K. (2021). Supine versus prone PCNL in lower calyceal stone: Comparative study in a tertiary care center. *Urologia, 88*(2), 148–152. doi:10.1177/0391560320962404 PMID:33028166

Cho, Y. I., & Kensey, K. R. (1991). Effects of the non-Newtonian viscosity of blood flows in a diseased arterial vessel, part-1: Steady flows. *Biorheology, 28*(3-4), 241–262. doi:10.3233/BIR-1991-283-415 PMID:1932716

Cilingir, A. C. (2010). Finite element analysis of the contact mechanics of ceramic-on-ceramic hip resurfacing prostheses. *Journal of Bionics Engineering, 7*(3), 244–253. doi:10.1016/S1672-6529(10)60247-8

Cipriano, C. A., Issack, P. S., Beksaç, B., Della Valle, A. G., Sculco, T. P., & Salvati, E. A. (2008). Metallosis after metal-on-polyethylene total hip arthroplasty. *American Journal of Orthopedics (Belle Mead, N.J.), 37*, E18–E25. PMID:18401490

Clark, J. M. (1990). The organisation of collagen fibrils in the superficial zones of articular cartilage. *Journal of Anatomy, 171*, 117. PMID:2081698

Clavica, F., Zhao, X., ElMahdy, M., Drake, M. J., Zhang, X., & Carugo, D. (2014). Investigating the flow dynamics in the obstructed and stented ureter by means of a biomimetic artificial model. *PLoS One, 9*(2), e87433. doi:10.1371/journal.pone.0087433 PMID:24498322

Colemont, L. J., & Camilleri, M. (1989, January). Chronic intestinal pseudo-obstruction: Diagnosis and treatment. *Mayo Clinic Proceedings, 64*(1), 60–70. doi:10.1016/S0025-6196(12)65304-X PMID:2642997

Colyer, S. L., Evans, M., Cosker, D. P., & Salo, A. I. T. (2018). A review of the evolution of vision-based motion analysis and the integration of advanced computer vision methods towards developing a markerless system. *Sports Medicine - Open, 4*(1), 24. doi:10.118640798-018-0139-y PMID:29869300

Croft, L., & Mofrad, M. (2010). *Computational Modeling of Aortic Heart Valves.* . doi:10.1007/978-90-481-3575-2_7

Dalstra, M., & Huiskes, R. (1995). Load transfer across the pelvic bone. *Journal of Biomechanics, 28*(6), 715–724. doi:10.1016/0021-9290(94)00125-N PMID:7601870

Daoud, A. I., Geissler, G. J., Wang, F., Saretsky, J., Daoud, Y. A., & Lieberman, D. E. (2012). Foot strike and injury rates in endurance runners: A retrospective study. *Medicine and Science in Sports and Exercise, 44*(7), 1325–1334. doi:10.1249/MSS.0b013e3182465115 PMID:22217561

Davis, R. B. (1988). Clinical Gait Analysis. *IEEE Engineering in Medicine and Biology Magazine, 7*(3), 35–40. doi:10.1109/51.7933 PMID:18244073

Deck, J. D. (1986). Endothelial cell orientation on aortic valve leaflets. *Cardiovascular Research, 20*(10), 760–767. doi:10.1093/cvr/20.10.760 PMID:3791342

Deetjen, M. E., Biewener, A. A., & Lentink, D. (2017). High-speed surface reconstruction of a flying bird using structured light. *The Journal of Experimental Biology, 220*(11), 1956–1961. doi:10.1242/jeb.149708 PMID:28348041

Demzik, A., Filippou, P., Chew, C., Deal, A., Mercer, E., Mahajan, S., Wallen, E. M., Tan, H. J., & Smith, A. B. (2021). Gender-Based Differences in Urology Residency Applicant Personal Statements. *Urology, 150*, 2–8. doi:10.1016/j.urology.2020.08.066 PMID:33035562

De, S., Autorino, R., Kim, F. J., Zargar, H., Laydner, H., Balsamo, R., Torricelli, F. C., Di Palma, C., Molina, W. R., Monga, M., & De Sio, M. (2015). Percutaneous nephrolithotomy versus retrograde intrarenal surgery: A systematic review and meta-analysis. *European Urology, 67*(1), 125–137. doi:10.1016/j.eururo.2014.07.003 PMID:25064687

Desai, A., Vafaee, T., Rooney, P., Kearney, J. N., Berry, H. E., Ingham, E., Fisher, J., & Jennings, L. M. (2018). In vitro biomechanical and hydrodynamic characterisation of decellularised human pulmonary and aortic roots. *Journal of the Mechanical Behavior of Biomedical Materials, 79*(September), 53–63. doi:10.1016/j.jmbbm.2017.09.019

Desai, M., Sun, Y., Buchholz, N., Fuller, A., Matsuda, T., Matlaga, B., Miller, N., Bolton, D., Alomar, M., & Ganpule, A. (2017). Treatment selection for urolithiasis: Percutaneous nephrolithomy, ureteroscopy, shock wave lithotripsy, and active monitoring. *World Journal of Urology, 35*(9), 1395–1399. doi:10.100700345-017-2030-8 PMID:28303335

Dhason, R., Roy, S., & Datta, S. (2020). A biomechanical study on the laminate stacking sequence in composite bone plates for vancouver femur B1 fracture fixation. *Computer Methods and Programs in Biomedicine, 196*, 105680. Advance online publication. doi:10.1016/j.cmpb.2020.105680 PMID:32763643

Dhawan, S. (1991). Bird flight. Sadhana. *Academy Proceedings in Engineering Sciences, 16*(4), 275–352.

Di Grazia, E., & La Rosa, P. (2013). Split-leg percutaneous nephrolithotomy: a safe and versatile technique. *Archivio italiano di urologia, andrologia: organo ufficiale (di) Societa italiana di ecografia urologica e nefrologica, 85*(2), 82–85.

Diegel, O., Nordin, A., & Motte, D. (2019). *A Practical Guide to Design for Additive Manufacturing* (1st ed.). Springer.

Diwan, S. S., Ravichandran, S., Govindarajan, R., & Narasimha, R. (2020). Understanding transmission dynamics of COVID-19-type infections by direct numerical simulations of cough/sneeze flows. *Transactions of the Indian National Academy of Engineering*, *5*(2), 255–261. doi:10.100741403-020-00106-w

Dixon, J. L., Stoops, J. D., Parker, J. L., Laughlin, M. H., Weisman, G. A., & Sturek, M. (1999). Dyslipidemia and vascular dysfunction in diabetic pigs fed an atherogenic diet. *Arteriosclerosis, Thrombosis, and Vascular Biology*, *19*(12), 2981–2992. doi:10.1161/01.ATV.19.12.2981 PMID:10591679

Dixon, S., Fleming, P., James, I., & Carré, M. (2015). *The Science and Engineering of Sport Surfaces*. Routledge. doi:10.4324/9780203133385

Draper, M. H. (1966). Avian Physiology. *Quarterly Journal of Experimental Physiology and Cognate Medical Sciences: Translation and Integration*, *51*(4), 381–382. doi:10.1113/expphysiol.1966.sp001873

Driscoll, M. (2019). The Impact of the Finite Element Method on Medical Device Design. *Journal of Medical and Biological Engineering*, *39*(2), 171–172. doi:10.100740846-018-0428-4

Driscoll, T. P., Nerurkar, N. L., Jacobs, N. T., Elliott, D. M., & Mauck, R. L. (2011). Fiber angle and aspect ratio influence the shear mechanics of oriented electrospun nanofibrous scaffolds. *Journal of the Mechanical Behavior of Biomedical Materials*, *4*(8), 1627–1636. doi:10.1016/j.jmbbm.2011.03.022 PMID:22098865

Dubov, A., Kim, S. Y. R., Shah, S., Schemitsch, E. H., Zdero, R., & Bougherara, H. (2011). The biomechanics of plate repair of periprosthetic femur fractures near the tip of a total hip implant: The effect of cable-screw position. *Proceedings of the Institution of Mechanical Engineers. Part H, Journal of Engineering in Medicine*, *225*(9), 857–865. doi:10.1177/0954411911410642 PMID:22070023

Dunlop, J. W., & Fratzl, P. (2010). Biological composites. *Annual Review of Materials Research*, *40*(1), 1–24. doi:10.1146/annurev-matsci-070909-104421

Dvořák, R. (2016). Aerodynamics of bird flight. *EPJ Web of Conferences*, *114*(01001).

Ebrahimi, H., Rabinovich, M., Vuleta, V., Zalcman, D., Shah, S., Dubov, A., Roy, K., Siddiqui, F. S., Schemitsch, E. H., Bougherara, H., & Zdero, R. (2012). Biomechanical properties of an intact, injured, repaired, and healed femur: An experimental and computational study. *Journal of the Mechanical Behavior of Biomedical Materials*, *16*(1), 121–135. doi:10.1016/j.jmbbm.2012.09.005 PMID:23182385

Edwards, W. B. (2018). Modelling Overuse Injuries in Sport as a Mechanical Fatigue Phenomenon. *Exercise and Sport Sciences Reviews*, *46*(4), 224–231. doi:10.1249/JES.0000000000000163 PMID:30001271

Eldredge, J. D., & Jones, A. R. (2019). Leading-edge vortices: Mechanics and modeling. *Annual Review of Fluid Mechanics*, *51*(1), 75–104. doi:10.1146/annurev-fluid-010518-040334

El-Hamamsy, I., Chester, A. H., & Yacoub, M. H. (2010). Cellular regulation of the structure and function of aortic valves. *Journal of Advanced Research*, *1*(1), 5–12. doi:10.1016/j.jare.2010.02.007

Engler, A. J., Sen, S., Sweeney, H. L., & Discher, D. E. (2006). Matrix elasticity directs stem cell lineage specification. *Cell*, *126*(4), 677–689. doi:10.1016/j.cell.2006.06.044 PMID:16923388

Erickson, R. H., & Kim, Y. S. (1990). Digestion and absorption of dietary protein. *Annual Review of Medicine*, *41*(1), 133–139. doi:10.1146/annurev.me.41.020190.001025 PMID:2184718

Essner, A., Sutton, K., & Wang, A. (2005). Hip simulator wear comparison of metal-on-metal, ceramic-on-ceramic and crosslinked UHMWPE bearings. *Wear*, *259*(7-12), 992–995. doi:10.1016/j.wear.2005.02.104

Farrington, T., Onambele-Pearson, G., Taylor, R. L., Earl, P., & Winwood, K. (2012). A review of facial protective equipment use in sport and the impact on injury incidence. *British Journal of Oral & Maxillofacial Surgery*, *50*(3), 233–238. doi:10.1016/j.bjoms.2010.11.020 PMID:21295384

Fauci, A. S., Lane, H. C., & Redfield, R. R. (2020). *Covid-19—navigating the uncharted.* Academic Press.

Federico, S., & Gasser, T. C. (2010). Nonlinear elasticity of biological tissues with statistical fibre orientation. *Journal of the Royal Society, Interface*, *7*(47), 955–966. doi:10.1098/rsif.2009.0502 PMID:20053655

Feng, Y., Marchal, T., Sperry, T., & Yi, H. (2020). Influence of wind and relative humidity on the social distancing effectiveness to prevent COVID-19 airborne transmission: A numerical study. *Journal of Aerosol Science*, *147*, 105585. doi:10.1016/j.jaerosci.2020.105585 PMID:32427227

Fernandes, P., Folgado, J., & Ruben, R. (2004). Shape optimization of a cementless hip stem for a minimum of interface stress and displacement. *Computer Methods in Biomechanics and Biomedical Engineering*, *7*(1), 51–61. doi:10.1080/10255840410001661637 PMID:14965880

Fernandez-Redondo, V., Gomez-Centeno, P., & Toribio, J. (1998). Chronic urticaria from a dental bridge. *Contact Dermatitis*, *38*(3), 178–179. doi:10.1111/j.1600-0536.1998.tb05696.x PMID:9536421

Finch, C. F., & Cook, J. (2014). Categorising sports injuries in epidemiological studies: The subsequent injury categorisation (SIC) model addresses multiple, recurrent and exacerbation injuries. *British Journal of Sports Medicine*, *48*(17), 1276–1280. doi:10.1136/bjsports-2012-091729 PMID:23501833

Fisher, J. (2011). Bioengineering reasons for the failure of metal-on-metal hip prostheses. *The Journal of Bone and Joint Surgery. British Volume*, *93*(8), 1001–1004. doi:10.1302/0301-620X.93B8.26936 PMID:21768619

Fitts, M. K., Pike, D. B., Anderson, K., & Shiu, Y. T. (2014). Hemodynamic Shear Stress and Endothelial Dysfunction in Hemodialysis Access. *The Open Urology & Nephrology Journal*, *7*(Suppl 1 M5), 33–44. doi:10.2174/1874303X01407010033

Fong, D. T. P., & Chan, Y. Y. (2010). The use of wearable inertial motion sensors in human lower limb biomechanics studies: A systematic review. *Sensors (Basel)*, *10*(12), 11556–11565. doi:10.3390101211556 PMID:22163542

Frantz, C., Stewart, K. M., & Weaver, V. M. (2010). The extracellular matrix at a glance. *Journal of Cell Science*, *123*(24), 4195–4200. doi:10.1242/jcs.023820 PMID:21123617

Frater, R. W. M., Gong, G., Hoffman, D., & Liao, K. (1992). Endothelial covering of biological artificial heart valves. *The Annals of Thoracic Surgery*, *53*(3), 371–372. doi:10.1016/0003-4975(92)90252-Y PMID:1540049

Fratzl, P. (2008). Collagen: structure and mechanics, an introduction. In *Collagen* (pp. 1–13). Springer. doi:10.1007/978-0-387-73906-9_1

Fry, D. L. (1968). Acute vascular endothelial changes associated with increased blood velocity gradients. *Circulation Research*, *12*(2), 165–197. doi:10.1161/01.RES.22.2.165 PMID:5639037

Fry, D. L. (1969). Certain histological and chemical responses of the vascular interface to acutely induced mechanical stress in the aorta of the dog. *Circulation Research*, *24*(1), 93–108. doi:10.1161/01.RES.24.1.93 PMID:5763742

Fung, Y. C. (1997). *Biomechanics: Circlulation* (2nd ed.). Springer. doi:10.1007/978-1-4757-2696-1

Fu, W., Yang, Z., Xie, Z., & Yan, H. (2017). Intravenous misplacement of the nephrostomy catheter following percutaneous nephrostolithotomy: Two case reports and literature review. *BMC Urology*, *17*(1), 43. doi:10.118612894-017-0233-3 PMID:28615052

Gabbard, S. L., & Lacy, B. E. (2013). Chronic intestinal pseudo-obstruction. *Nutrition in Clinical Practice*, *28*(3), 307–316. doi:10.1177/0884533613485904 PMID:23612903

Gadzhiev, N., Malkhasyan, V., Akopyan, G., Petrov, S., Jefferson, F., & Okhunov, Z. (2020). Percutaneous nephrolithotomy for staghorn calculi: Troubleshooting and managing complications. *Asian Journal of Urology*, *7*(2), 139–148. doi:10.1016/j.ajur.2019.10.004 PMID:32257807

Galbusera, F. (2018). The spine: its evolution, function, and shape. In *Biomechanics of the Spine* (pp. 3–9). Academic Press. doi:10.1016/B978-0-12-812851-0.00001-X

Gamal, W., Moursy, E., Hussein, M., Mmdouh, A., Hammady, A., & Aldahshoury, M. (2015). Supine pediatric percutaneous nephrolithotomy (PCNL). *Journal of Pediatric Urology*, *11*(2), 78.e1–78.e785. doi:10.1016/j.jpurol.2014.10.012 PMID:25819602

Ganley, K. J., & Powers, C. M. (2004). Determination of lower extremity anthropometric parameters using dual energy X-ray absorptiometry: The influence on net joint moments during gait. *Clinical Biomechanics (Bristol, Avon)*, *19*(1), 50–56. doi:10.1016/j.clinbiomech.2003.08.002 PMID:14659930

Ganong, W. F. (2001). *Review of Medical Physiology*. Appleton.

Garg, S., & Pant, M. (2017). Meshfree Methods: A Comprehensive Review of Applications. *International Journal of Computational Methods*, *15*(04), 1830001. Advance online publication. doi:10.1142/S0219876218300015

Garg, V., Muth, A. N., Ransom, J. F., Schluterman, M. K., Barnes, R., King, I. N., Grossfeld, P. D., & Srivastava, D. (2005). Mutations in NOTCH1 cause aortic valve disease. *Nature*, *437*(7056), 270–274. doi:10.1038/nature03940 PMID:16025100

Gearity, B. T., & Murray, M. A. (2011). Athletes' experiences of the psychological effects of poor coaching. *Psychology of Sport and Exercise*, *12*(3), 213–221. doi:10.1016/j.psychsport.2010.11.004

Gee, A., Bougherara, H., Schemitsch, E. H., & Zdero, R. (2021). Biomechanical design using in-vitro finite element modeling of distal femur fracture plates made from semi-rigid materials versus traditional metals for post-operative toe-touch weight-bearing. *Medical Engineering & Physics*, *87*, 95–103. doi:10.1016/j.medengphy.2020.11.015 PMID:33461680

Gee, C. J. (2010). How does sporsportschology actually improve athletic performance? A framework to facilitate athletes' and coaches' understanding. *Behavior Modification*, *34*(5), 386–402. doi:10.1177/0145445510383525 PMID:20935240

Geiringer, S. R. (1985). The biomechanics of running: Implications for the prevention of foot injuries. *Sports Medicine (Auckland, N.Z.)*, *2*(2), 144–153. doi:10.2165/00007256-198502020-00006 PMID:2860714

Gentleman, E., Lay, A. N., Dickerson, D. A., Nauman, E. A., Livesay, G. A., & Dee, K. C. (2003). Mechanical characterization of collagen fibers and scaffolds for tissue engineering. *Biomaterials*, *24*(21), 3805–3813. doi:10.1016/S0142-9612(03)00206-0 PMID:12818553

Gerrity, R. G., Natarajan, R., Nadler, J. L., & Kimsey, T. (2001). Diabetes-Induced Accelerated Atherosclerosis in Swine. *Diabetes*, *50*(7), 1654–1665. doi:10.2337/diabetes.50.7.1654 PMID:11423488

Gessner, F. B. (1973). Haemodynamic theories of atherogenesis. *Circulation Research*, *3*(3), 259–266. doi:10.1161/01.RES.33.3.259

Getgood, A., Hoshino, Y., & Roessler, P. P. (2019). Biomechanics of musculoskeletal injuries. *The Sports Medicine Physician*, 27-35.

Ghosh, K., Pan, Z., Guan, E., Ge, S., Liu, Y., Nakamura, T., Ren, X.-D., Rafailovich, M., & Clark, R. A. (2007). Cell adaptation to a physiologically relevant ECM mimic with different viscoelastic properties. *Biomaterials*, *28*(4), 671–679. doi:10.1016/j.biomaterials.2006.09.038 PMID:17049594

Ghosh, R., & Gupta, S. (2014). Bone remodelling around cementless composite acetabular components: The effects of implant geometry and implant–bone interfacial conditions. *Journal of the Mechanical Behavior of Biomedical Materials*, *32*, 257–269. doi:10.1016/j.jmbbm.2014.01.010 PMID:24508712

Ghosh, R., Gupta, S., Dickinson, A., & Browne, M. (2013a). Experimental validation of numerically predicted strain and micromotion in intact and implanted composite hemi-pelvises. *Proceedings of the Institution of Mechanical Engineers. Part H, Journal of Engineering in Medicine*, *227*(2), 162–174. doi:10.1177/0954411912461238 PMID:23513987

Ghosh, R., Mukherjee, K., & Gupta, S. (2013b). Bone remodelling around uncemented metallic and ceramic acetabular components. *Proceedings of the Institution of Mechanical Engineers. Part H, Journal of Engineering in Medicine*, *227*(5), 490–502. doi:10.1177/0954411913478703 PMID:23637259

Ghosh, R., Pal, B., Ghosh, D., & Gupta, S. (2015). Finite element analysis of a hemi-pelvis: The effect of inclusion of cartilage layer on acetabular stresses and strain. *Computer Methods in Biomechanics and Biomedical Engineering*, *18*(7), 697–710. doi:10.1080/10255842.2013.843674 PMID:24156480

Gilbert, T. W., Wognum, S., Joyce, E. M., Freytes, D. O., Sacks, M. S., & Badylak, S. F. (2008). Collagen fiber alignment and biaxial mechanical behavior of porcine urinary bladder derived extracellular matrix. *Biomaterials*, *29*(36), 4775–4782. doi:10.1016/j.biomaterials.2008.08.022 PMID:18801572

Gittes, F., & MacKintosh, F. C. (1998). Dynamic shear modulus of a semiflexible polymer network. *Physical Review. E*, *58*(2), R1241–R1244. doi:10.1103/PhysRevE.58.R1241

Giusti, G., & De Lisa, A. (2020). PCNL in the prone position VS PCNL in the modified supine Double-S position: Is there a better position? A prospective randomized trial. *Urolithiasis*, *48*(1), 63–69. doi:10.100700240-018-1088-0 PMID:30456414

Gloeckner, D. G., Bihir, K. L., & Sacks, M. S. (1999). Effects of Mechanical Fatigue on the Bending Properties of the Porcine Bioprosthetic Heart Valve. *ASAIO Journal (American Society for Artificial Internal Organs)*, *45*(1), 59–63. doi:10.1097/00002480-199901000-00014 PMID:9952009

Goldberg, J. M., & Cullen, K. E. (2011). Vestibular control of the head: Possible functions of the vestibulocollic reflex. *Experimental Brain Research*, *210*(3), 331–345. doi:10.100700221-011-2611-5 PMID:21442224

Gómez-Blanco, J. C., Martínez-Reina, F. J., Cruz, D., Pagador, J. B., Sánchez-Margallo, F. M., & Soria, F. (2016). Fluid Structural Analysis of Urine Flow in a Stented Ureter. *Computational and Mathematical Methods in Medicine*, *5710798*, 1–7. Advance online publication. doi:10.1155/2016/5710798 PMID:27127535

Goswami, P., Mandal, D. K., Manna, N. K., & Chakrabarti, S. (2015a). Wall Shear Stress Characteristics for the Progression of the Disease, Atherosclerosis. *Journal of the Institution of Engineers (India): Series C*, *96*(3), 311-323.

Goswami, P., Mandal, D. K., Manna, N. K., & Chakrabarti, S. (2014). Study on the Effect of Steady, Simple Pulsatile and Physiological Pulsatile Flows through a Stenosed Artery. *Heat and Mass Transfer*, *50*(10), 1343–1352. doi:10.100700231-014-1334-0

Goswami, P., Mandal, D. K., Manna, N. K., & Chakrabarti, S. (2015b). Analysis of Steady and Physiological Pulsatile Flow Characteristics in an Artery with Various Percentages of Restrictions. *International Journal of Fluid Mechanics Research*, *42*(3), 260–280. doi:10.1615/InterJFluidMechRes.v42.i3.60

Goswami, P., Mandal, D. K., Manna, N. K., & Chakrabarti, S. (2016). Analysis of Wall Shear Parameters of Physiological Pulsating Flow through Mild and Severe Arterial Stenosis and Correlation to Atherosclerosis. *International Journal of Science and Technology*, *2*(3), 40–54.

Goswami, P., Mandal, D. K., Manna, N. K., & Chakrabarti, S. (2019). Numerical Investigation of Fluid Mechanical Factors of Realistic Pulsatile Flow through Constricted artery and Various Aspects of Plaque Deposition. *Journal of Mechanical Engineering Science*, *13*(3), 5306–5322. doi:10.15282/jmes.13.3.2019.07.0433

Grant, T. M., Thompson, M. S., Urban, J., & Yu, J. (2013). Elastic fibres are broadly distributed in tendon and highly localized around tenocytes. *Journal of Anatomy*, *222*(6), 573–579. doi:10.1111/joa.12048 PMID:23587025

Gronley, J. K., & Perry, J. (1984). Gait analysis techniques. *Physical Therapy*, *64*(12), 1831–1838. doi:10.1093/ptj/64.12.1831 PMID:6505028

Guisasola, I., James, I., Stiles, V., & Dixon, S. (2010). Dynamic behaviour of soils used for natural turf sports surfaces. *Sports Engineering*, *12*(3), 111–122. doi:10.100712283-010-0036-1

Gupta, A. K., & Agrawal, S. P. (2015). Computational modeling and analysis of the hydrodynamic parameters of blood through stenotic artery. *Procedia Computer Science*, *57*, 403–410. doi:10.1016/j.procs.2015.07.355

Gupta, Y., Iyer, R., Dommeti, V. K., Nutu, E., Rana, M., Merdji, A., & Roy, S. (2021). Design of dental implant using design of experiment and topology optimization: A finite element analysis study. *Proceedings of the Institution of Mechanical Engineers. Part H, Journal of Engineering in Medicine*, *235*(2), 157–166. doi:10.1177/0954411920967146 PMID:33094686

Hachinski, V. (2008). A historical review of gait analysis. *Neurosciences*, *13*(4), 460. PMID:21063384

Hadi, M. F., Sander, E. A., & Barocas, V. H. (2012). *Multiscale model predicts tissue-level failure from collagen fiber-level damage*. Academic Press.

Hadjipanayi, E., Mudera, V., & Brown, R. A. (2009a). Close dependence of fibroblast proliferation on collagen scaffold matrix stiffness. *Journal of Tissue Engineering and Regenerative Medicine*, *3*(2), 77–84. doi:10.1002/term.136 PMID:19051218

Hadjipanayi, E., Mudera, V., & Brown, R. A. (2009b). Guiding cell migration in 3D: A collagen matrix with graded directional stiffness. *Cell Motility and the Cytoskeleton*, *66*(3), 121–128. doi:10.1002/cm.20331 PMID:19170223

Hadjizadeh, A., Ajji, A., & Bureau, M. N. (2011). Nano/micro electro-spun polyethylene terephthalate fibrous mat preparation and characterization. *Journal of the Mechanical Behavior of Biomedical Materials*, *4*(3), 340–351. doi:10.1016/j.jmbbm.2010.10.014 PMID:21316622

Haider, M., Benedict, B., Kim, E., & Guilak, F. (2010). *Computational Modeling of Cell Mechanics in Articular Cartilage*. . doi:10.1007/978-90-481-3575-2_11

Han, J., Hui, Z., Tian, F., & Chen, G. (2021). Review on bio-inspired flight systems and bionic aerodynamics. *Chinese Journal of Aeronautics*, *34*(7), 170–186. doi:10.1016/j.cja.2020.03.036

Han, L., Grodzinsky, A. J., & Ortiz, C. (2011). Nanomechanics of the cartilage extracellular matrix. *Annual Review of Materials Research*, *41*(1), 133–168. doi:10.1146/annurev-matsci-062910-100431 PMID:22792042

Haque, A., & Dickman, J. D. (2005). Vestibular gaze stabilization: Different behavioral strategies for arboreal and terrestrial avians. *Journal of Neurophysiology*, *93*(3), 1165–1173. doi:10.1152/jn.00966.2004 PMID:15525803

Harris, G. F., & Wertsch, J. J. (1994). Procedures for gait analysis. *Archives of Physical Medicine and Rehabilitation, 75*(2), 216–225. doi:10.1016/0003-9993(94)90399-9 PMID:8311681

Harris, W. H. (2012). Edge loading has a paradoxical effect on wear in metal-on-polyethylene total hip arthroplasties. *Clinical Orthopaedics and Related Research, 470*(11), 3077–3082. doi:10.100711999-012-2330-7 PMID:22644421

Hasan, A., Ragaert, K., Swieszkowski, W., Selimović, Š., Paul, A., Camci-Unal, G., Mofrad, M. R. K., & Khademhosseini, A. (2014). Biomechanical properties of native and tissue engineered heart valve constructs. *Journal of Biomechanics, 47*(9), 1949–1963. doi:10.1016/j.jbiomech.2013.09.023 PMID:24290137

Hauser, R. A., & Dolan, E. E. (2011). Ligament Injury and Healing : An Overview of current clinical concepts. *Journal of Prolotherapy, 3*, 836–846.

Heers, A. M., Rankin, J. W., & Hutchinson, J. R. (2018). Building a Bird: Musculoskeletal Modeling and Simulation of Wing-Assisted Incline Running During Avian Ontogeny. *Frontiers in Bioengineering and Biotechnology, 6*, 140. Advance online publication. doi:10.3389/fbioe.2018.00140 PMID:30406089

Heiderscheit, B. C. (2014). Increasing running step rate reduces patellofemoral joint forces. *Medicine and Science in Sports and Exercise*, 557–564. PMID:23917470

Heim, M., Römer, L., & Scheibel, T. (2010). Hierarchical structures made of proteins. The complex architecture of spider webs and their constituent silk proteins. *Chemical Society Reviews, 39*(1), 156–164. doi:10.1039/B813273A PMID:20023846

He, X., Xie, D., Du, C., Zhu, W., Li, W., Wang, K., Li, Y., Lu, H., & Guo, F. (2015). Improved nephrostomy tube can reduce percutaneous nephrolithotomy postoperative bleeding. *International Journal of Clinical and Experimental Medicine, 8*(3), 4243–4249. PMID:26064336

Higginson, B. K. (2009). Methods of running gait analysis. *Current Sports Medicine Reports, 8*(3), 136–141. doi:10.1249/JSR.0b013e3181a6187a PMID:19436169

Holzapfel, G. A., Gasser, T. C., & Ogden, R. W. (2000). A new constitutive framework for arterial wall mechanics and a comparative study of material models. *Journal of Elasticity and the Physical Science of Solids, 61*(1), 1-48.

Hong, Y., Huber, A., Takanari, K., Amoroso, N. J., Hashizume, R., Badylak, S. F., & Wagner, W. R. (2011). Mechanical properties and in vivo behavior of a biodegradable synthetic polymer microfiberâ€“extracellular matrix hydrogel biohybrid scaffold. *Biomaterials, 32*(13), 3387–3394. doi:10.1016/j.biomaterials.2011.01.025 PMID:21303718

Hooton, D., Lentle, R., Monro, J., Wickham, M., & Simpson, R. (2015). The secretion and action of brush border enzymes in the mammalian small intestine. *Reviews of Physiology, Biochemistry and Pharmacology, 168*, 59–118. doi:10.1007/112_2015_24 PMID:26345415

Hornick, J. (2015). *3D printing will rock the world.* Paseo Publishing.

Hosseini, S. R., Mohseni, M. G., Aghamir, S., & Rezaei, H. (2019). Effect of Irrigation Solution Temperature on Complication of Percutaneous Nephrolithotomy: A Randomized Clinical Trial. *Urology Journal, 16*(6), 525–529. PMID:30882166

Hua, X., Wang, L., Al-Hajjar, M., Jin, Z., Wilcox, R. K., & Fisher, J. (2014). Experimental validation of finite element modelling of a modular metal-on-polyethylene total hip replacement. *Proceedings of the Institution of Mechanical Engineers. Part H, Journal of Engineering in Medicine, 228*(7), 682–692. doi:10.1177/0954411914541830 PMID:24963036

Huizinga, J. D., Chen, J. H., Zhu, Y. F., Pawelka, A., McGinn, R. J., Bardakjian, B. L., & Chen, D. (2014). The origin of segmentation motor activity in the intestine. *Nature Communications, 5*(1), 1–11. doi:10.1038/ncomms4326 PMID:24561718

Hunt, D. M., Carvalho, L. S., Cowing, J. A., & Davies, W. L. (2009). Evolution and spectral tuning of visual pigments in birds and mammals. *Philosophical Transactions of the Royal Society of London. Series B, Biological Sciences, 364*(1531), 2941–2955. doi:10.1098/rstb.2009.0044 PMID:19720655

Ichaoui, H., Samet, A., Ben Hadjalouane, H., Hermi, A., Hedhli, H., Bakir, M. A., Khiari, R., & Ghozzi, S. (2019). Percutaneous Nephrolithotomy (PCNL): Standard Technique Versus Tubeless - 125 Procedures. *Cureus, 11*(3), e4251. doi:10.7759/cureus.4251 PMID:31131174

Innocenti, B. (2018). Biomechanics: A fundamental tool with a long history (and even longer future!). *Muscles, Ligaments and Tendons Journal, 7*(4), 491–492. doi:10.32098/mltj.04.2017.02 PMID:29721449

Ishikawa, T., Guimaraes, L. F. R., Oshima, S., & Yamane, R. (1998). Effect of non-Newtonian property of blood on flow through a stenosed tube. *Fluid Dynamics Research, 22*(5), 251–264. doi:10.1016/S0169-5983(97)00041-5

Islamov, M., Sypabekova, M., Kanayeva, D., & Rojas-Solórzano, L. (2017). CFD Modeling of Chamber Filling in a Micro-Biosensor for Protein Detection. *Biosensors (Basel), 7*(4), 45. doi:10.3390/bios7040045 PMID:28972568

ISO/ASTM 52900:2015 Additive manufacturing — General principles — Terminology. https://www.iso.org/obp/ui/#iso:std:iso-astm:52900:ed-2:v1:en

Izol, V., Deger, M., Akdogan, N., Ok, F., Bayazit, Y., & Aridogan, I. A. (2021). The Effect of Percutaneous Nephrolithotomy on the Estimated Glomerular Filtration Rate in Patients with Chronic Kidney Disease. *Journal of Endourology, 35*(5), 583–588. doi:10.1089/end.2020.0512 PMID:33054416

Jastifer, J. R., McNitt, A. S., Mack, C. D., Kent, R. W., McCullough, K. A., Coughlin, M. J., & Anderson, R. B. (2019). Synthetic Turf: History, Design, Maintenance, and Athlete Safety. *Sports Health, 11*(1), 84–90. doi:10.1177/1941738118793378 PMID:30096021

Jena, S., Arunachalam, T., & Panda, S. K. (2020). Experimental and numerical investigation of a polypropylene orthotic device for assistance in level ground walking. *Proceedings of the Institution of Mechanical Engineers. Part H, Journal of Engineering in Medicine, 234*(4), 356–369. doi:10.1177/0954411919894091 PMID:31854229

Jena, S., Sakhare, G. M., Panda, S. K., & Thirugnanam, A. (2021). Implementation of Multiple Regression Technique for Detection of Gait Asymmetry Using Experimental Gait Data. *Journal of Medical and Biological Engineering, 41*(1), 1–10. doi:10.100740846-020-00533-8

Kabla, A., & Mahadevan, L. (2007). Nonlinear mechanics of soft fibrous networks. *Journal of the Royal Society, Interface, 4*(12), 99–106. doi:10.1098/rsif.2006.0151 PMID:17015287

Kallidonis, P., Panagopoulos, V., Kyriazis, I., & Liatsikos, E. (2016). Complications of percutaneous nephrolithotomy: Classification, management, and prevention. *Current Opinion in Urology, 26*(1), 88–94. doi:10.1097/MOU.0000000000000232 PMID:26555687

Kärrholm, J., Anderber, C., Snorrason, F., Thanner, J., Langeland, N., Malchau, H., & Herberts, P. (2002). Evaluation of a femoral stem with reduced stiffness: A randomized study with use of radiostereometry and bone densitometry. *JBJS, 84*(9), 1651–1658. doi:10.2106/00004623-200209000-00020 PMID:12208924

Katkat, D., Bulut, Y., Demir, M., & Akar, S. (2009). Effects of different sports surfaces on muscle performance. *Biology of Sport, 26*(3), 285–296. doi:10.5604/20831862.894793

Keegan, K. (2011). Gait analysis. *Clinical Veterinary Advisor: The Horse*, (February), 736–740.

Keith, D. A., Paz, A., Gallop, P. M., & Glimcher, M. J. (1977). *Histologic and biochemical identification and characterization of an elastin in cartilage*. Academic Press.

Ker, R. F. (1999). The design of soft collagenous load-bearing tissues. *The Journal of Experimental Biology, 202*(23), 3315–3324. doi:10.1242/jeb.202.23.3315 PMID:10562514

Khalifa, A. M. A., & Giddens, D. P. (1981). Characterization and evolution of poststenotic flow disturbances. *Journal of Biomechanics, 14*(5), 279–296. doi:10.1016/0021-9290(81)90038-5 PMID:7263720

Kharb, A., Saini, V., Jain, Y., & Dhiman, S. (2011). A review of gait cycle and its parameters. *International Journal of Computational Engineering & Management, 13*, 78–83.

Kim, C. H., Chung, D. Y., Rha, K. H., Lee, J. Y., & Lee, S. H. (2020). Effectiveness of Percutaneous Nephrolithotomy, Retrograde Intrarenal Surgery, and Extracorporeal Shock Wave Lithotripsy for Treatment of Renal Stones: A Systematic Review and Meta-Analysis. *Medicina (Kaunas, Lithuania), 57*(1), 26. doi:10.3390/medicina57010026 PMID:33396839

Kim, J. H., Kim, S. H., & Chang, S. H. (2011). Estimation of the movement of the inter-fragmentary gap of a fractured human femur in the presence of a composite bone plate. *Journal of Composite Materials, 45*(14), 1491–1498. doi:10.1177/0021998310383730

Kim, K. W., Choi, Y. H., Lee, S. B., Baba, Y., Kim, H. H., & Suh, S. H. (2017). Analysis of Urine Flow in Three Different Ureter Models. *Computational and Mathematical Methods in Medicine, 5172641*, 1–11. Advance online publication. doi:10.1155/2017/5172641 PMID:28659992

King, C. E., Toskes, P. P., & Watkins, J. B. (1984). Breath tests in the diagnosis of small intestine bacterial overgrowth. *CRC Critical Reviews in Clinical Laboratory Sciences, 21*(3), 269–281. doi:10.3109/10408368409165785 PMID:6439469

Kinn, A. C., & Lykkeskov-Andersen, H. (2002). Impact on ureteral peristalsis in a stented ureter. An experimental study in the pig. *Urological Research, 30*(4), 213–218. doi:10.100700240-002-0258-1 PMID:12202937

Kirac, M., Bozkurt, Ö. F., Tunc, L., Guneri, C., Unsal, A., & Biri, H. (2013). *Comparison of retrograde intrarenal surgery and mini-percutaneous nephrolithotomy in management of lower-pole renal stones with a diameter of smaller than 15 mm*. Academic Press.

Klaschik, E., Nauck, F., & Ostgathe, C. (2003). Constipation—Modern laxative therapy. *Supportive Care in Cancer, 11*(11), 679–685. doi:10.100700520-003-0525-x PMID:14505158

Knezevic, V., Sim, A. J., Borg, T. K., & Holmes, J. W. (2002). Isotonic biaxial loading of fibroblast-populated collagen gels: A versatile, low-cost system for the study of mechanobiology. *Biomechanics and Modeling in Mechanobiology, 1*(1), 59–67. doi:10.100710237-002-0005-0 PMID:14586707

Knudson, D. (2007). Qualitative biomechanical principles for application in coaching. *Sports Biomechanics, 6*(1), 109–118. doi:10.1080/14763140601062567 PMID:17542182

Koch, T. M., Reddy, B. D., Zilla, P., & Franz, T. (2010). Aortic valve leaflet mechanical properties facilitate diastolic valve function. *Computer Methods in Biomechanics and Biomedical Engineering, 13*(2), 225–234. doi:10.1080/10255840903120160 PMID:19657802

Konstantinidis, L., Hauschild, O., Beckmann, N. A., Hirschmüller, A., Südkamp, N. P., & Helwig, P. (2010). Treatment of periprosthetic femoral fractures with two different minimal invasive angle-stable plates: Biomechanical comparison studies on cadaveric bones. *Injury, 41*(12), 1256–1261. doi:10.1016/j.injury.2010.05.007 PMID:21288467

Kopperdahl, D. L., & Keaveny, T. M. (1998). Yield strain behavior of trabecular bone. *Journal of Biomechanics, 31*(7), 601–608. doi:10.1016/S0021-9290(98)00057-8 PMID:9796682

Kopperdahl, D. L., Morgan, E. F., & Keaveny, T. M. (2002). Quantitative computed tomography estimates of the mechanical properties of human vertebral trabecular bone. *Journal of Orthopaedic Research*, *20*(4), 801–805. doi:10.1016/S0736-0266(01)00185-1 PMID:12168670

Korossis, S. (2018). Structure-Function Relationship of Heart Valves in Health and Disease. In *Structural Insufficiency Anomalies in Cardiac Valves* (pp. 1–38). InTech. doi:10.5772/intechopen.78280

Kraaij, G., Zadpoor, A.A., Tuijthof, G.J., Dankelman, J., Nelissen, R.G., & Valstar, E.R. (2014). Mechanical properties of human bone–implant interface tissue in aseptically loose hip implants. *Journal of the Mechanical Behavior of Biomedical Materials, 38*, 59-68.

Ku, D. N. (1997). Blood flow in arteries. *Annual Review of Fluid Mechanics*, *29*(1), 399–434. doi:10.1146/annurev.fluid.29.1.399

Ku, D. N., Giddens, D. P., Zarins, C. K., & Glagoy, S. (1985). Pulsatile flow and atherosclerosis in the human carotid bifurcation—Positive correlation between plaque location and low and oscillating shear– stress. *Arteriosclerosis (Dallas, Tex.)*, *5*(3), 293–302. doi:10.1161/01.ATV.5.3.293 PMID:3994585

Kumar, P., Bach, C., Kachrilas, S., Papatsoris, A. G., Buchholz, N., & Masood, J. (2012). Supine percutaneous nephrolithotomy (PCNL): 'in vogue' but in which position? *BJU International*, *110*(11c, 11 Pt C), E1018–E1021. doi:10.1111/j.1464-410X.2012.11188.x PMID:22564784

Lacroix, D., Chateau, A., Ginebra, M. P., & Planell, J. A. (2006). Micro-finite element models of bone tissue-engineering scaffolds. *Biomaterials*, *27*(30), 5326–5334. doi:10.1016/j.biomaterials.2006.06.009 PMID:16824593

Lahme, S., Bichler, K. H., Strohmaier, W. L., & Götz, T. (2001). Minimally invasive PCNL in patients with renal pelvic and calyceal stones. *European Urology*, *40*(6), 619–624. doi:10.1159/000049847 PMID:11805407

Langton, D., Jameson, S., Joyce, T., Gandhi, J., Sidaginamale, R., Mereddy, P., Lord, J., & Nargol, A. (2011). Accelerating failure rate of the ASR total hip replacement. *The Journal of Bone and Joint Surgery*, *93*(8), 1011–1016. doi:10.1302/0301-620X.93B8.26040 PMID:21768621

Latif, N., Quillon, A., Sarathchandra, P., McCormack, A., Lozanoski, A., Yacoub, M. H., & Chester, A. H. (2015). Modulation of Human Valve Interstitial Cell Phenotype and Function Using a Fibroblast Growth Factor 2 Formulation. *PLoS One*, *10*(6), e0127844. doi:10.1371/journal.pone.0127844 PMID:26042674

Leary, M. (2019). *Design for Additive Manufacturing* (1st ed.). Elsevier.

Lee, J. M., Courtman, D. W., & Boughner, D. R. (1984). The glutaraldehyde-stabilized porcine aortic valve xenograft. I. Tensile viscoelastic properties of the fresh leaflet material. *Journal of Biomedical Materials Research*, *18*(1), 61–77. doi:10.1002/jbm.820180108 PMID:6699033

Lee, J. S., & Fung, Y. C. (1970). Flow in locally constricted tubes at low Reynolds number. *Journal of Applied Mechanics*, *37*(1), 9–16. doi:10.1115/1.3408496

Lee, K. W., & Xu, X. Y. (2002). Modelling of flow and wall behaviour in a mildly stenosed tube. *Medical Engineering & Physics*, *24*(9), 575–586. doi:10.1016/S1350-4533(02)00048-6 PMID:12376044

Lehmann, W., Rupprecht, M., Hellmers, N., Sellenschloh, K., Briem, D., Püschel, K., Amling, M., Morlock, M., & Rueger, J. M. (2010). Biomechanical evaluation of peri- and interprosthetic fractures of the femur. *The Journal of Trauma Injury Infection and Critical Care*, *68*(6), 1459–1463. doi:10.1097/TA.0b013e3181bb8d89 PMID:20093986

Lenz, M., Perren, S. M., Gueorguiev, B., Höntzsch, D., & Windolf, M. (2013). Mechanical behavior of fixation components for periprosthetic fracture surgery. *Clinical Biomechanics (Bristol, Avon)*, *28*(9–10), 988–993. doi:10.1016/j.clinbiomech.2013.09.005 PMID:24080369

Lenz, M., Perren, S. M., Gueorguiev, B., Richards, R. G., Hofmann, G. O., Fernandez Dell'Oca, A., Höntzsch, D., & Windolf, M. (2014). A biomechanical study on proximal plate fixation techniques in periprosthetic femur fractures. *Injury*, *45*(Suppl. 1), S71–S75. Advance online publication. doi:10.1016/j.injury.2013.10.027 PMID:24252576

Leonhäuser, D., Kranz, J., Leidolf, R., Arndt, P., Schwantes, U., Geyer, J., & Grosse, J. O. (2019). Expression of components of the urothelial cholinergic system in bladder and cultivated primary urothelial cells of the pig. *BMC Urology*, *19*(1), 62. doi:10.118612894-019-0495-z PMID:31288793

Leonidou, A., Moazen, M., Lepetsos, P., Graham, S. M., Macheras, G. A., & Tsiridis, E. (2015). The biomechanical effect of bone quality and fracture topography on locking plate fixation in periprosthetic femoral fractures. *Injury*, *46*(2), 213–217. doi:10.1016/j.injury.2014.10.060 PMID:25467710

Lerman, D. A., Prasad, S., & Alotti, N. (2015). Calcific Aortic Valve Disease: Molecular Mechanisms And Therapeutic Approaches. *European Cardiology Review*, *10*(2), 108. doi:10.15420/ecr.2015.10.2.108 PMID:27274771

Le, T. T., Andreadakis, Z., Kumar, A., Román, R. G., Tollefsen, S., Saville, M., & Mayhew, S. (2020). The COVID-19 vaccine development landscape. *Nature Reviews. Drug Discovery*, *19*(5), 305–306. doi:10.1038/d41573-020-00073-5 PMID:32273591

Levin, R. J. (1994). Digestion and absorption of carbohydrates—From molecules and membranes to humans. *The American Journal of Clinical Nutrition*, *59*(3), 690S–698S. doi:10.1093/ajcn/59.3.690S PMID:8116552

Li, Q., Ichim, I., Loughran, J., Li, W., Swain, M., & Kieser, J. (n.d.). Numerical simulation of crack formation in all ceramic dental bridge. Key Engineering Materials.

Liao, J., Joyce, E. M., & Sacks, M. S. (2008). Effects of decellularization on the mechanical and structural properties of the porcine aortic valve leaflet. *Biomaterials*, *29*(8), 1065–1074. doi:10.1016/j.biomaterials.2007.11.007 PMID:18096223

Li, B., Zhu, Q., Guo, S., Yang, F., Li, Y., Zhu, Z., Chen, S., Song, R., & Li, Y. (2020). Design and experiment of a bionic flapping wing mechanism with flapping–twist–swing motion based on a single rotation. *AIP Advances*, *10*(6), 065018. doi:10.1063/5.0008792

Li, D., Zhao, S., Da Ronch, A., Xiang, J., Drofelnik, J., Li, Y., Zhang, L., Wu, Y., Kintscher, M., Monner, H. P., Rudenko, A., Guo, S., Yin, W., Kirn, J., Storm, S., & De Breuker, R. (2018). A review of modelling and analysis of morphing wings. *Progress in Aerospace Sciences*, *100*, 46–62. doi:10.1016/j.paerosci.2018.06.002

Li, J., Luo, X. Y., & Kuang, Z. B. (2001). A nonlinear anisotropic model for porcine aortic heart valves. *Journal of Biomechanics*, *34*(10), 1279–1289. doi:10.1016/S0021-9290(01)00092-6 PMID:11522307

Lin, C.-L., Wang, J.-C., Ramp, L. C., & Liu, P.-R. (2008). Biomechanical response of implant systems placed in the maxillary posterior region under various conditions of angulation, bone density, and loading. *The International Journal of Oral & Maxillofacial Implants*, 23. PMID:18416413

Lin, D., Li, Q., Li, W., & Swain, M. (2009). Dental implant induced bone remodeling and associated algorithms. *Journal of the Mechanical Behavior of Biomedical Materials*, *2*(5), 410–432. doi:10.1016/j.jmbbm.2008.11.007 PMID:19627848

Lindman, B. R., Clavel, M.-A., Mathieu, P., Iung, B., Lancellotti, P., Otto, C. M., & Pibarot, P. (2016). Calcific aortic stenosis. *Nature Reviews. Disease Primers*, *2*(1), 16006. doi:10.1038/nrdp.2016.6 PMID:27188578

Lin-Lin, Z., Hui, G., & Chui-Jie, W. (2016). Three-dimensional numerical simulation of a bird model in unsteady flight. *Computational Mechanics*, *58*(1), 1-11.

Lin, T. W., Cardenas, L., & Soslowsky, L. J. (2004). Biomechanics of tendon injury and repair. *Journal of Biomechanics*, *37*(6), 865–877. doi:10.1016/j.jbiomech.2003.11.005 PMID:15111074

Lipsky, M. J., Shapiro, E. Y., Cha, D. Y., & Gupta, M. (2013). Modified-PCNL without modified instruments: A description of technique. *Journal of Endourology*, *27*(6), 684–687. doi:10.1089/end.2012.0604 PMID:23268559

Li, Q., Wan, L., Liu, S., Li, M., Chen, L., Hou, Z., & Zhang, W. (2020). Clinical efficacy of enhanced recovery after surgery in percutaneous nephrolithotripsy: A randomized controlled trial. *BMC Urology*, *20*(1), 162. doi:10.118612894-020-00728-w PMID:33081762

Liu, G-T., Wang, X-J., Ai, B-Q., & Liu, L-G. (2004). Numerical study of pulsating flow through a tapered artery with stenosis. *Chinese Journal of Physics*, *42*(4-1), 401-409.

Liu, A. C., Joag, V. R., & Gotlieb, A. I. (2007). The Emerging Role of Valve Interstitial Cell Phenotypes in Regulating Heart Valve Pathobiology. *American Journal of Pathology*, *171*(5), 1407–1418. doi:10.2353/ajpath.2007.070251 PMID:17823281

Liu, H., He, S., Shen, L., & Hong, J. (2021). Simulation-based study of COVID-19 outbreak associated with air-conditioning in a restaurant. *Physics of Fluids*, *33*(2), 023301. doi:10.1063/5.0040188 PMID:33746488

Li, W., Swain, M. V., Li, Q., & Steven, G. P. (2005). Towards automated 3D finite element modeling of direct fiber reinforced composite dental bridge. *Journal of Biomedical Materials Research. Part B, Applied Biomaterials*, *74*(1), 520–528. doi:10.1002/jbm.b.30233 PMID:15912531

Li, W., Swain, M., Li, Q., Ironside, J., & Steven, G. (2004a). Fibre reinforced composite dental bridge. Part II: Numerical investigation. *Biomaterials*, *25*(20), 4995–5001. doi:10.1016/j.biomaterials.2004.01.011 PMID:15109861

Li, W., Swain, M., Li, Q., Ironside, J., & Steven, G. (2004b). Fibre reinforced composite dental bridge.: Part I: experimental investigation. *Biomaterials*, *25*(20), 4987–4993. doi:10.1016/j.biomaterials.2004.01.010 PMID:15109860

Li, X., Long, Q., Chen, X., He, D., & He, H. (2014). Real-time ultrasound-guided PCNL using a novel SonixGPS needle tracking system. *Urolithiasis*, *42*(4), 341–346. doi:10.100700240-014-0671-2 PMID:24965272

Lo, C. M., Wang, H. B., Dembo, M., & Wang, Y. L. (2000). Cell movement is guided by the rigidity of the substrate. *Biophysical Journal*, *79*(1), 144–152. doi:10.1016/S0006-3495(00)76279-5 PMID:10866943

Lodish, H., Berk, A., Kaiser, C. A., Kaiser, C., Krieger, M., Scott, M. P., & Matsudaira, P. (2008). *Molecular cell biology*. Macmillan.

Longo, U. G., Ronga, M., & Maffulli, N. (2009). Acute ruptures of the Achilles tendon. *Sports Medicine and Arthroscopy Review*, *17*(2), 127–138. doi:10.1097/JSA.0b013e3181a3d767 PMID:19440140

Long, Q., Xu, X. Y., Ramnarine, K. V., & Hoskins, P. (2001). Numerical investigation of physiologically realistic pulsatile flow through arterial stenosis. *Journal of Biomechanics*, *34*(10), 1229–1242. doi:10.1016/S0021-9290(01)00100-2 PMID:11522303

Lotfi, M., Hamblin, M. R., & Rezaei, N. (2020). COVID-19: Transmission, prevention, and potential therapeutic opportunities. *Clinica Chimica Acta*, *508*, 254–266. doi:10.1016/j.cca.2020.05.044 PMID:32474009

Lovekamp, J. J., Simionescu, D. T., Mercuri, J. J., Zubiate, B., Sacks, M. S., & Vyavahare, N. R. (2006). Stability and function of glycosaminoglycans in porcine bioprosthetic heart valves. *Biomaterials, 27*(8), 1507–1518. doi:10.1016/j.biomaterials.2005.08.003 PMID:16144707

Lycke, R., Walls, M., & Calve, S. (2019). Computational Modeling of Developing Cartilage Using Experimentally Derived Geometries and Compressive Moduli. *Journal of Biomechanical Engineering, 141*(8), 0810021–0810028. Advance online publication. doi:10.1115/1.4043208 PMID:30874718

MacDonald, D. A. (1979). On steady flow through modelled vascular stenosis. *Journal of Biomechanics, 12*(1), 13–20. doi:10.1016/0021-9290(79)90004-6 PMID:762177

Majumder, S., Roychowdhury, A., & Pal, S. (2007). Simulation of hip fracture in sideways fall using a 3D finite element model of pelvis–femur–soft tissue complex with simplified representation of whole body. *Medical Engineering & Physics, 29*(10), 1167–1178. doi:10.1016/j.medengphy.2006.11.001 PMID:17270483

Malandrino, A. (2018). Intervertebral disc. In *Biomechanics of the Spine* (pp. 89–103). Academic Press.

Mamun, K., Ali, M., & Akhter, M. N. (2016). Physiological non-Newtonian blood flow through single stenosed artery. *Theoretical and Applied Mechanics, 43*(1), 99–115. doi:10.2298/TAM160322006M

Mandal, D. K., & Chakrabarti, S. (2007a). Effect of Restriction and Reynolds Number on the Pressure of Blood of a Stenotic Artery. *International Journal of Fluid Mechanics Research, 34*(2), 159–178. doi:10.1615/InterJFluidMechRes.v34.i2.50

Mandal, D. K., & Chakrabarti, S. (2007b). Two Dimensional Simulation of Steady Blood Flow through a Stenosed Coronary Artery. *International Journal of Dynamics of Fluids, 3*(2), 187–209.

Mandal, D. K., & Chakrabarti, S. (2007c). Study of Pressure Drop and Flow Characteristics across Rectangular Stenotic Models. *International Journal of Fluid Mechanics Research, 34*(5), 434–461. doi:10.1615/InterJFluidMechRes.v34.i5.40

Mandal, D. K., & Chakrabarti, S. (2008). Effect of Stricture Length on Reattachment Point and Wall Shear Stress through a Stenosed Coronary Artery. *International Journal of Fluid Mechanics Research, 35*(2), 188–202. doi:10.1615/InterJFluidMechRes.v35.i2.70

Mandal, D. K., & Chakrabarti, S. (2009). Effect of Stenosis Length on Flow Characteristics across Rectangular Stenotic Models. *International Journal of Fluid Mechanics, 1*(1), 29–39.

Mandal, D. K., & Chakrabarti, S. (2010). Study on the effect of Different Shaped Stenoses on Blood flow through Coronary Artery. *International Journal of Biomedical Engineering and Technology, 4*(1), 1–17. doi:10.1504/IJBET.2010.034274

Mandal, D. K., Manna, N. K., Bandyopadhyay, S., Biswas, B. P., & Chakrabarti, S. (2011b). A Numerical Study on the Performance of a Sudden Expansion with Multisteps as a Diffuser. *International Journal of Applied Mechanics, 3*(4), 779–802. doi:10.1142/S1758825111001238

Mandal, D. K., Manna, N. K., & Chakrabarti, S. (2010a). Numerical Study of Blood Flow Through Different Double Bell Shaped Stenosed Coronary Artery During the Progression of the Disease, Atherosclerosis. *International Journal of Numerical Methods for Heat & Fluid Flow, 20*(6), 670–698. doi:10.1108/09615531011056827

Mandal, D. K., Manna, N. K., & Chakrabarti, S. (2010b). A numerical model study of steady flow through bell-shaped stenoses with and without asymmetry. *International Journal of Experimental and Computational Biomechanics, 1*(3), 306–331. doi:10.1504/IJECB.2010.035263

Mandal, D. K., Manna, N. K., & Chakrabarti, S. (2011a). Influence of Primary Stenosis on Secondary One and Vice Versa in case of Double Stenoses. *Journal of Applied Fluid Mechanics, 4*(4), 31–42.

Manea, A., Bran, S., Dinu, C., Rotaru, H., Barbur, I., Crisan, B., Armencea, G., Onisor, F., Lazar, M., Ostas, D., Baciut, M., Vacaras, S., Mitre, I., Crisan, L., Muresan, O., Roman, R., & Baciut, G. (2019). Principles of biomechanics in oral implantology. *Medicine and Pharmacy Reports, 92*(Suppl No 3), S14–S19. doi:10.15386/mpr-1512 PMID:31989104

Manickam, P. S., Ghosh, G., & Roy, S. (2022). Optimization of Bone Graft Shapes of S-Type Cervical Cage through Genetic Algorithm. *International Journal for Multiscale Computational Engineering, 20*(1), 55–68. doi:10.1615/IntJ-MultCompEng.2021039717

Manickam, P. S., & Roy, S. (2021). The biomechanical effects of S-type dynamic cage using Ti and PEEK for ACDF surgery on cervical spine varying loads. *The International Journal of Artificial Organs, 44*(10), 748–755. doi:10.1177/03913988211039525 PMID:34387526

Manickam, P. S., & Roy, S. (2021). The biomechanical study of cervical spine: A Finite Element Analysis. *The International Journal of Artificial Organs*. Advance online publication. doi:10.1177/0391398821995495 PMID:33645324

Manickam, P. S., Roy, S., & Shetty, G. M. (2021). Biomechanical evaluation of a novel S-type, dynamic zero-profile cage design for anterior cervical discectomy and fusion with variations in bone graft shape: A finite element analysis. *World Neurosurgery, 154*, e199–e214. doi:10.1016/j.wneu.2021.07.013 PMID:34246827

Marey, E. J. (1890). *Physiologie du mouvement: vol des oiseaux, 207. G.* Masson. doi:10.5962/bhl.title.115247

Martin, C., & Sun, W. (2012). Biomechanical characterization of aortic valve tissue in humans and common animal models. *Journal of Biomedical Materials Research. Part A, 100A*(6), 1591–1599. doi:10.1002/jbm.a.34099 PMID:22447518

Martin, R. B. (1972). The effects of geometric feedback in the development of osteoporosis. *Journal of Biomechanics, 5*(5), 447–455. doi:10.1016/0021-9290(72)90003-6 PMID:4667271

Martin, R. B. (1983). Porosity and specific surface of bone. *Critical Reviews in Biomedical Engineering, 10*, 179–222. PMID:6368124

Massoud, S. N., Hunter, J. B., Holdsworth, B. J., Wallace, W. A., & Juliusson, R. (1997). Early femoral loosening in one design of cemented hip replacement. *The Journal of Bone and Joint Surgery. British Volume, 79*(4), 603–608. doi:10.1302/0301-620X.79B4.0790603 PMID:9250746

Matloubieh, J. E., Eghbali, M., & Abraham, N. (2020). Strategies to Encourage Medical Student Interest in Urology. *Current Urology Reports, 21*(10), 34. doi:10.100711934-020-00984-1 PMID:32767185

Matsumoto, Y., Adams, V., Jacob, S., Mangner, N., Schuler, G., & Linke, A. (2010). Regular Exercise Training Prevents Aortic Valve Disease in Low-Density Lipoprotein–Receptor–Deficient Mice. *Circulation, 121*(6), 759–767. doi:10.1161/CIRCULATIONAHA.109.892224 PMID:20124122

Mattheck, C., Vorberg, U., & Kranz, C. (1990). Effects of hollow shaft endoprosthesis on stress distribution in cortical bone. *Biomedizinische Technik. Biomedical Engineering, 35*, 316–319. doi:10.1515/bmte.1990.35.12.316 PMID:2078647

Mauck, R. L., Baker, B. M., Nerurkar, N. L., Burdick, J. A., Li, W. J., Tuan, R. S., & Elliott, D. M. (2009). Engineering on the straight and narrow: The mechanics of nanofibrous assemblies for fiber-reinforced tissue regeneration. *Tissue Engineering. Part B, Reviews, 15*(2), 171–193. doi:10.1089/ten.teb.2008.0652 PMID:19207040

Maurer, G. (2006). Aortic regurgitation. *Heart (British Cardiac Society), 92*(7), 994–1000. doi:10.1136/hrt.2004.042614 PMID:16775114

Ma, X., Schickel, M. E., Stevenson, M. D., Sarang-Sieminski, A. L., Gooch, K. J., Ghadiali, S. N., & Hart, R. T. (2013). Fibers in the extracellular matrix enable long-range stress transmission between cells. *Biophysical Journal, 104*(7), 1410–1418. doi:10.1016/j.bpj.2013.02.017 PMID:23561517

McDonald, D. A. (1974). *Blood Flow in Arteries*. Camelot.

McIntosh, S. H., Agrawal, S. K., & Khan, Z. (2006). Design of a mechanism for biaxial rotation of a wing for a hovering vehicle. *IEEE/ASME Transactions on Mechatronics*, *11*(2), 145–153. doi:10.1109/TMECH.2006.871089

Mears, A. C., Osei-Owusu, P., Harland, A. R., Owen, A., & Roberts, J. R. (2018). Perceived links between playing surfaces and injury: A worldwide study of elite association football players. *Sports Medicine - Open*, *4*(1), 40. doi:10.118640798-018-0155-y PMID:30128862

Mech, D. J., & Rizvi, M. S. (2021). *In-silico modeling of the micromechanics of fibrous scaffolds and stiffness sensing by cells.* arXiv preprint arXiv:2112.06064.

Menzel, H.-J. (1986). Biomechanics of javelin throwing. *Biomechanics of Sports*, *1*(3), 85–98.

Mercer, R. R., & Crapo, J. D. (1990). Spatial distribution of collagen and elastin fibers in the lungs. *Journal of Applied Physiology*, *69*(2), 756–765. doi:10.1152/jappl.1990.69.2.756 PMID:2228886

Merryman, W. D., Bieniek, P. D., Guilak, F., & Sacks, M. S. (2009). Viscoelastic Properties of the Aortic Valve Interstitial Cell. *Journal of Biomechanical Engineering*, *131*(4), 1–7. doi:10.1115/1.3049821 PMID:19275434

Merryman, W. D., Shadow Huang, H. Y., Schoen, F. J., & Sacks, M. S. (2006). The effects of cellular contraction on aortic valve leaflet flexural stiffness. *Journal of Biomechanics*, *39*(1), 88–96. doi:10.1016/j.jbiomech.2004.11.008 PMID:16271591

Miller, M., & Thompson, S. (2014). Anatomy and Biomechanics of the Knee. *Operative Techniques in Sports Medicine*, *11*(3), 172–186.

Mills, C. J., Gabe, I. T., Gault, J. H., Mason, D. T., Ross, J., Braunwald, E., & Shillingford, J. P. (1970). Pressure-flow relationships and vascular impedance in man. *Cardiovascular Research*, *4*(4), 405–417. doi:10.1093/cvr/4.4.405 PMID:5533085

Mills, R., Hildenbrandt, H., Taylor, G. K., & Hemelrijk, C. K. (2018). Physics-based simulations of aerial attacks by peregrine falcons reveal that stooping at high speed maximizes catch success against agile prey. *PLoS Computational Biology*, *14*(4), e1006044. doi:10.1371/journal.pcbi.1006044 PMID:29649207

Minns, R. J., & Steven, F. S. (1977). The collagen fibril organization in human articular cartilage. *Journal of Anatomy*, *123*(Pt 2), 437. PMID:870478

Mirnajafi, A., Raymer, J. M., McClure, L. R., & Sacks, M. S. (2006). The flexural rigidity of the aortic valve leaflet in the commissural region. *Journal of Biomechanics*, *39*(16), 2966–2973. doi:10.1016/j.jbiomech.2005.10.026 PMID:16360160

Mirzendehdel, A. M., & Suresh, K. (2017). *A Hands-on Introduction to Topology Optimization* (1st ed.). Createspace Independent Pub.

Misfeld, M., & Sievers, H.-H. (2007). Heart valve macro- and microstructure. *Philosophical Transactions of the Royal Society of London. Series B, Biological Sciences*, *362*(1484), 1421–1436. doi:10.1098/rstb.2007.2125 PMID:17581807

Mishra, E., Jena, S., Bhoi, C., Arunachalam, T., & Panda, S. K. (2019). Effect of high heel gait on hip and knee-ankle-foot rollover characteristics while walking over inclined surfaces— A pilot study. *The Foot*, *40*, 8–13. doi:10.1016/j.foot.2019.03.004 PMID:30981083

Misra, J. C., & Shit, G. C. (2006). Blood flow through arteries in a pathological state: A theoretical study. *International Journal of Engineering Science*, *44*(10), 662–671. doi:10.1016/j.ijengsci.2005.12.011

Mithieux, S. M., & Weiss, A. S. (2005). Elastin. *Advances in Protein Chemistry, 70*, 437–461. doi:10.1016/S0065-3233(05)70013-9 PMID:15837523

Mithoefer, K., McAdams, T. R., Scopp, J. M., & Mandelbaum, B. R. (2009). Emerging options for treatment of articular cartilage injury in the athlete. *Clinics in Sports Medicine, 28*(1), 25–40. doi:10.1016/j.csm.2008.09.001 PMID:19064163

Mittal, R., Ni, R., & Seo, J. H. (2020). The flow physics of COVID-19. *Journal of Fluid Mechanics*, 894.

Moazen, M., Mak, J. H., Etchels, L. W., Jin, Z., Wilcox, R. K., Jones, A. C., & Tsiridis, E. (2013). The effect of fracture stability on the performance of locking plate fixation in periprosthetic femoral fractures. *The Journal of Arthroplasty, 28*(9), 1589–1595. doi:10.1016/j.arth.2013.03.022 PMID:23642449

Modi, P. K., Kaufman, S. R., Caram, M. V., Ellimoottil, C., Shahinian, V. B., & Hollenbeck, B. K. (2019). Impact of Medicare Office Visit Payment Reform on Urologic Practices. *Urology, 126*, 83–88. doi:10.1016/j.urology.2019.01.013 PMID:30682462

Morio, C., Lake, M. J., Gueguen, N., Rao, G., & Baly, L. (2009). The influence of footwear on foot motion during walking and running. *Journal of Biomechanics, 42*(13), 2081–2088. doi:10.1016/j.jbiomech.2009.06.015 PMID:19643421

Mostafaei, A., Elliott, A. M., Barnes, J. E., Li, F., Tan, W., Cramer, C. L., Nandwana, P., & Chmielus, M. (2021). Binder jet 3D printing—Process parameters, materials, properties, modeling, and challenges. *Progress in Materials Science, 119*, 100707.

Muheim, R. (2011). Behavioural and physiological mechanisms of polarized light sensitivity in birds. *Philosophical Transactions of the Royal Society of London. Series B, Biological Sciences, 366*(1565), 763–771. doi:10.1098/rstb.2010.0196 PMID:21282180

Muir, H., Bullough, P., & Maroudas, A. (1970). The distribution of collagen in human articular cartilage with some of its physiological implications. *The Journal of Bone and Joint Surgery. British Volume, 52*(3), 554–563. doi:10.1302/0301-620X.52B3.554 PMID:4247851

Muiznieks, L. D., & Keeley, F. W. (2013). Molecular assembly and mechanical properties of the extracellular matrix: A fibrous protein perspective. Biochimica et Biophysica Acta (BBA)-. *Molecular Basis of Disease, 1832*(7), 866–875. doi:10.1016/j.bbadis.2012.11.022 PMID:23220448

Mukherjee, K., & Gupta, S. (2016). The effects of musculoskeletal loading regimes on numerical evaluations of acetabular component. *Proceedings of the Institution of Mechanical Engineers. Part H, Journal of Engineering in Medicine, 230*(10), 918–929. doi:10.1177/0954411916661368 PMID:27475907

Murakumo, M., Ushiki, T., Abe, K., Matsumura, K., Shinno, Y., & Koyanagi, T. (1995). Three-dimensional arrangement of collagen and elastin fibers in the human urinary bladder: A scanning electron microscopic study. *The Journal of Urology, 154*(1), 251–256. doi:10.1016/S0022-5347(01)67289-6 PMID:7776441

Muratoglu, O. (2001). A novel method of crosslinking UHMWPE to improve wear, reduce oxidation and retain mechanical properties. *The Journal of Bone and Joint Surgery. British Volume, 83*, 447.

Muratoglu, O. K., Bragdon, C. R., O'Connor, D. O., Jasty, M., Harris, W. H., Gul, R., & McGarry, F. (1999). Unified wear model for highly crosslinked ultra-high molecular weight polyethylenes (UHMWPE). *Biomaterials, 20*(16), 1463–1470. doi:10.1016/S0142-9612(99)00039-3 PMID:10458559

Muratoglu, O. K., Bragdon, C. R., O'Connor, D., Perinchief, R. S., Estok, D. M. II, Jasty, M., & Harris, W. H. (2001). Larger diameter femoral heads used in conjunction with a highly cross-linked ultra–high molecular weight polyethylene: A new concept. *The Journal of Arthroplasty, 16*(8), 24–30. doi:10.1054/arth.2001.28376 PMID:11742447

Muro-de-la-Herran, A., García-Zapirain, B., & Méndez-Zorrilla, A. (2014). Gait analysis methods: An overview of wearable and non-wearable systems, highlighting clinical applications. *Sensors (Switzerland)*, *14*(2), 3362–3394. doi:10.3390140203362 PMID:24556672

Murrell, G. A. C., Maddali, S., Horovitz, L., Oakley, S. P., & Warren, R. F. (2001). The effects of time course after anterior cruciate ligament injury in correlation with meniscal and cartilage loss. *The American Journal of Sports Medicine*, *29*(1), 9–14. doi:10.1177/03635465010290012001 PMID:11206263

Mutuku, J. K., Hou, W. C., & Chen, W. H. (2020). An overview of experiments and numerical simulations on airflow and aerosols deposition in human airways and the role of bioaerosol motion in COVID-19 transmission. *Aerosol and Air Quality Research*, *20*(6), 1172–1196. doi:10.4209/aaqr.2020.04.0185

Nabawy, M. R., & Crowther, W. J. (2017). The role of the leading edge vortex in lift augmentation of steadily revolving wings: A change in perspective. *Journal of the Royal Society, Interface*, *14*(132), 159. doi:10.1098/rsif.2017.0159 PMID:28747395

Nan, Y., Peng, B., Chen, Y., & McGlinchey, D. (2019). From studying real hummingbirds to designing hummingbird-like robots—A literature review. *IEEE Access: Practical Innovations, Open Solutions*, *7*, 131785–131804. doi:10.1109/ACCESS.2019.2939491

Natali, A.N., Carniel, E.L., & Pavan, P.G. (2010). Modelling of mandible bone properties in the numerical analysis of oral implant biomechanics. *Computer Methods and Programs in Biomedicine, 100*, 158-165.

Nath, S., Bodhak, S., & Basu, B. (2007). Tribological investigation of novel HDPE-HAp-Al2O3 hybrid biocomposites against steel under dry and simulated body fluid condition. *Journal of Biomedical Materials Research. Part A*, *83*(1), 191–208. doi:10.1002/jbm.a.31203 PMID:17397040

Nath, S., Bodhak, S., & Basu, B. (2009). HDPE-Al2O3-HAp composites for biomedical applications: Processing and characterizations. *Journal of Biomedical Materials Research. Part B, Applied Biomaterials*, *88*(1), 1–11. doi:10.1002/jbm.b.31050 PMID:18338785

Ndaïrou, F., Area, I., Nieto, J. J., & Torres, D. F. (2020). Mathematical modeling of COVID-19 transmission dynamics with a case study of Wuhan. *Chaos, Solitons, and Fractals*, *135*, 109846. doi:10.1016/j.chaos.2020.109846 PMID:32341628

Nerurkar, N. L., Elliott, D. M., & Mauck, R. L. (2007). Mechanics of oriented electrospun nanofibrous scaffolds for annulus fibrosus tissue engineering. *Journal of Orthopaedic Research*, *25*(8), 1018–1028. doi:10.1002/jor.20384 PMID:17457824

Newell, N., Little, J. P., Christou, A., Adams, M. A., Adam, C. J., & Masouros, S. D. (2017). Biomechanics of the human intervertebral disc: A review of testing techniques and results. *Journal of the Mechanical Behavior of Biomedical Materials*, *69*, 420–434. doi:10.1016/j.jmbbm.2017.01.037 PMID:28262607

Ng, M. R., Besser, A., Danuser, G., & Brugge, J. S. (2012). Substrate stiffness regulates cadherin-dependent collective migration through myosin-II contractility. *The Journal of Cell Biology*, *199*(3), 545–563. doi:10.1083/jcb.201207148 PMID:23091067

Nigg, B. M., Nurse, M. A., & Stefanyshyn, D. J. (1999). Shoe inserts and orthotics for sport and physical activities. *Medicine and Science in Sports and Exercise*, *31*(Supplement), 31. doi:10.1097/00005768-199907001-00003 PMID:10416543

Nikić, P., Durutović, O., Kajmaković, B., Nale, D., Bumbaširević, U., Radovanović, M., Milenković-Petronić, D., & Džamić, Z. (2014). Complications associated with percutaneous nephrolitholapaxy (PCNL)—Our experience and literature review. *Acta Chirurgica Iugoslavica*, *61*(1), 51–56. doi:10.2298/ACI1401051N PMID:25782226

Norman, T. L., Shultz, T., Noble, G., Gruen, T., & Blaha, J. (2013). Bone creep and short and long term subsidence after cemented stem total hip arthroplasty (THA). *Journal of Biomechanics*, *46*(5), 949–955. doi:10.1016/j.jbiomech.2012.12.010 PMID:23357700

Nourissat, G., Berenbaum, F., & Duprez, D. (2015). Tendon injury: From biology to tendon repair. *Nature Reviews. Rheumatology*, *11*(4), 223–233. doi:10.1038/nrrheum.2015.26 PMID:25734975

Nouri, Z., Norouzi, N., Norouzi, N., Ataei, E., & Azizi, S. (2021). Virologic microparticle fluid mechanics simulation: COVID-19 transmission inside an elevator space. *International Journal of Computational Materials Science and Engineering*, *10*(02), 2150007. doi:10.1142/S204768412150007X

Nurse, M. A., Hulliger, M., Wakeling, J. M., Nigg, B. M., & Stefanyshyn, D. J. (2005). Changing the texture of footwear can alter gait patterns. *Journal of Electromyography and Kinesiology*, *15*(5), 496–506. doi:10.1016/j.jelekin.2004.12.003 PMID:15935961

Nutt, J. G., Marsden, C. D., & Thompson, P. D. (1993). Human walking and higher-level gait disorders, particularly in the elderly. *Neurology*, *43*(2), 268–279. doi:10.1212/WNL.43.2.268 PMID:8437689

O'Hara, A. M., & Shanahan, F. (2006). The gut flora as a forgotten organ. *EMBO Reports*, *7*(7), 688–693. doi:10.1038j.embor.7400731 PMID:16819463

Ojha, M., Cobbold, C., Johnston, K. W., & Hummel, R. L. (1989). Pulsatile flow through constricted tubes: An experimental investigation using photochromic tracer methods. *Journal of Fluid Mechanics*, *203*, 173–197. doi:10.1017/S0022112089001424

Oliver, R., Wells, H., Traxer, O., Knoll, T., Aboumarzouk, O., Biyani, C. S., Somani, B. K., & Group, Y. A. U. (2018). Ureteric stents on extraction strings: A systematic review of literature. *Urolithiasis*, *46*(2), 129–136. doi:10.100700240-016-0898-1 PMID:27324264

Ostrom, J. H. (1979). Bird Flight: How Did It Begin? Did birds begin to fly" from the trees down" or" from the ground up"? Reexamination of Archaeopteryx adds plausibility to an" up from the ground" origin of avian flight. *American Scientist*, *67*(1), 46–56. PMID:434589

Otto, C. M. (2002). Calcification of bicuspid aortic valves. *Heart (British Cardiac Society)*, *88*(4), 321–322. doi:10.1136/heart.88.4.321 PMID:12231576

Oxland, T. R. (2015). A history of spine biomechanics. *Der Unfallchirurg*, *118*(1), 80–92. doi:10.100700113-015-0087-7 PMID:26526281

Oxland, T. R. (2016). Fundamental biomechanics of the spine—What we have learned in the past 25 years and future directions. *Journal of Biomechanics*, *49*(6), 817–832. doi:10.1016/j.jbiomech.2015.10.035 PMID:26706717

Pal, A., & Brasseur, J. G. (2002). The mechanical advantage of local longitudinal shortening on peristaltic transport. *Journal of Biomechanical Engineering*, *124*(1), 94–100. doi:10.1115/1.1427700 PMID:11871611

Park, J. Y., Park, K., & Jeong, S. J. (2020). History of *Investigative and Clinical Urology* and an analysis of published articles. *Investigative and Clinical Urology*, *61*(Suppl 1), S64–S69. doi:10.4111/icu.2020.61.S1.S64 PMID:32055756

Patankar, S. V. (1980). *Numerical heat transfer and fluid flow*. Hemisphere Publication.

Pelit, E. S., Kati, B., Çanakci, C., Sağir, S., & Çiftçi, H. (2017). Outcomes of miniaturized percutaneous nephrolitotomy in infants: Single centre experience. *International Braz J Urol: Official Journal of the Brazilian Society of Urology*, *43*(5), 932–938.

Peng, S., Chen, Q., & Liu, E. (2020). The role of computational fluid dynamics tools on investigation of pathogen transmission: Prevention and control. *The Science of the Total Environment*, *746*, 142090. doi:10.1016/j.scitotenv.2020.142090 PMID:33027870

Perona, P. G., Lawrence, J., Paprosky, W. G., Patwardhan, A. G., & Sartori, M. (1992). Acetabular micromotion as a measure of initial implant stability in primary hip arthroplasty: An in vitro comparison of different methods of intial acetabular component fixation. *The Journal of Arthroplasty*, *7*(4), 537–547. doi:10.1016/S0883-5403(06)80076-8 PMID:1479374

Perry, J. (1992). *Gait analysis normal and pathological function*. SLACK Incorporated.

Perttunen, J., & Heinonen, A. (2000). Biomechanical loading in the triple jump. *Journal of Sports Sciences*, *18*(5), 363–370. doi:10.1080/026404100402421 PMID:10855682

Pettersson, G. B., Crucean, A. C., Savage, R., Halley, C. M., Grimm, R. A., Svensson, L. G., Naficy, S., Gillinov, A. M., Feng, J., & Blackstone, E. H. (2008). Toward Predictable Repair of Regurgitant Aortic Valves. A Systematic Morphology-Directed Approach to Bicommissural Repair. *Journal of the American College of Cardiology*, *52*(1), 40–49. doi:10.1016/j.jacc.2008.01.073 PMID:18582633

Pewowaruk, R., Rutkowski, D., Hernando, D., Kumapayi, B. B., Bushman, W., & Roldán-Alzate, A. (2020). A pilot study of bladder voiding with real-time MRI and computational fluid dynamics. *PLoS One*, *15*(11), e0238404. doi:10.1371/journal.pone.0238404 PMID:33211706

Phan, C. T., & Tso, P. (2001). Intestinal lipid absorption and transport. *Frontiers in Bioscience*, *6*(5), D299–D319.

Pietrabissa, R., Contro, R., Quaglini, V., Soncini, M., Gionso, L., & Simion, M. (2000). Experimental and computational approach for the evaluation of the biomechanical effects of dental bridge misfit. *Journal of Biomechanics*, *33*(11), 1489–1495. doi:10.1016/S0021-9290(00)00089-0 PMID:10940408

Porras, A. M., van Engeland, N. C. A., Marchbanks, E., McCormack, A., Bouten, C. V. C., Yacoub, M. H., Latif, N., & Masters, K. S. (2017). Robust Generation of Quiescent Porcine Valvular Interstitial Cell Cultures. *Journal of the American Heart Association*, *6*(3). Advance online publication. doi:10.1161/JAHA.116.005041 PMID:28292746

Prager-Khoutorsky, M., Lichtenstein, A., Krishnan, R., Rajendran, K., Mayo, A., Kam, Z., Geiger, B., & Bershadsky, A. D. (2011). Fibroblast polarization is a matrix-rigidity-dependent process controlled by focal adhesion mechanosensing. *Nature Cell Biology*, *13*(12), 1457–1465. doi:10.1038/ncb2370 PMID:22081092

Prakash, C., Kumar, R., & Mittal, N. (2018). Recent developments in human gait research: Parameters, approaches, applications, machine learning techniques, datasets and challenges. *Artificial Intelligence Review*, *49*(1), 1–40. doi:10.100710462-016-9514-6

Prasad, M. K., Varshney, R. K., Jain, P., Choudhary, A. K., Khare, A., & Jheetay, G. S. (2020). Postoperative analgesic efficacy of fluoroscopy-guided erector spinae plane block after percutaneous nephrolithotomy (PCNL): A randomized controlled study. *Saudi Journal of Anaesthesia*, *14*(4), 480–486. doi:10.4103ja.SJA_26_20 PMID:33447190

Prochazka, A., & Ellaway, P. (2012). Sensory systems in the control of movement. *Comprehensive Physiology*, *2*(4), 2615–2627. doi:10.1002/cphy.c100086 PMID:23720260

Pu, J., & Komvopoulos, K. (2014). Mechanical properties of electrospun bilayer fibrous membranes as potential scaffolds for tissue engineering. *Acta Biomaterialia*, *10*(6), 2718–2726. doi:10.1016/j.actbio.2013.12.060 PMID:24434536

Quinn, G., Studart, A.R., Hebert, C., VerHoef, J., & Arola, D. (2010). Fatigue of zirconia and dental bridge geometry: Design implications. *Dental Materials, 26*, 1133-1136.

Rabby, M. S., Shupti, S. P., & Molla, M. M. (2014). Pulsatile non-Newtonian laminar blood flows through arterial double stenosis. *Journal of Fluids*, *2014*, 1–13. doi:10.1155/2014/757902

Ragaert, K., De Somer, F., Somers, P., De Baere, I., Cardon, L., & Degrieck, J. (2012). Flexural mechanical properties of porcine aortic heart valve leaflets. *Journal of the Mechanical Behavior of Biomedical Materials*, *13*, 78–84. doi:10.1016/j.jmbbm.2012.04.009 PMID:22842278

Raharja, P., Atmoko, W., Rasyid, N., & Birowo, P. (2019). Safety and Effectiveness of Externalized Ureteral Catheter in Tubeless Percutaneous Nephrolithotomy. *Urology Journal*, *17*(5), 456–461. PMID:31422576

Rajamannan, N. M. (2011). Calcific Aortic Valve Disease: Cellular Origins of Valve Calcification. *Arteriosclerosis, Thrombosis, and Vascular Biology*, *31*(12), 2777–2778. doi:10.1161/ATVBAHA.111.237610 PMID:22096095

Ramasamy, J. G., & Akkus, O. (2007). Local variations in the micromechanical properties of mouse femur: The involvement of collagen fiber orientation and mineralization. *Journal of Biomechanics*, *40*(4), 910–918. doi:10.1016/j.jbiomech.2006.03.002 PMID:16678186

Rassoli, A., Shafigh, M., Seddighi, A., Seddighi, A., Daneshparvar, H., & Fatouraee, N. (2014). Biaxial mechanical properties of human ureter under tension. *Urology Journal*, *11*(3), 1678–1686. PMID:25015616

Rayner, J. M., Viscardi, P. W., Ward, S., & Speakman, J. R. (2001). Aerodynamics and energetics of intermittent flight in birds. *American Zoologist*, *41*(2), 188–204. doi:10.1093/icb/41.2.188

Rayz, V., & Berger, S. (2010). *Computational Modeling of Vascular Hemodynamics*. . doi:10.1007/978-90-481-3575-2_5

Reba, I. (1966). Applications of the Coanda effect. *Scientific American*, *214*(6), 84–93. doi:10.1038cientificamerican0666-84

Reinschmidt, C., & Nigg, B. M. (2000). Current Issues in the Design of Running and Court Shoes. *Sportverletzung Sportschaden*, *14*(3), 71–81. doi:10.1055-2000-7866 PMID:11081243

Resorlu, B., Unsal, A., Ziypak, T., Diri, A., Atis, G., Guven, S., Sancaktutar, A. A., Tepeler, A., Bozkurt, O. F., & Oztuna, D. (2013). Comparison of retrograde intrarenal surgery, shockwave lithotripsy, and percutaneous nephrolithotomy for treatment of medium-sized radiolucent renal stones. *World Journal of Urology*, *31*(6), 1581–1586. doi:10.100700345-012-0991-1 PMID:23179732

Richardson, L. F., & Gaunt, J. A. (1927). The deferred approach to the limit.Part I. Single lattice. Part II. Interpenetrating lattices. *Philosophical Transactions of the Royal Society of London. Series A, Containing Papers of a Mathematical or Physical Character*, *226*, 299–361.

Rieger, J. S., Jaeger, S., Schuld, C., Kretzer, J. P., & Bitsch, R. G. (2013). A vibrational technique for diagnosing loosened total hip endoprostheses: An experimental sawbone study. *Medical Engineering & Physics*, *35*(3), 329–337. doi:10.1016/j.medengphy.2012.05.007 PMID:22673003

Ritty, T. M., Ditsios, K., & Starcher, B. C. (2002). Distribution of the elastic fiber and associated proteins in flexor tendon reflects function. *The Anatomical Record: An Official Publication of the American Association of Anatomists*, *268*(4), 430–440. doi:10.1002/ar.10175 PMID:12420291

Rizvi, M. S., Kumar, P., Katti, D. S., & Pal, A. (2012). Mathematical model of mechanical behavior of micro/nano-fibrous materials designed for extracellular matrix substitutes. *Acta Biomaterialia*, *8*(11), 4111–4122. doi:10.1016/j.actbio.2012.07.025 PMID:22842037

Rizvi, M. S., & Pal, A. (2014). Statistical model for the mechanical behavior of the tissue engineering non-woven fibrous matrices under large deformation. *Journal of the Mechanical Behavior of Biomedical Materials*, *37*, 235–250. doi:10.1016/j.jmbbm.2014.05.026 PMID:24956158

Rizvi, M. S., Pal, A., & Das, S. L. (2016). Structure-induced nonlinear viscoelasticity of non-woven fibrous matrices. *Biomechanics and Modeling in Mechanobiology*, *15*(6), 1641–1654. doi:10.100710237-016-0788-z PMID:27090523

Roache, P. J. (1994). Perspective: A method for uniform reporting of grid refinement studies. *Journal of Fluids Engineering*, *116*(3), 405–441. doi:10.1115/1.2910291

Roy, S., Das, M., Chakraborty, P., Biswas, J.K., Chatterjee, S., Khutia, N., & Saha, S., & RoyChowdhury, A. (2017). Optimal selection of dental implant for different bone conditions based on the mechanical response. *Acta of Bioengineering and Biomechanics*, 19. PMID:28869633

Roy, S., Panda, D., Khutia, N., & Chowdhury, A. R. (2014). Pore geometry optimization of titanium (Ti6Al4V) alloy, for its application in the fabrication of customized hip implants. *International Journal of Biomaterials*, *2014*, 1–12. doi:10.1155/2014/313975 PMID:25400663

Ruben, R. B., Fernandes, P. R., & Folgado, J. (2012). On the optimal shape of hip implants. *Journal of Biomechanics*, *45*(2), 239–246. doi:10.1016/j.jbiomech.2011.10.038 PMID:22115063

Ruben, R. B., Folgado, J., & Fernandes, P. R. (2007). Three-dimensional shape optimization of hip prostheses using a multicriteria formulation. *Structural and Multidisciplinary Optimization*, *34*(3), 261–275. doi:10.100700158-006-0072-4

Rutkovskiy, A., Malashicheva, A., Sullivan, G., Bogdanova, M., Kostareva, A., Stensløkken, K. O., Fiane, A., & Vaage, J. (2017). Valve interstitial cells: The key to understanding the pathophysiology of heart valve calcification. *Journal of the American Heart Association*, *6*(9), 1–23. doi:10.1161/JAHA.117.006339 PMID:28912209

Sabnis, R. B., Chhabra, J. S., Ganpule, A. P., Abrol, S., & Desai, M. R. (2014). Current role of PCNL in pediatric urolithiasis. *Current Urology Reports*, *15*(7), 423. doi:10.100711934-014-0423-4 PMID:24898187

Sachs, G. (2005). Minimum shear wind strength required for dynamic soaring of albatrosses. *The Ibis*, *147*(1), 1–10. doi:10.1111/j.1474-919x.2004.00295.x

Sadun, A. S., Jalani, J., & Sukor, J. A. (2016). Force Sensing Resistor (FSR): a brief overview and the low-cost sensor for active compliance control. *First International Workshop on Pattern Recognition.*

Saha, N., Dubey, A. K., & Basu, B. (2012). Cellular proliferation, cellular viability, and biocompatibility of HA-ZnO composites. *Journal of Biomedical Materials Research. Part B, Applied Biomaterials*, *100*(1), 256–264. doi:10.1002/jbm.b.31948 PMID:22102555

Saha, N., Keskinbora, K., Suvaci, E., & Basu, B. (2010). Sintering, microstructure, mechanical, and antimicrobial properties of HAp-ZnO biocomposites. *Journal of Biomedical Materials Research. Part B, Applied Biomaterials*, *95*(2), 430–440. doi:10.1002/jbm.b.31734 PMID:20878929

Samiezadeh, S., Tavakkoli Avval, P., Fawaz, Z., & Bougherara, H. (2015). On optimization of a composite bone plate using the selective stress shielding approach. *Journal of the Mechanical Behavior of Biomedical Materials*, *42*, 138–153. doi:10.1016/j.jmbbm.2014.11.015 PMID:25482217

Sander, E. A., Stylianopoulos, T., Tranquillo, R. T., & Barocas, V. H. (2009). Image-based multiscale modeling predicts tissue-level and network-level fiber reorganization in stretched cell-compacted collagen gels. *Proceedings of the National Academy of Sciences of the United States of America*, *106*(42), 17675–17680. doi:10.1073/pnas.0903716106 PMID:19805118

Santra, S., Mandal, D. K., & Chakrabarti, S (2020).A New Approach for Assessing the LDL Species in Arterial Wall Layers. *Journal of The Institution of Engineers (India): Series C, 101*(1), 13–24.

Santra, S., Mandal, D. K., & Chakrabarti, S. (2017). Assessment of Accumulation Rate of LDL Species in Arterial Wall layers under Hypertension and Hyperlipidemia Conditions. *International Journal of Fluid Mechanics Research, 44*(1), 79–92. doi:10.1615/InterJFluidMechRes.2017016597

Santra, S., Mandal, D. K., & Chakrabarti, S. (2018). Effect of pulsatile blood flow on LDL transport in arterial layers. *Progress in Computational Fluid Dynamics, 18*(3), 177–187. doi:10.1504/PCFD.2018.091746

Saunders, J. B., Inman, V. T., & Eberhart, H. D. (1953). Six Determinants of Gait. *Journal of Bone and Joint Surgery, 35*(3), 543–558. doi:10.2106/00004623-195335030-00003

Saussine, C., Lechevallier, E., & Traxer, O. (2008). Les variantes techniques de la NLPC (PCNL: technical variations). *Progres en Urologie: Journal de l'Association Francaise d'Urologie et de la Societe Francaise d'Urologie, 18*(12), 897–900.

Sawaya, F., & Søndergaard, L. (2018). Aortic Regurgitation. In Diagnosis and Management of Adult Congenital Heart Disease (3rd ed., pp. 387–394). Elsevier. doi:10.1016/B978-0-7020-6929-1.00037-X

Saxod, R. (1996). Ontogeny of the cutaneous sensory organs. *Microscopy Research and Technique, 34*(4), 313–333. doi:10.1002/(SICI)1097-0029(19960701)34:4<313::AID-JEMT4>3.0.CO;2-P PMID:8807616

Schepsis, A. A., Jones, H., & Haas, A. L. (2002). Achilles tendon disorders in athletes. *The American Journal of Sports Medicine, 30*(2), 287–305. doi:10.1177/03635465020300022501 PMID:11912103

Schipper, H., Kranenbarg, S., van Leeuwen, J., Gijsen, M., Haazelager, M., & van Turnhout, M. (2008). Quantitative description of collagen structure in the articular cartilage of the young and adult equine distal metacarpus. *Animal Biology (Leiden, Netherlands), 58*(4), 353–370. doi:10.1163/157075608X383674

Schlichting, H., & Gersten, K. (2016). *Boundary-layer theory*. Springer.

Schmidt, J., & Hackenbroch, M. H. (1994). The Cenos hollow stem in total hip arthroplasty: First experiences in a prospective study. *Archives of Orthopaedic and Trauma Surgery, 113*(3), 117–120. doi:10.1007/BF00441616 PMID:8054230

Schoukens, G. (2009). *Developments in textile sports surfaces*. Advances in Carpet Manufacture.

Scott, D. R., Marcus, E. A., Weeks, D. L., & Sachs, G. (2002). Mechanisms of acid resistance due to the urease system of Helicobacter pylori. *Gastroenterology, 123*(1), 187–195. doi:10.1053/gast.2002.34218 PMID:12105847

Seel, T., Raisch, J., & Schauer, T. (2014). IMU-Based Joint Angle Measurement for Gait Analysis. *Sensors (Basel), 14*(4), 6891–6909. doi:10.3390140406891 PMID:24743160

Sender, R., Fuchs, S., & Milo, R. (2016). Revised estimates for the number of human and bacteria cells in the body. *PLoS Biology, 14*(8), e1002533. doi:10.1371/journal.pbio.1002533 PMID:27541692

Setti, L., Passarini, F., De Gennaro, G., Barbieri, P., Perrone, M. G., Borelli, M., ... Miani, A. (2020). *Airborne transmission route of COVID-19: why 2 meters/6 feet of inter-personal distance could not be enough*. Academic Press.

Shafaghi, A. H., Rokhsar Talabazar, F., Koşar, A., & Ghorbani, M. (2020). On the effect of the respiratory droplet generation condition on COVID-19 transmission. *Fluids, 5*(3), 113. doi:10.3390/fluids5030113

Shah, S., Kim, S. Y. R., Dubov, A., Schemitsch, E. H., Bougherara, H., & Zdero, R. (2011). The biomechanics of plate fixation of periprosthetic femoral fractures near the tip of a total hip implant: Cables, screws, or both? *Proceedings of the Institution of Mechanical Engineers. Part H, Journal of Engineering in Medicine, 225*(9), 845–856. doi:10.1177/0954411911413060 PMID:22070022

Shao, S., Zhou, D., He, R., Li, J., Zou, S., Mallery, K., Kumar, S., Yang, S., & Hong, J. (2021). Risk assessment of airborne transmission of COVID-19 by asymptomatic individuals under different practical settings. *Journal of Aerosol Science*, *151*, 105661. doi:10.1016/j.jaerosci.2020.105661 PMID:32968325

Shapiro, I. M., & Risbud, M. V. (2014). Introduction to the structure, function, and comparative anatomy of the vertebrae and the intervertebral disc. In *The intervertebral disc* (pp. 3–15). Springer. doi:10.1007/978-3-7091-1535-0_1

Sharabi, M., Wilke, H. J., & Haj-Ali, R. (2018). The Vertebral Bone. In *Biomechanics of the Spine* (pp. 71–87). Academic Press.

Sherwin, S. J., & Blackburn, H. M. (2005). Three-dimensional instabilities and transition of steady and pulsatile axisymmetric stenotic flows. *Journal of Fluid Mechanics*, *533*, 297–327. doi:10.1017/S0022112005004271

Shi, L., & Fok, A.S. (2009). Structural optimization of the fibre-reinforced composite substructure in a three-unit dental bridge. *Dental Materials, 25*, 791-801.

Shilo, Y., Modai, J., Leibovici, D., Dror, I., & Berkowitz, B. (2021). Comparative study of renal drainage with different ureteral stents subject to extrinsic ureteral obstruction using an in vitro ureter-stent model. *BMC Urology*, *21*(1), 100. doi:10.118612894-021-00865-w PMID:34261481

Sider, K. L., Blaser, M. C., & Simmons, C. A. (2011). Animal Models of Calcific Aortic Valve Disease. *International Journal of Inflammation, 2011*(Ldl), 1–18. doi:10.4061/2011/364310

Sigmund, O., & Bendsoe, M. P. (2011). *Topology Optimization* (2nd ed.). Springer.

Simon, S. R. (1993). Gait Analysis, Normal and Pathological Function. *The Journal of Bone & Joint Surgery*, *75*(3), 476–477. doi:10.2106/00004623-199303000-00027

Sinclair, J., Edmundson, C., Brooks, D., & Hobbs, S. (2011). Evaluation of kinematic methods of identifying gait events during running. *International Journal of Sports*, *05*(03), 188–192.

Siopack, J. S., Jergesen, H. E., & Francisco, S. (1995). Conferences and Reviews Total Hip Arthroplasty. *The Western Journal of Medicine*, *162*(2), 243–249. PMID:7725707

Sjodin, B. (2016). What's The Difference Between FEM, FDM, and FVM? In S. Mraz, R. Begg, & M. McBurnett (Eds.), *Machine Design*. Endeavor Business Media, LLC.

Slonaker, M., & Goswami, T. (2004). Review of wear mechanisms in hip implants: Paper II–ceramics IG004712. *Materials & Design*, *25*(5), 395–405. doi:10.1016/j.matdes.2003.11.011

Smith, M. (2005). A review of the initial management of soft tissue sports injuries. *Journal of Orthopaedic Nursing*, *9*(2), 103–107. doi:10.1016/j.joon.2004.09.011

Smith, N. F. (1972). Bernoulli and Newton in fluid mechanics. *The Physics Teacher*, *10*(8), 451–455. doi:10.1119/1.2352317

Smoljanovic, T., Bojanic, I., Hannafin, J. A., Hren, D., Delimar, D., & Pecina, M. (2009). Traumatic and overuse injuries among international elite junior rowers. *The American Journal of Sports Medicine*, *37*(6), 1193–1199. doi:10.1177/0363546508331205 PMID:19299531

Sokolis, D. P. (2012). Multiaxial mechanical behaviour of the passive ureteral wall: Experimental study and mathematical characterisation. *Computer Methods in Biomechanics and Biomedical Engineering*, *15*(11), 1145–1156. doi:10.1080/10255842.2011.581237 PMID:21660781

Sourial, M. W., Francois, N., Box, G. N., & Knudsen, B. E. (2019). Supracostal access tubeless percutaneous nephrolithotomy: Minimizing complications. *World Journal of Urology, 37*(7), 1429–1433. doi:10.100700345-018-2518-x PMID:30302593

Southgate, D. A. (1995). Digestion and metabolism of sugars. *The American Journal of Clinical Nutrition, 62*(1), 203S–210S. doi:10.1093/ajcn/62.1.203S PMID:7598078

Spiller, R. (2006). Role of motility in chronic diarrhoea. *Neurogastroenterology and Motility, 18*(12), 1045–1055. doi:10.1111/j.1365-2982.2006.00836.x PMID:17109687

Srisubat, A., Potisat, S., Lojanapiwat, B., Setthawong, V., & Laopaiboon, M. (2009). Extracorporeal shock wave lithotripsy (ESWL) versus percutaneous nephrolithotomy (PCNL) or retrograde intrarenal surgery (RIRS) for kidney stones. *Cochrane Database of Systematic Reviews*, (4), CD007044. PMID:19821393

Steck, R., Tami, A., Sidler, H-J., Anderson, E., & Niederer, P. (2010). *Computational Modeling of Extravascular Flow in Bone*. . doi:10.1007/978-90-481-3575-2_10

Stella, J. A., Liao, J., & Sacks, M. S. (2007). Time-dependent biaxial mechanical behavior of the aortic heart valve leaflet. *Journal of Biomechanics, 40*(14), 3169–3177. doi:10.1016/j.jbiomech.2007.04.001 PMID:17570376

Stella, J. A., & Sacks, M. S. (2007). On the Biaxial Mechanical Properties of the Layers of the Aortic Valve Leaflet. *Journal of Biomechanical Engineering, 129*(5), 757–766. doi:10.1115/1.2768111 PMID:17887902

Stella, J. A., Wagner, W. R., & Sacks, M. S. (2010). Scaleâ€ dependent fiber kinematics of elastomeric electrospun scaffolds for soft tissue engineering. *Journal of Biomedical Materials Research Part A: An Official Journal of The Society for Biomaterials, The Japanese Society for Biomaterials, and The Australian Society for Biomaterials and the Korean Society for Biomaterials, 93*(3), 1032–1042. PMID:19753623

Stern, F., Wilson, R. V., Coleman, H. W., & Paterson, E. G. (2001). Comprehensive approach to verication and validation of cfd simulations part 1: Methodology and procedures. *Journal of Fluids Engineering, 123*(4), 793–802. doi:10.1115/1.1412235

Stiehm, M., Wüstenhagen, C., Siewert, S., Grabow, N., & Schmitz, K. (2017). Numerical simulation of pulsatile flow through a coronary nozzle model based on FDA's benchmark geometry. *Current Directions in Biomedical Engineering, 3*(2), 775–778. doi:10.1515/cdbme-2017-0163

Stiles, V. H., James, I. T., Dixon, S. J., & Guisasola, I. N. (2009). Natural turf surfaces: The case for continued research. *Sports Medicine (Auckland, N.Z.), 39*(1), 65–84. doi:10.2165/00007256-200939010-00005 PMID:19093696

Stops, A., Wilcox, R., & Jin, Z. (2012). Computational modelling of the natural hip: A review of finite element and multibody simulations. *Computer Methods in Biomechanics and Biomedical Engineering, 15*(9), 963–979. doi:10.1080/10255842.2011.567983 PMID:21574077

Stradins, P., Lacis, R., Ozolanta, I., Purina, B., Ose, V., Feldmane, L., & Kasyanov, V. (2004). Comparison of biomechanical and structural properties between human aortic and pulmonary valve. *European Journal of Cardio-Thoracic Surgery, 26*(3), 634–639. doi:10.1016/j.ejcts.2004.05.043 PMID:15302062

Strocchi, R., De Pasquale, V., Gubellini, P., Facchini, A., Marcacci, M., Buda, R., ... Ruggeri, A. (1992). The human anterior cruciate ligament: Histological and ultrastructural observations. *Journal of Anatomy, 180*(Pt 3), 515. PMID:1487443

Stylianopoulos, T., & Barocas, V. H. (2007). Volume-averaging theory for the study of the mechanics of collagen networks. *Computer Methods in Applied Mechanics and Engineering, 196*(31-32), 2981–2990. doi:10.1016/j.cma.2006.06.019

Stylianopoulos, T., Bashur, C. A., Goldstein, A. S., Guelcher, S. A., & Barocas, V. H. (2008). Computational predictions of the tensile properties of electrospun fibre meshes: Effect of fibre diameter and fibre orientation. *Journal of the Mechanical Behavior of Biomedical Materials*, *1*(4), 326–335. doi:10.1016/j.jmbbm.2008.01.003 PMID:19627797

Sucosky, P., Balachandran, K., Elhammali, A., Jo, H., & Yoganathan, A. P. (2009). Altered shear stress stimulates upregulation of endothelial VCAM-1 and ICAM-1 in a BMP-4- and TGF-β1-dependent pathway. *Arteriosclerosis, Thrombosis, and Vascular Biology*, *29*(2), 254–260. doi:10.1161/ATVBAHA.108.176347 PMID:19023092

Sugarbaker, D. J., Rattan, S., & Goyal, R. K. (1984a). Mechanical and electrical activity of esophageal smooth muscle during peristalsis. *American Journal of Physiology. Gastrointestinal and Liver Physiology*, *246*(2), G145–G150. doi:10.1152/ajpgi.1984.246.2.G145 PMID:6696111

Sugarbaker, D. J., Rattan, S., & Goyal, R. K. (1984b). Swallowing induces sequential activation of esophageal longitudinal smooth muscle. *American Journal of Physiology. Gastrointestinal and Liver Physiology*, *247*(5), G515–G519. doi:10.1152/ajpgi.1984.247.5.G515 PMID:6496741

Surer, E., & Kose, A. (2011). Methods and technologies for Gait analysis. *Computer Analysis of Human Behavior,* 105-123.

Surer, E., & Kose, A. (2011). Method and technologies for gait anlaysis. In A. A. Salah & T. Gevers (Eds.), *Computer Analysis of Human Behavior* (pp. 105–123). Springer. doi:10.1007/978-0-85729-994-9_5

Sweeting, K., & Mock, M. (2007, June). Gait and posture - assessment in general practice. *Australian Family Physician*, *36*(6), 398–401, 404–405. PMID:17565395

Tabassi, N. C. B., & Garnero, P. (2007). Monitoring cartilage turnover. *Current Rheumatology Reports*, *9*(1), 16–24. doi:10.100711926-007-0017-y PMID:17437662

Taddei, F., Pancanti, A., & Viceconti, M. (2004). An improved method for the automatic mapping of computed tomography numbers onto finite element models. *Medical Engineering & Physics*, *26*(1), 61–69. doi:10.1016/S1350-4533(03)00138-3 PMID:14644599

Taguchi, M., Yoshida, K., Sugi, M., Kinoshita, H., & Matsuda, T. (2018). Simplified method using kidney / ureter / bladder x-ray to determine the appropriate length of ureteral stents. *International braz j urol: official journal of the Brazilian Society of Urology, 44*(6), 1224–1233. doi:10.1590/S1677-5538.IBJU.2017.0620

Takahashi, T. (2011). Flow behavior of digesta and the absorption of nutrients in the gastrointestine. *Journal of Nutritional Science and Vitaminology*, *57*(4), 265–273. doi:10.3177/jnsv.57.265 PMID:22041908

Takeda, H., Nakagawa, T., Nakamura, K., & Engebretsen, L. (2011). Prevention and management of knee osteoarthritis and knee cartilage injury in sports. *British Journal of Sports Medicine*, *45*(4), 304–309. doi:10.1136/bjsm.2010.082321 PMID:21357577

Tan, E. P. S., & Lim, C. T. (2006). Mechanical characterization of nanofibersâ€“a review. *Composites Science and Technology*, *66*(9), 1102–1111. doi:10.1016/j.compscitech.2005.10.003

Taylor, C. A., Hughes, T. J. R., & Zarins, C. K. (1998). Finite element modeling of blood flow in arteries. *Computer Methods in Applied Mechanics and Engineering*, *158*(1-2), 155–196. doi:10.1016/S0045-7825(98)80008-X

Taylor, P. M., Batten, P., Brand, N. J., Thomas, P. S., & Yacoub, M. H. (2003). The cardiac valve interstitial cell. *The International Journal of Biochemistry & Cell Biology*, *35*(2), 113–118. doi:10.1016/S1357-2725(02)00100-0 PMID:12479860

ten Broeke, R. H., Tarala, M., Arts, J. J., Janssen, D. W., Verdonschot, N., & Geesink, R. G. (2014). Improving periprosthetic bone adaptation around cementless hip stems: A clinical and finite element study. *Medical Engineering & Physics*, *36*(3), 345–353. doi:10.1016/j.medengphy.2013.12.006 PMID:24378381

Terkawi,, M.A., Matsumae,, G., Shimizu,, T., Takahashi,, D., Kadoya,, K., & Iwasaki,, N. (2022). Interplay between Inflammation and Pathological Bone Resorption: Insights into Recent Mechanisms and Pathways in Related Diseases for Future Perspectives. Int J Mol Sci., 23(3). doi:10.3390/ijms23031786 PMID:35163708

Tesio, L., & Rota, V. (2019). The Motion of Body Center of Mass During Walking: A Review Oriented to Clinical Applications. *Frontiers in Neurology*, *10*, 1–22. doi:10.3389/fneur.2019.00999 PMID:31616361

Thacker, S. B., Stroup, D. F., Branche, C. M., Gilchrist, J., Goodman, R. A., & Kelling, E. P. (2003). Prevention of knee injuries in sports: A systematic review of the literature. *The Journal of Sports Medicine and Physical Fitness*, *43*(2), 165–179. PMID:12853898

Thelin, N., Holmberg, S., & Thelin, A. (2006). Knee injuries account for the sports-related increased risk of knee osteoarthritis. *Scandinavian Journal of Medicine & Science in Sports*, *16*(5), 329–333. doi:10.1111/j.1600-0838.2005.00497.x PMID:16978252

Thomas, A. L. (1997). On the tails of birds. *Bioscience*, 215–225.

Tkatchenko, T. V., Moreno-Rodriguez, R. A., Conway, S. J., Molkentin, J. D., Markwald, R. R., & Tkatchenko, A. V. (2009). Lack of periostin leads to suppression of Notch1 signaling and calcific aortic valve disease. *Physiological Genomics*, *39*(3), 160–168. doi:10.1152/physiolgenomics.00078.2009 PMID:19723774

Tobalske, B. W. (2007). Biomechanics of bird flight. *The Journal of Experimental Biology*, *210*(18), 3135–3146. doi:10.1242/jeb.000273 PMID:17766290

Tobalske, B. W., & Dial, K. (1996). Flight kinematics of black-billed magpies and pigeons over a wide range of speeds. *The Journal of Experimental Biology*, *199*(2), 263–280. doi:10.1242/jeb.199.2.263 PMID:9317775

Toniollo, M. B., Macedo, A. P., Palhares, D., Calefi, P. L., Sorgini, D. B., & Mattos, M. G. C. d. (2012). Morse taper implants at different bone levels: A finite element analysis of stress distribution. *Brazilian Journal of Oral Sciences*, *11*, 440–444.

Tripathi, G., & Basu, B. (2014). In vitro osteogenic cell proliferation, mineralization, and in vivo osseointegration of injection molded high-density polyethylene-based hybrid composites in rabbit animal model. *Journal of Biomaterials Applications*, *29*(1), 142–157. doi:10.1177/0885328214520805 PMID:24452882

Tripathi, G., Dubey, A. K., & Basu, B. (2012). Evaluation of physico-mechanical properties and in vitro biocompatibility of compression molded HDPE based biocomposites with HA/Al2O3 ceramic fillers and titanate coupling agents. *Journal of Applied Polymer Science*, *124*(4), 3051–3063. doi:10.1002/app.35339

Tripathi, G., Gough, J. E., Dinda, A., & Basu, B. (2013). In vitro cytotoxicity and in vivo osseointergration properties of compression-molded HDPE-HA-Al2O3 hybrid biocomposites. *Journal of Biomedical Materials Research. Part A*, *101*(6), 1539–1549. doi:10.1002/jbm.a.34452 PMID:23065866

Tsang, H. G., Cui, L., Farquharson, C., Corcoran, B. M., Summers, K. M., & Macrae, V. E. (2018). Exploiting novel valve interstitial cell lines to study calcific aortic valve disease. *Molecular Medicine Reports*, *17*(2), 2100–2106. doi:10.3892/mmr.2017.8163 PMID:29207136

Tseng. (2011). Elastic fibers in the aortic valve spongiosa: A fresh perspective on its structure and role in overall tissue function. *Acta Biomaterialia, 7*(5), 2101–2108. doi:10.1016/j.actbio.2011.01.022

Tseng, B., Balaoing, L. R., Grigoryan, B., Raphael, R. M., Killian, T. C., Souza, G. R., & Grande-Allen, K. J. (2014). A three-dimensional co-culture model of the aortic valve using magnetic levitation. *Acta Biomaterialia*, *10*(1), 173–182. doi:10.1016/j.actbio.2013.09.003 PMID:24036238

Tseng, K., Kim, E. J., Connell, P. S., Ayoub, S., Shah, J. V., & Grande-Allen, K. J. (2013). The Tensile and Viscoelastic Properties of Aortic Valve Leaflets Treated with a Hyaluronidase Gradient. *Cardiovascular Engineering and Technology*, *4*(2), 151–160. doi:10.100713239-013-0122-1

Tu, C., Deville, M., Dheur, L., & Vanderschuren, L. (1992). Finite element simulation of pulsatile flow through arterial stenosis. *Journal of Biomechanics*, *25*(10), 1141–1152. doi:10.1016/0021-9290(92)90070-H PMID:1400514

Tucker, V. A. (1992). Pitching equilibrium, wing span and tail span in a gliding Harris' hawk, *Parabuteo unicinctus. The Journal of Experimental Biology*, *165*(1), 21–41. doi:10.1242/jeb.165.1.21

Turkyilmaz, I., Tözüm, T., & Tumer, C. (2007). Bone density assessments of oral implant sites using computerized tomography. *Journal of Oral Rehabilitation*, *34*(4), 267–272. doi:10.1111/j.1365-2842.2006.01689.x PMID:17371564

Tutty, O. R. (1992). Pulsatile flow in a constricted channel. *Journal of Biomechanical Engineering*, *114*(1), 50–54. doi:10.1115/1.2895449 PMID:1491586

Twist, C., & Eston, R. (2005). The effects of exercise-induced muscle damage on maximal intensity intermittent exercise performance. *European Journal of Applied Physiology*, *94*(5–6), 652–658. doi:10.100700421-005-1357-9 PMID:15887020

Tyler, T. F., Silvers, H. J., Gerhardt, M. B., & Nicholas, S. J. (2010). Groin injuries in sports medicine. *Sports Health*, *2*(3), 231–236. doi:10.1177/1941738110366820 PMID:23015943

Usyk, T., & McCulloch, A. (2003). Computational Methods for Soft Tissue Biomechanics. *Biomech. Soft Tissue Cardiovasc. Syst.*, *441*, 273–342. Advance online publication. doi:10.1007/978-3-7091-2736-0_7

Vafaee, T., Thomas, D., Desai, A., Jennings, L. M., Berry, H., Rooney, P., Kearney, J., Fisher, J., & Ingham, E. (2018). Decellularization of human donor aortic and pulmonary valved conduits using low concentration sodium dodecyl sulfate. *Journal of Tissue Engineering and Regenerative Medicine*, *12*(2), 841–853. doi:10.1002/term.2391 PMID:27943656

Varghese, S. S., & Frankel, S. H. (2003). Numerical modeling of pulsatile turbulent flow in stenotic vessels. *Journal of Biomechanical Engineering*, *125*(4), 445–460. doi:10.1115/1.1589774 PMID:12968569

Velavan, T. P., & Meyer, C. G. (2020). The COVID-19 epidemic. *Tropical Medicine & International Health*, *25*(3), 278–280. doi:10.1111/tmi.13383 PMID:32052514

Vesely, I. (1997). The role of elastin in aortic valve mechanics. *Journal of Biomechanics*, *31*(2), 115–123. doi:10.1016/S0021-9290(97)00122-X PMID:9593204

Videler, J. J. (2006). *Avian flight*. Oxford University Press. doi:10.1093/acprof:oso/9780199299928.001.0001

Virulsri, C., Tangpornprasert, P., & Romtrairat, P. (2015). Femoral hip prosthesis design for Thais using multi-objective shape optimization. *Materials & Design*, *68*, 1–7. doi:10.1016/j.matdes.2014.11.027

Vormittag, K., Calonje, R., & Briner, W. W. (2009). Foot and ankle injuries in the barefoot sports. *Current Sports Medicine Reports*, *8*(5), 262–266. doi:10.1249/JSR.0b013e3181b9e3be PMID:19741354

Wade, K. (2018). Vertebral Endplates. In *Biomechanics of the Spine* (pp. 125–140). Academic Press.

Wagner, P., Olsson, H., Ranstam, J., Robertsson, O., Zheng, M. H., & Lidgren, L. (2012). Metal-on-metal joint bearings and hematopoetic malignancy: A review. *Acta Orthopaedica*, *83*(6), 553–558. doi:10.3109/17453674.2012.747055 PMID:23140092

Walker, A., & Subic, A. (2013). Advances in design and materials for indoor sports surfaces. *Advanced Materials Research*, *633*, 47–61. doi:10.4028/www.scientific.net/AMR.633.47

Wallman, J. O. S. H., Velez, J. O. S. E., Weinstein, B., & Green, A. E. (1982). Avian vestibuloocular reflex: Adaptive plasticity and developmental changes. *Journal of Neurophysiology, 48*(4), 952–967. doi:10.1152/jn.1982.48.4.952 PMID:6982960

Walsh, S. A., Iwaniuk, A. N., Knoll, M. A., Bourdon, E., Barrett, P. M., Milner, A. C., Nudds, R. L., Abel, R. L., & Dello Sterpaio, P. (2013). Avian cerebellar floccular fossa size is not a proxy for flying ability in birds. *PLoS One, 8*(6), e67176. doi:10.1371/journal.pone.0067176 PMID:23825638

Wang, Y., Brasseur, J., Banco, G., Webb, A., Ailiani, A., & Neuberger, T. (2010). *Development of a Lattice-Boltzmann Method for Multiscale Transport and Absorption with Application to Intestinal Function.* doi:10.1007/978-90-481-3575-2_3

Wang, G., Wang, D., Mao, J., Lin, Y., Yin, Z., Wang, B., He, Y., & Sun, S. (2016). Three dimensional finite-element analysis of treating Vancouver B1 periprosthetic femoral fractures with three kinds of internal fixation. *International Journal of Clinical and Experimental Medicine, 9*(4), 7557–7564.

Wang, H., Abhilash, A. S., Chen, C. S., Wells, R. G., & Shenoy, V. B. (2014). Long-range force transmission in fibrous matrices enabled by tension-driven alignment of fibers. *Biophysical Journal, 107*(11), 2592–2603. doi:10.1016/j.bpj.2014.09.044 PMID:25468338

Wang, J., Bai, Y., Yin, S., Cui, J., Tang, Y., Wang, Z., Chen, B., Li, H., Wei, W., & Wang, J. (2020). Risk factors for deterioration of renal function after percutaneous nephrolithotomy in solitary kidney patients with staghorn calculi. *Translational Andrology and Urology, 9*(5), 2022–2030. doi:10.21037/tau-20-916 PMID:33209666

Wang, X. (2006). From Immersed Boundary Method to Immersed Continuum Methods. *International Journal for Multiscale Computational Engineering, 4*(1), 127–146. doi:10.1615/IntJMultCompEng.v4.i1.90

Warashina, H., Sakano, S., Kitamura, S., Yamauchi, K.-I., Yamaguchi, J., Ishiguro, N., & Hasegawa, Y. (2003). Biological reaction to alumina, zirconia, titanium and polyethylene particles implanted onto murine calvaria. *Biomaterials, 24*(21), 3655–3661. doi:10.1016/S0142-9612(03)00120-0 PMID:12818536

Warburton, A., Girdler, S. J., Mikhail, C. M., Ahn, A., & Cho, S. K. (2020). Biomaterials in spinal implants: A review. *Neurospine, 17*(1), 101–110. doi:10.14245/ns.1938296.148 PMID:31694360

Warrick, D. R., Bundle, M. W., & Dial, K. P. (2002). Bird maneuvering flight: Blurred bodies, clear heads. *Integrative and Comparative Biology, 42*(1), 141–148. doi:10.1093/icb/42.1.141 PMID:21708703

Waters, R. L., & Mulroy, S. (1999, July). (199 9). The energy expenditure of normal and pathologic gait. *Gait & Posture, 9*(3), 207–231. doi:10.1016/S0966-6362(99)00009-0 PMID:10575082

Weinans, H., Huiskes, R., Van Rietbergen, B., Sumner, D., Turner, T., & Galante, J. (1993). Adaptive bone remodeling around bonded noncemented total hip arthroplasty: A comparison between animal experiments and computer simulation. *Journal of Orthopaedic Research, 11*(4), 500–513. doi:10.1002/jor.1100110405 PMID:8340823

Weinberg, E. J., & Kaazempur Mofrad, M. R. (2007). Transient, Three-dimensional, Multiscale Simulations of the Human Aortic Valve. *Cardiovascular Engineering (Dordrecht, Netherlands), 7*(4), 140–155. doi:10.100710558-007-9038-4 PMID:18026835

Weinberg, E. J., & Kaazempur Mofrad, M. R. (2008). A multiscale computational comparison of the bicuspid and tricuspid aortic valves in relation to calcific aortic stenosis. *Journal of Biomechanics, 41*(16), 3482–3487. doi:10.1016/j.jbiomech.2008.08.006 PMID:18996528

Weis-Fogh, T. (1972). Energetics of hovering flight in hummingbirds and in Drosophila. *The Journal of Experimental Biology, 56*(1), 79–104. doi:10.1242/jeb.56.1.79

Weiss, C., Rosenberg, L., & Helfet, A. J. (1968). An ultrastructural study of normal young adult human articular cartilage. *JBJS*, *50*(4), 663–674. doi:10.2106/00004623-196850040-00002 PMID:5658553

Weiss, K., & Whatman, C. (2015). Biomechanics associated with patellofemoral pain and ACL injuries in sports. *Sports Medicine (Auckland, N.Z.)*, *45*(9), 1325–1337. doi:10.100740279-015-0353-4 PMID:26130304

Wei, W., & Yalong, L. (2021). Study on treatment and rehabilitation training of ligament injury of javelin throwers based on sports biomechanics. *Measurement: Journal of the International Measurement Confederation*, *171*, 108757. doi:10.1016/j.measurement.2020.108757

Weston, M. W., LaBorde, D. V., & Yoganathan, A. P. (1999). Estimation of the Shear Stress on the Surface of an Aortic Valve Leaflet. *Annals of Biomedical Engineering*, *27*(4), 572–579. doi:10.1114/1.199 PMID:10468241

Weston, M. W., & Yoganathan, A. P. (2001). Biosynthetic Activity in Heart Valve Leaflets in Response to In Vitro Flow Environments. *Annals of Biomedical Engineering*, *29*(9), 752–763. doi:10.1114/1.1397794 PMID:11599583

Wheeler, A. P., Morad, S., Buchholz, N., & Knight, M. M. (2012). The shape of the urine stream—From biophysics to diagnostics. *PLoS One*, *7*(10), e47133. doi:10.1371/journal.pone.0047133 PMID:23091609

Whitehead, W. E., Engel, B. T., & Schuster, M. M. (1980). Irritable bowel syndrome. *Digestive Diseases and Sciences*, *25*(6), 404–413. doi:10.1007/BF01395503 PMID:7379673

Whittle, M. (1985). Applications of gait analysis. In Gait Analysis (pp. 177–193). doi:10.1016/B978-075068883-3.50010-6

Wilcox, D. C. (2006). *Turbulence modeling for CFD* (3rd ed.). DCW Industries, Inc.

Wilhelm, K., Fritsche, H. M., & Netsch, C. (2015). Perkutane Steintherapie heute: Standard-, Mini-, Micro-, Ultramini-PCNL (Percutaneous Stone Treatment Today: Standard-, Mini-, Micro-, Ultramini-PCNL). *Aktuelle Urologie*, *46*(4), 297–302. doi:10.1055-0035-1555863 PMID:26227130

Winter, D. A., Quanbury, A. O., & Reimer, G. D. (1976). Analysis of instantaneous energy of normal gait. *Journal of Biomechanics*, *9*(4), 253–257. doi:10.1016/0021-9290(76)90011-7 PMID:1262360

Wiwatanapataphee, B., Poltem, D., Wu, Y. H., & Lenbury, Y. (2006). Simulation of pulsatile flow of blood in stenosed coronary artery bypass with graft. *Mathematical Biosciences and Engineering*, *3*(2), 71–383. PMID:20361829

Womersley, J. R. (1955). Method for the calculation of velocity, rate of flow, and viscous drag in arteries when the pressure gradient is known. *The Journal of Physiology*, *127*(3), 553–563. doi:10.1113/jphysiol.1955.sp005276 PMID:14368548

Wood, J. D. (2004). Peristalsis. Encyclopedia of Gastroenterology, 164-165.

Wootton, D. M., & Ku, D. N. (1999). Fluid mechanics of vascular systems, diseases, and thrombosis. *Annual Review of Biomedical Engineering*, *1*(1), 299–329. doi:10.1146/annurev.bioeng.1.1.299 PMID:11701491

Yahia, L. H., & Drouin, G. (1989). Microscopical investigation of canine anterior cruciate ligament and patellar tendon: Collagen fascicle morphology and architecture. *Journal of Orthopaedic Research*, *7*(2), 243–251. doi:10.1002/jor.1100070212 PMID:2918423

Yamako, G., Chosa, E., Totoribe, K., Hanada, S., Masahashi, N., Yamada, N., & Itoi, E. (2014a). In-vitro biomechanical evaluation of stress shielding and initial stability of a low-modulus hip stem made of β type Ti-33.6 Nb-4Sn alloy. *Medical Engineering & Physics*, *36*(12), 1665–1671. doi:10.1016/j.medengphy.2014.09.002 PMID:25282098

Yamako, G., Chosa, E., Zhao, X., Totoribe, K., Watanabe, S., Sakamoto, T., & Nakane, N. (2014b). Load-transfer analysis after insertion of cementless anatomical femoral stem using pre-and post-operative CT images based patient-specific finite element analysis. *Medical Engineering & Physics*, *36*(6), 694–700. doi:10.1016/j.medengphy.2014.02.018 PMID:24629623

Yang, C.-T., Wei, H.-W., Kao, H.-C., & Cheng, C.-K. (2009). Design and test of hip stem for medullary revascularization. *Medical Engineering & Physics*, *31*(8), 994–1001. doi:10.1016/j.medengphy.2009.06.001 PMID:19581119

Yan, J., Qiang, L., Gao, Y., Cui, X., Zhou, H., Zhong, S., Wang, Q., & Wang, H. (2012). Effect of fiber alignment in electrospun scaffolds on keratocytes and corneal epithelial cells behavior. *Journal of Biomedical Materials Research. Part A*, *100*(2), 527–535. doi:10.1002/jbm.a.33301 PMID:22140085

Yeung, T., Georges, P. C., Flanagan, L. A., Marg, B., Ortiz, M., Funaki, M., Zahir, N., Ming, W., Weaver, V., & Janmey, P. A. (2005). Effects of substrate stiffness on cell morphology, cytoskeletal structure, and adhesion. *Cell Motility and the Cytoskeleton*, *60*(1), 24–34. doi:10.1002/cm.20041 PMID:15573414

Yin, F. C., & Fung, Y. C. (1971). Mechanical properties of isolated mammalian ureteral segments. *The American Journal of Physiology*, *221*(5), 1484–1493. doi:10.1152/ajplegacy.1971.221.5.1484 PMID:5124294

Yoganandan, N., Kumaresan, S., & Pintar, F. A. (2000). Geometric and mechanical properties of human cervical spine ligaments. *Journal of Biomechanical Engineering*, *122*(6), 623–629. doi:10.1115/1.1322034 PMID:11192384

Young, D. F., & Tsai, F. Y. (1973). Flow characteristics in models of arterial stenoses-I. Steady flow. *Journal of Biomechanics*, *6*(4), 395–410. doi:10.1016/0021-9290(73)90099-7 PMID:4732939

Yperman, J., De Visscher, G., Holvoet, P., & Flameng, W. (2004). Molecular and functional characterization of ovine cardiac valve-derived interstitial cells in primary isolates and cultures. *Tissue Engineering*, *10*(9–10), 1368–1375. doi:10.1089/ten.2004.10.1368 PMID:15588397

Yu, J. M., Leu, T. S., & Miau, J. J. (2017). Investigation of reduced frequency and freestream turbulence effects on dynamic stall of a pitching airfoil. *Journal of Visualization / the Visualization Society of Japan*, *20*(1), 31–44. doi:10.100712650-016-0366-6

Yuri, P., Hariwibowo, R., Soeroharjo, I., Danarto, R., Hendri, A. Z., Brodjonegoro, S. R., Rasyid, N., Birowo, P., & Widyahening, I. S. (2018). Meta-analysis of Optimal Management of Lower Pole Stone of 10 - 20 mm: Flexible Ureteroscopy (FURS) versus Extracorporeal Shock Wave Lithotripsy (ESWL) versus Percutaneus Nephrolithotomy (PCNL). *Acta Medica Indonesiana*, *50*(1), 18–25. PMID:29686172

Zabirnyk, A., Perez, M. del M., Blasco, M., Stensløkken, K., Ferrer, M. D., Salcedo, C., & Vaage, J. (2020). A Novel Ex Vivo Model of Aortic Valve Calcification. A Preliminary Report. *Frontiers in Pharmacology*, *11*(December), 1–7. doi:10.3389/fphar.2020.568764 PMID:33390945

Zago, M., Luzzago, M., Marangoni, T., Cecco, D. M., Tarabini, M., & Galli, M. (2020). 3D tracking of human motion using visual skeletonization and stereoscopic vision. *Frontiers in Bioengineering and Biotechnology*, *8*, 1–11. doi:10.3389/fbioe.2020.00181 PMID:32195243

Zakikhani, P., Ho, R., Wang, W., & Li, Z. (2019). Biomechanical assessment of aortic valve stenosis: Advantages and limitations. *Medicine in Novel Technology and Devices*, *2*(September), 100009. doi:10.1016/j.medntd.2019.100009

Zeidler, S. (2020). *Additive Manufacturing in Orthopedics*. https://www.sme.org/technologies/articles/2020/may/additive-manufacturing-in-orthopedics/

Zemel, A., Rehfeldt, F., Brown, A. E. X., Discher, D. E., & Safran, S. A. (2010). Optimal matrix rigidity for stress-fibre polarization in stem cells. *Nature Physics*, *6*(6), 468–473. doi:10.1038/nphys1613 PMID:20563235

Zendehbudi, G. R., & Moayeri, M. S. (1999). Comparison of physiological and simple pulsatile flows through stenosed arteries. *Journal of Biomechanics*, *32*(9), 959–969. doi:10.1016/S0021-9290(99)00053-6 PMID:10460133

Zhang, B., Liu, S., Liu, Y., Wu, B., Zhang, X., Wang, X., Liang, X., Cao, X., Wang, D., & Wu, C. L. (2021). Novel CFD modeling approaches to assessing urine flow in prostatic urethra after transurethral surgery. *Scientific Reports*, *11*(1), 663. doi:10.103841598-020-79505-6 PMID:33436678

Zhang, W., Titze, M., Cappi, B., Wirtz, D., Telle, R., & Fischer, H. (2010). Improved mechanical long-term reliability of hip resurfacing prostheses by using silicon nitride. *Journal of Materials Science. Materials in Medicine*, *21*(11), 3049–3057. doi:10.100710856-010-4144-z PMID:20725769

Zheng, J., Pan, J., Qin, Y., Huang, J., Luo, Y., Gao, X., & Zhou, X. (2015). Role for intravesical prostatic protrusion in lower urinary tract symptom: A fluid structural interaction analysis study. *BMC Urology*, *15*(1), 86. doi:10.118612894-015-0081-y PMID:26285823

Zhou, J., Hu, S., Ding, J., Xu, J., Shi, J., & Dong, N. (2013). Tissue engineering of heart valves: PEGylation of decellularized porcine aortic valve as a scaffold for in vitro recellularization. *Biomedical Engineering Online*, *12*(1), 87. doi:10.1186/1475-925X-12-87 PMID:24006837

Zhu, Y., Imbrie-Moore, A. M., Paulsen, M. J., Priromprintr, B., Park, M. H., Wang, H., Lucian, H. J., Farry, J. M., & Woo, Y. J. (2021). A Novel Aortic Regurgitation Model from Cusp Prolapse with Hemodynamic Validation Using an Ex Vivo Left Heart Simulator. *Journal of Cardiovascular Translational Research*, *14*(2), 283–289. doi:10.100712265-020-10038-z PMID:32495264

Zoetendal, E. G., Raes, J., Van Den Bogert, B., Arumugam, M., Booijink, C. C., Troost, F. J., & Kleerebezem, M. (2012). The human small intestinal microbiota is driven by rapid uptake and conversion of simple carbohydrates. *The ISME Journal*, *6*(7), 1415–1426. doi:10.1038/ismej.2011.212 PMID:22258098

Zuo, Y. Y., Uspal, W. E., & Wei, T. (2020). Airborne transmission of COVID-19: Aerosol dispersion, lung deposition, and virus-receptor interactions. *ACS Nano*, *14*(12), 16502–16524. doi:10.1021/acsnano.0c08484 PMID:33236896

About the Contributors

Pritam Pain has completed his B.Tech degree from WBUT in 2014 and then completed his M.Tech degree from Haldia Institute of Technology in 2016. He is currently working as an assistant professor in Haldia Institute of Technology. He had several journal papers and book chapters regarding Non-Traditional machining. He has been appointed as a reviewer in many journals. His main interest of research is in nature-inspired modern optimization algorithms.

Sreerup Banerjee is presently working as Associate Professor in the department of Mechanical Engineering at Haldia Institute of Technology, Haldia, West Bengal, India. Areas of expertise include biomechanics and medical image processing. Have 4.5 years of experience as Assistant Professor in NIT Agartala, and 1-year experience in VIT University Bhopal. Obtained bachelor's degree in Mechanical Engineering from Haldia Institute of Technology, master's degree in Biomedical Engineering from Jadavpur University and PhD degree from IIT Kanpur.

Goutam Kumar Bose is currently working as HOD and Professor in Mechanical Engineering Department, Haldia Institute of Technology, Haldia, India. He obtained his PhD in Production Engineering from the Jadavpur University, Kolkata, India. He has obtained a Master's in Engineering in Mechanical Engineering from the Bengal Engineering & Science University, Shibpur, India. He has worked as an Assistant Professor in the Department of Mechanical Engineering at the College of Engineering & Management, Kolaghat for ten years. He was an Engineer in R & D Centre of M/s Hindustan Motors Ltd. West Bengal, India. His active areas of interest are Metal Cutting, Non- conventional machining and Industrial and Production Management. He has published research papers in journals of international repute. He has attended several international conferences in India and abroad.

* * *

Thirugnanam Arunachalam is working as an Associate Professor, in the Department of Biotechnology and Medical Engineering, National Institute of Technology Rourkela, India. His current research area includes Biomechanics, Biotransport and Biomaterials. Prior to joining NIT Rourkela, he worked in the Industry as Deputy Manager in the Quality Control department in Aerospace and Defense projects.

Ravi Kant Avvari is an Assistant Professor at Biotechnology and Medical Engineering, NIT Rourkela, India. He has completed B.Tech. in Electrical & Electronics Engineering from GEC Thrissur and Ph.D. in Biomechanics from IIT Kanpur followed by a post-doctoral research in the area of Nanotechnology from Computational Nanoscience Lab at IIT Kanpur. His research interest includes biomechanics, biofluid mechanics, intestinal digestion, bionics, controls modeling, and nanotechnology.

Anju R. Babu, Assistant Professor in the Department of Biotechnology and Medical Engineering, National Institute of Technology Rourkela, received her Ph.D. from the Indian Institute of Science, Bangalore, India. She did her post-doctoral research at the University of Oklahoma and the Technical University of Graz. She did her M.Tech in Nanomedical Sciences from Amrita Center for Nano-sciences, and B.Tech in Biomedical engineering from Sahrdaya College of Engineering and Technology, Kerala, India. Her area of research focuses on investigating the structure-property relationship of soft biological tissues like the aorta, valves, and skin.

Mirza Khalid Baig is an Assistant Professor at Department of Biotechnology and Medical Engineering, National Institute of Technology (NIT), Rourkela. Previously, he was a postdoctoral research associate at the Centre for Bio-Inspired Technology (CBIT), Department of Electrical and Electronic Engineering (EE), Imperial College London. He received his Ph.D. (2018) from the CBIT, Department of EE, Imperial College London under the supervision of Prof. Christofer Toumazou, focused on developing system-on-chip for closed-loop neuromodulation therapy for managing obesity through Vagus Nerve Stimulation (VNS). Currently, his research interests are focused on developing efficient, small-size, implantable, sensors, and closed-loop devices for 'electroceutical' applications.

Ranjit Barua is a Senior Researchers at Centre for Healthcare Science and Technology, IIEST-SHIBPUR. Core research area Bio-Mechanical System, 3D Bioprinting, Tissue Engineering.

Priyobroto Basu is an M.Tech final year student studying Biomedical Engineering in NIT Rourkela.

Jetal Bhanarkar is mechanical graduate and currently pursuing post-graduation in Biomedical Engineering. Her interests lie in exploring the biomedical field from an engineering perspective.

Nirmalendu Biswas did Master in Mechanical Engineering (Specialization in Thermal) in 2008 from Jadavpur University. He has successfully completed doctoral programme and is awarded Ph.D degree in Engineering (Mechanical) from the Department of Mechanical Engineering, Jadavpur University, Kolkata, India in 2016. His research interests includes heat transfer, single phase and multi phase flow, free and forced convection, Numerical and Experimental analysis. He has published number of papers in International Journals and Proceedings regarding improvement on thermal performance of system.

Subhomoy Chatterjee completed PhD on medical devices from Indian Institute of Engineering Science and Technology, Shibpur (IIESTS), W. Bengal on January, 2017. His research was focused on biomechanics of prosthetic total hip replacement, based on computational (finite element) simulations and response of living tissues (bones) to the altered loading environment due to implantation. He performed his post doctoral research on medical devices, focusing on analytical biomechanics including response of prosthetic acetabular socket in total hip replacement and of various designs of dental implants in Indian Institute of Science (IISc), Bangalore. His Under-graduate and Post-graduate education were on Prosthetics & Orthotics. Currently he is working as Assistant Professor (Prosthetics & Orthotics) in Schieffelin Institute of Health Research and Leprosy Centre in Vellore, Tamil Nadu.

Hrijuta Datta is presently serving as Assistant Professor in the Department of Bioengineering at National Institute of Technology, Agartala since July, 2015. She is presently pursuing her Ph.D. from

National Institute of Technology, Patna. Her area of research work in Ph.D. is digital image processing. She has completed her M.Tech in Biomedical Engineering from SRM Institute Of Technology, Tamil Nadu in 2015. Hrijuta's research interest includes biomechanics and medical image analysis. She had worked on the biomechanics behind rehabilitation of frozen shoulder and computational processing of diseased retina.

Shubhabrata Datta, presently Research Professor in the Department of Mechanical Engineering, and Coordinator of Centre for Composites and Advanced Materials at SRM Institute of Science and Technology, Kattankulathur, Tamil Nadu, did his Bachelors, Masters and PhD in Engineering from Indian Institute of Engineering Science and Technology, Shibpur, India (previously known as B.E. College Shibpur) in the field of Metallurgical and Materials Engineering. Prof. Datta has 30 years of teaching and research experience. His research interest is in the domain of materials informatics and composites. Prof. Datta has published more than 150 papers in journals and peer-reviewed conference proceedings, with one authored and three edited books. Eleven of his graduate students have been conferred with PhD degree. He was bestowed with the Exchange Scientist Award from Royal Academy of Engineering, UK and worked in the University of Sheffield, UK. He also worked Dept of Materials Science and Engineering, Helsinki University of Technology, Finland, Dept of Materials Science and Engineering, Iowa State University, Ames, USA and Heat Engineering Lab, Dept of Chemical Engineering, Åbo Akademi University, Finland as Visiting Scientist. He is a Fellow of Institution of Engineers (India), Associate Editor, Journal of the Institution of Engineers (India): Series D, and editorial board member of several international journals. Prof. Datta has featured in the list of "World Ranking of Top 2% Scientists" in 2021 database (Published by Stanford University & Elsevier).

Thirumalai Deepak is pursuing a Ph.D. degree in the Department of Biotechnology and Medical Engineering, National Institute of Technology, Rourkela, Odisha. He has been working in the research area of tissue engineering in Soft Tissue Biomechanics Laboratory under the guidance of Dr. Anju R.Babu. He post-graduated in M.Tech, Biotechnology from Sri Venkateswara College of Engineering (affiliation to Anna University), Chennai, Tamilnadu, India, and graduated B.Tech, Biotechnology from Sree Sastha Institute of Engineering and Technology (affiliation to Anna University), Chennai, Tamilnadu, India.

Raja Dhason is currently working as an Assistant Professor in Department of Mechanical Engineering in SRM IST.

Partha Goswami has completed his Masters in applied mechanics (Fluid) from Bengal Engineering and Science University, Shibpur, West Bengal in the year 2007. He did his PhD in Mechanical Engineering in the year of 2018 from the Jadavpur University. He has fifteen years industrial experience in thermal power plant. His research areas are heat transfer, power plant, computational fluid dynamics and bio-fluid dynamics.

Tharani Kumaran is currently a research scholar in the Department of Biotechnology and Medical Engineering at National Institute of Technology Rourkela, Odisha, India. Her area of interest includes Biomechanics and Biomaterials. She is currently working on developing a indigenous force plate for gait profiling of locomotive syndrome.

Dipak Mandal has completed his Masters in Heat Power Engineering from Bengal Engineering and Science University, Shibpur, West Bengal in the year 1999. He did his PhD in the year of 2010 from the same University. He has ten years industrial experience in thermal power plant. Now he is working as a associate professor at the College of Engineering and Management, Kolaghat. His research areas are computational fluid dynamics, bio-fluid dynamics, nanoparticles, heat transfer, MHD, porous cavity, renewable energy and etc. He has published more than 50 journal papers in peer reviewed journals.

Nirmal Manna has completed his Masters in Heat Power Engineering from Bengal Engineering College (DU) presently known as IIEST, Shibpur, West Bengal in 1996. He did Ph.D. in 2005 from Jadavpur University, Kolkata, West Bengal. He has four years of industrial experience on turnkey projects of ash handling systems of thermal power plants before in academics. Now he is a full professor in the Mechanical Engineering Department of Jadavpur University. His research interests are Computational Fluid Dynamics (CFD), development of CFD codes, Eulerian-Eulerian approach of multiphase flow simulation, CFD-based investigation of multiphysics problems on heat and fluid flow.

Dhruba Jyoti Mech is currently pursuing a Doctor of Philosophy in Biomedical Engineering from the Indian Institute of Technology, Hyderabad (IITH). His area of interest is computational modeling, biomechanics, and finite element analysis. He has completed his post-graduation from the Indian Institute of Engineering Science and Technology, Shibpur (IIESTs) in Biomedical Engineering and graduated from Assam Down Town University (ADTU) in Mechanical Engineering.

Yugesh Ramdhun came from Mauritius and completed his High School Certificate from Mahatma Gandhi Institute Moka with distinction and an Oustanding Cambridge Learner Award. for the highest marks in Cambridge AS level Biology in 2018, together with other high ranks in his field. He is currently doing his B.Tech in Biomedical Engineering at the National Institute of Technology Rourkela and doing his minor course in Mechanical Engineering. He has also won the Academic Excellence Award in his first year of B.Tech (2019-2020), and being currently in his third year of B.Tech; he is still one of the best students in his class. He is presently working in preliminary research at the Biomechanics and Biomaterials Laboratory of the Biotechnology and Medical Engineering Department. After graduation, Yugesh plans to continue his postgraduation and pursue a career in the biomechanics field.

Sandipan Roy is associated with SRM Institute of Science and Technology Tamil Nadu as Research Assistant Professor. His core areas of research include biomaterials, bio-tribology, biomechanics, and finite element analysis.

Shammodip Roy has more than 12 years of orthopaedic implant design and medical device R&D experience, in designing, developing and innovating across orthopedic, neuro, spine and GI specialties. As a technical leader he has launched multiple medical devices globally, working with FDA and other regulatory agencies, holds several design patents and has authored research papers with clinicians around the world, including leading surgeons from HSS New York, Mayo Clinic, Cleveland Clinic and ETH Zurich. He has a Masters degree in Engineering from IIT Kanpur (Biomechanics major, mathematical modeling) with a University Gold Medal for his graduate thesis, and is currently at Stryker Orthopedics R&D, setting technical direction and driving new product development solutions in joint arthroplasty, for globally dispersed R&D teams. He has 10+ years experience in the inter-disciplinary fields of biomechanics, medical image processing, surgical robotics and additive manufacturing, and has clinical networks in India and abroad. Outside of corporate responsibilities, he is also engaged in academic and research collaborations globally in the field of biomechanics and artificial intelligence.

Monisha Gowri S. is currently a Research scholar in the Department of Biotechnology and Medical Engineering, National Institute of Technology Rourkela, India. She is working on developing an indigenous KSN for comprehensive gait profiling, in a project funded by the Department of Science and Technology, GoI. Her research area includes Biomechanics and Gait Analysis.

Anupama Usha Rani Siddhanathi is a graduate in MBBS and has more than 10 years of experience. Currently she is working at CWS hospital, Rourkela, Odisha.

Rashi Singh is a final year undergraduate B.Tech. in Biotechnology Engineering at NIT Rourkela. Her research interest includes Microbiology and Biomechanics.

Moutoshi Singha Roy is presently working as Assistant Professor in the department of Mechanical Engineering at National Institute of Technology Agartala, India. Mrs. Roy is pursuing her PhD on material science and engineering from Tripura University, India. She has completed Master's degree on Production engineering with a specialization on Computer Integrated Manufacturing from NIT Agartala, and Bachelor's degree in Mechanical Engineering from Tripura Institute of Technology (TIT), India. Her research interest involves on materials, composites, tribology, and biomechanics.

Patina Yamini is currently pursuing M.Tech in the Department of Biotechnology and Medical Engineering, National Institute of Technology, Rourkela, Odisha. She has been working in the research area of tissue engineering in Soft Tissue Biomechanics Laboratory under the guidance of Dr. Anju R. Babu. She graduated B.E, Electrical and Electronics Engineering from Roland Institute of Technology, Berhampur, Odisha.

Index

A

additive manufacturing 48-52, 54-59, 62-64
aerodynamics 223-224, 230, 240-242
altitude 235, 237, 244-246, 253, 255
ANSYS 16, 22, 29-31, 47, 61, 154, 159
Antibody tests 258
Antigen tests 258
arterial diseases 153, 155, 169
atherosclerosis 148, 153-156, 160, 169, 175-177

B

bevel needle 210, 215, 221
biomechanical properties 91, 137, 141-142, 144, 146, 149
Biomechanics 1-7, 9-14, 16, 18-20, 25, 35-36, 39-45, 65-66, 82-84, 86, 91-92, 94, 97-101, 106-107, 113-118, 120, 131, 133, 135, 137, 143, 145, 149-152, 174-179, 195, 207, 216, 223-224, 238, 243
bone condition 16, 30-31, 34, 36, 38, 40
bone density 16, 31-34, 42, 44
bone remodeling 16, 42, 45
bone remodelling 19, 30-34, 38, 41, 46-47
boundary conditions 10, 16, 20, 23, 25, 28, 60-62, 88, 90, 125, 153, 159, 163, 185-186, 215

C

Cages materials 94
cervical spine 94-99
CFD 10, 153-155, 159, 163-164, 167, 169, 178, 198, 200-202, 206, 208-209
collagen 95-96, 122, 128, 131-139, 145
composite bone plate 88, 92
computational biomechanics 1-2, 5, 7, 9, 11-13, 177
Computational Fluid Dynamics 9-10, 174, 177, 200-202, 205, 207, 209, 223, 256
Computer Solution 153
coronavirus 244-248, 253, 255, 258
COVID-19 244-248, 253, 255-259
COVID-19 vaccines 258

D

Dental screw 16
diameter 23, 42, 57, 129, 135, 155, 158-160, 171, 183-184, 201-202, 210, 219, 226, 244-245, 247-248, 253, 255
digestion 179-182, 186-187, 189-193, 195-197
discrete fiber model 121, 129
DJ stent 198, 201-204, 208, 210, 214, 221
dynamic cage 94, 97-98

E

Element (in FEM) 46
EMG markers 65

F

FDM 16, 44, 47-49, 52-55
FEA 7, 39, 62, 88, 94
FEM 6-8, 16-19, 38-39, 44, 46-47, 60, 62, 199, 210, 212, 215, 221
fermentation 179, 186, 192, 194
fiber material 88, 123, 127
fibrous scaffold 121-122, 125-131
flight mechanism 223, 240
Fluid Mechanics 153, 155, 175-178, 224, 241, 243, 256
force plate 4-5, 81, 100, 114-115
force platform 65, 81, 113

G

gait abnormalities 65-66, 73
gait analysis 65-66, 73, 76, 78-79, 81-86, 105, 114-115, 117-119

H

Haemodynamics 153
hematuria 198, 204, 208

I

Imaging Test 258
implant 16-19, 22-23, 29-31, 34-36, 38-47, 49-50, 60, 63, 88-89, 91-92, 97-98, 201
injury prevention 65, 82, 100, 107

K

kidney stone 198, 208, 211-212, 217, 222
kinematics 2, 65-66, 69, 71, 77, 82-83, 100-101, 106, 113-115, 135, 223-224, 230, 232-235, 237, 239-241, 243, 245
kinetics 2, 65-66, 71, 73, 79, 83, 100-101, 113, 115, 190, 223

M

maneuverability 223, 240
mass 5, 8, 10, 21, 32, 61-62, 66, 72-73, 80, 86-87, 106, 109, 155-156, 164, 174-175, 229, 231-232, 234-235, 245-248, 250
Mathematical modeling 121, 182, 256
medical reconstruction 22
Mesh (in FEM) 46
meshing 17-18, 20, 24, 30, 36, 89-90
micromechanics 121, 134
MIMICS 22-23, 25, 29, 90
MIS 210, 213, 221
motion analysis 65, 76-78, 84, 113, 116
musculoskeletal 2, 7, 42, 65-66, 74, 83-84, 105-107, 115, 117, 223, 233, 241

N

Node (in FEM) 46

P

PCNL 210-222
Performance improvement 100, 107
Periprosthetic Femoral Fracture 88-89
peristalsis 179-184, 186-187, 192, 194-195, 197, 199-200, 207
Personal Protective Equipment(PPE) 258
polymer 37, 44, 115, 131-133, 189, 221
Polymerase chain reaction (PCR) 258
pulsatile flow 153-156, 159-161, 163, 165-169, 171-172, 175-178

R

range 5, 7-9, 22, 48, 56-57, 65-66, 81, 87, 89-90, 94, 96, 101, 103, 108, 112-113, 115, 126, 128, 160, 165, 173, 230-231, 237-238, 240, 243-244, 247-250, 253-255
Range of Motion 66, 87, 94, 96
Rapid Diagnostic Test (RDT) 259
regurgitation 137-138, 145-147, 150, 152
renal stone 202, 210-211, 213, 221-222
respiratory droplet 244-249, 252-255, 257

S

size 17, 23-24, 39, 52-59, 61, 90, 128-129, 160, 164-165, 167, 181, 201, 205, 211, 222, 231-232, 234, 236, 238, 243-245, 247, 253, 255
SLA 48-49, 53-54, 56
small intestine 9, 179-196
sniff tests 259
sports biomechanics 4, 7, 100-101, 107, 113, 115-116, 118, 120
sports surfaces 101-102, 117-120
statistical modeling 127, 129
stenosis 94, 137-140, 145-146, 149, 152, 154-156, 158-160, 167-169, 171, 174-178, 198, 201
Stiffness Matrix 6, 26-28
stiffness sensing 121, 134
Stokes 244-246
streamline 60, 154, 156, 167-171, 186, 228
stress shielding 17, 19, 45, 50, 88, 91-92

T

terminal velocity 244-245, 247, 249, 253, 255
total hip replacement 17, 22, 41, 46-47
trajectory 3, 69, 187, 231, 244-245, 254-255

U

Urology 134, 198, 201, 206-208, 211-212, 217-220

UTI 198, 208

V

Valvular endothelial cells 137-139, 148
Valvular interstitial cells 137-138, 140

W

wall pressure 156, 159, 165, 168, 172-173
wall shear stress 154-156, 165, 167-169, 173, 175-176, 188

www.ingramcontent.com/pod-product-compliance
Lightning Source LLC
LaVergne TN
LVHW062144300926
849161LV00010B/2620